AF353113

Recent Advances in Underwater Signal Processing

Recent Advances in Underwater Signal Processing

Editors

Haixin Sun
Xuebo Zhang

MDPI • Basel • Beijing • Wuhan • Barcelona • Belgrade • Manchester • Tokyo • Cluj • Tianjin

Editors
Haixin Sun
Xiamen University
Xiamen, China

Xuebo Zhang
Whale Wave Technology Inc.
Kunming, China

Editorial Office
MDPI
St. Alban-Anlage 66
4052 Basel, Switzerland

This is a reprint of articles from the Special Issue published online in the open access journal *Sensors* (ISSN 1424-8220) (available at: https://www.mdpi.com/journal/sensors/special_issues/raudsp_sensors).

For citation purposes, cite each article independently as indicated on the article page online and as indicated below:

LastName, A.A.; LastName, B.B.; LastName, C.C. Article Title. *Journal Name* **Year**, *Volume Number*, Page Range.

ISBN 978-3-0365-8134-7 (Hbk)
ISBN 978-3-0365-8135-4 (PDF)

Contents

About the Editors

Haixin Sun

Dr. Sun received his Ph.D. in Signal and Information Processing from the Institute of Acoustics, Chinese Academy of Sciences, in 2006. Since 2006, he has been with the Department of Information and Communication Engineering, School of Information Technology, Xiamen University, China. He has presided over 3 National Natural Science Foundation projects,1 innovation Special Zone Key Project and 2 general projects, 1 National Key Research and Development Program Project and 1 Provincial Marine Economic Development Key Project. He is the author and co-author of more than 100 publications. He is also published a monograph in 2023.

Dr. Sun is an active Member of the IEEE. He is a Guest Associate Editor of Frontiers Marine Science. He served as a reviewer for IEEE TIE, IEEE JOE, JASA, Applied Acoustic, China Ocean Engineering, and Acoustics, and other internationally renowned acoustic and Marine journals. He is also a letter review expert of the National Natural Science Foundation of China, a member of Xiamen Communication Society, a member of Xiamen Institute of Electronics and so on.

Xuebo Zhang

Dr. Zhang (Senior Member, IEEE) received his B.Eng degree in electronic engineering and the Ph.D. degree in underwater acoustic engineering in 2009 and 2014, respectively. Currently, he has been with Whale Wave Technology Inc., Kunming, China. Up till now, he has authored and coauthored more than 70 papers, 10 Chinese Patents and a monograph.

Dr. Zhang is currently served as the Technology Committee Co-Chair for IEEE Oceanic Engineering Society. He has been the Associate Editor for IEEE Access. Besides, he has become the Editorial Board Member for Journal of Electronics & Information Technology and Current Engineering Letters and Reviews, and he has also been the Topical Advisory Member for Journal of Imaging, Journal of Marine Science and Engineering and Remote Sensing. Additionally, he is further the Section Editor for Recent Patents on Engineering. He has served as the Session Chair, Regional Chair, Track Chair, Publicity Chair, Invited Speaker and Special Session Chair for IEEE ICCC, ICCCS, ICIVC, ICCT, ICCSN, WAIE, IGARSS 2023 and so on. He received 'Outstanding Reviewer'and 'Outstanding Editorial Board Member' for Journal of Electronics & Information Technology, Science and Technology Progress First Prize, etc. Dr. Zhang is the Lead Guest Editor for Special Issues of Advances in Mechanical Engineering, Journal of Marine Science and Engineering, Electronics Letters, Recent Patents on Engineering, Frontiers in Marine Science and International Journal of Distributed Sensor Networks. Besides, he is also the Guest Editor for Special Issues of Wireless Communications & Mobile Computing, Journal of Electronics & Information Technology and MDPI Sensors.

 *sensors*

Editorial

Recent Advances in Underwater Signal Processing

Xuebo Zhang [1] **and Haixin Sun** [2,*]

1 Whale Wave Technology Inc., Kunming 650200, China; xby_zhang@nwnu.edu.cn
2 School of Informatics, Xiamen University, Xiamen 361005, China
* Correspondence: hxsun@xmu.edu.cn

Citation: Zhang, X.; Sun, H. Recent Advances in Underwater Signal Processing. *Sensors* **2023**, *23*, 5777. https://doi.org/10.3390/s23135777

Received: 10 June 2023
Accepted: 19 June 2023
Published: 21 June 2023

The ocean, covering 71% of the Earth's surface, is integral to human life. To solve its mysteries, equipment such as sonar and radar have emerged to perform topography, underwater communication, target detection, positioning, imaging, and ocean monitoring. Recent advances in signal processing and electronic technology have propelled new theories, mechanisms, and processing technologies for underwater equipment to a new stage.

This Special Issue aims to highlight recent advancements, developments, and applications in underwater signal processing methodologies, including characterization, simulation, real data processing, as well as applications to underwater engineering.

The editorial of this Special Issue introduces a total of 12 articles, which are divided into four types: (1) optimization of ship navigation, (2) underwater acoustic communication, (3) underwater acoustic signal recognition, and (4) underwater detection and positioning. The specific breakdown is as follows: two articles are introduced for the first type, four articles for the second type, three articles for the third type, and three articles for the fourth type.

1. Optimization of Ship Navigation

Ship navigation optimization has received much attention in recent years due to the increasing demand for ocean resources and space, presenting various challenges to ocean management and security. The need to improve ship navigation safety and detect ship operating trajectories is growing. In this issue, two articles present valuable methods and ideas.

In [1], researchers discuss the significance of atomic interference gravimeters for underwater navigation assistance. The instrument requires high precision to measure weak gravitational signals. To ensure accuracy, vibration isolation is essential in reducing external interference. The article reviews three vibration isolation methods: passive isolation, active isolation, and vibration compensation. It also highlights the direction of vibration compensation improvement as a future development trend.

In [2], researchers address the storage, management, analysis, and mining of ship target data. It designs and develops the overall structure and functional modules of the Ship Trajectory Data Management and Analysis System (STDMAS), proposing a ship identification method based on motion characteristics. The system is user-friendly, easy to maintain, and expandable, meeting the actual needs of ocean target data management, analysis, and mining. However, the current processing capacity for AIS data is limited. Future research can utilize big data algorithms and cloud computing architecture to improve the efficiency of processing massive data.

2. Underwater Acoustic Communication

Underwater acoustic communication utilizes sound waves to propagate through water and has various applications, including ocean detection, underwater sensing, and underwater operations. In recent years, underwater acoustic communication technology has gained increasing attention and research. In this article, we introduce four new methods to enhance the effectiveness of waterborne acoustic communication from different perspectives.

The article [3] explores the impact of temporal and spatial fluctuations of the ocean acoustic field on waterborne acoustic communication. Researchers have theoretically derived the fluctuation of signal intensity concerning changes in horizontal distance, signal frequency, bandwidth, and deployment depth, which were further verified through simulation and analysis experiments in the Yellow Sea acoustic field. The experimental results showed that a vertical array should be used for reception in shallow-water acoustic communication to improve the signal-to-noise ratio and system reliability. In addition to the acoustic field, corresponding achievements have been made in underwater signal sampling and transmission. Ref. [4] proposes a signal transmission and reception method for sonobuoys based on an autoencoder. By using an autoencoder at the transmission and reception ends, signal compression and restoration can be efficiently achieved, reducing the impact of environmental noise and improving the reliability of signal transmission. Furthermore, the article [5] designs a hardware/software platform for measuring channels and testing transmission technology under actual conditions in the ultrasonic band, allowing for the analysis and design of new underwater communication system solutions.

The overview article [6] takes the routing protocol as a starting point and discusses its various aspects, such as the concept and causes of void regions and the main challenges researchers face when designing routing protocols. The most advanced void avoidance protocol using OR technology is studied in-depth. We believe that these research results will provide robust support for applying and promoting waterborne acoustic communication technology.

3. Underwater Acoustic Signal Recognition

We have achieved significant progress not only in underwater acoustic communication but also in underwater acoustic signal recognition. This has become a prominent research field, with applications in underwater communication, ocean exploration, sonar detection, and other areas.

In [7], researchers proposed an EVMD algorithm that overcomes the accuracy limitations of VMD execution in the field of underwater acoustic communication. Simulation and experimental results show that this method has a recognition rate superior to traditional ship-radiated noise feature extraction methods, with a recognition rate of up to 96.6667%. However, the paper only discusses VMD mode numbers, and future work will consider optimizing the number of models and quadratic penalty terms for achieving higher decomposition accuracy.

Reducing noise and efficiently acquiring target underwater acoustic signals by sensors are equally crucial in signal recognition. In [8], the researchers proposed a method using Hidden Markov Models (HMM) to detect sequence acoustic data without separate training data. The stability and accuracy of detecting signals of interest (SOI) were improved using genetic algorithms and multiple measurements. Therefore, the multi-measurement GA-HMM exhibited excellent performance in both passive and active acoustic data.

Similarly, we have made significant progress in image transmission recognition. In [9], a normalization-based adaptive modulator (INAM) was proposed, which amplifies pixel deviations through adaptive predictive modulation factors. INAM was introduced into the learning of image-adaptive 3D LUT for underwater image enhancement, achieving good results.

4. Underwater Detection and Positioning

In addition to the previously mentioned fields, the importance of underwater detection and positioning technology in ocean science cannot be understated. Nowadays, this technology has been extensively utilized in ocean exploration, ocean resource development, underwater safety monitoring, and other areas. However, the complex marine environment can significantly impact the accuracy of underwater detection, and developing this technology entails overcoming substantial challenges. This issue of the *Sensors* journal presents three articles on underwater detection and positioning that are of significant importance for advancing this technology.

To address the issue of low signal-to-noise ratios in the underwater environment, which poses challenges for active sonar in detecting, tracking, and identifying underwater targets, the article [10] proposes a Tacotron model-based deep neural network (DNN) approach for active sonar signal synthesis. This method applies the Tacotron model to sonar signal synthesis and successfully synthesizes data that are almost identical to the data used for training, as confirmed by spectral comparison, attention result inspection, and MOS testing.

To improve the tracking performance of low SRR underwater targets, the article [11] proposes a particle filtering track-before-detect algorithm based on the knowledge-aided (KA-PF-TBD) algorithm. This method maximizes the utilization of prior information on the underwater diver target and establishes a set of multi-directional motion models to address the mismatch between conventional model sets and actual target motion.

In [12], researchers conducted a full-scale experiment simulating the underwater localization of magnetic sensors. Two natural computing algorithms were utilized to solve the signals generated by the known ferromagnetic object trajectory, successfully determining the positions of eight magnetometers. The methods performed exceptionally well, particularly in the multi-target version, accurately determining the position of the sensor with a relative error of 1% to 3%. The near sensor and the far sensor exhibited absolute errors of between 20 and 35 cm, respectively. These three papers illustrate the continual development and innovation in underwater detection and positioning technology.

5. Conclusions

The theme of this Special Issue focuses on underwater signal and ocean signal processing. This Special Issue highlights 12 articles that can be divided into four categories: optimization of ship navigation [1,2], underwater acoustic communication [3–6], underwater acoustic signal recognition [7–9], and underwater detection and positioning [10–12]. In addition to traditional underwater acoustic signals, research objects also include underwater sensors, underwater environments, ships, underwater images, etc. Therefore, in the field of underwater acoustic communication, in addition to traditional signal processing and analysis methods, there are many related technologies and applications worthy of research. In addition, in terms of algorithm design, the development of artificial intelligence algorithms also provides new solutions to analyze and process underwater signals. Signal processing algorithms that combine artificial intelligence algorithms with underwater signal processing technology will be a very important development trend in the future.

Conflicts of Interest: The authors declare no conflict of interest.

References

1. Gong, W.; Li, A.; Huang, C.; Che, H.; Feng, C.; Qin, F. Effects and Prospects of the Vibration Isolation Methods for an Atomic Interference Gravimeter. *Sensors* **2022**, *22*, 583. [CrossRef] [PubMed]
2. Feng, C.; Fu, B.; Luo, Y.; Li, H. The Design and Development of a Ship Trajectory Data Management and Analysis System Based on AIS. *Sensors* **2022**, *22*, 310. [CrossRef] [PubMed]
3. Lv, Z.; Du, L.; Li, H.; Wang, L.; Qin, J.; Yang, M.; Ren, C. Influence of Temporal and Spatial Fluctuations of the Shallow Sea Acoustic Field on Underwater Acoustic Communication. *Sensors* **2022**, *22*, 5795. [CrossRef] [PubMed]
4. Park, J.; Seok, J.; Hong, J. Autoencoder-Based Signal Modulation and Demodulation Methods for Sonobuoy Signal Transmission and Reception. *Sensors* **2022**, *22*, 6510. [CrossRef] [PubMed]
5. Fernández-Plazaola, U.; López-Fernández, J.; Martos-Naya, E.; Paris, J.F.; Cañete, F.J. HW/SW Platform for Measurement and Evaluation of Ultrasonic Underwater Communications. *Sensors* **2022**, *22*, 6514. [CrossRef] [PubMed]
6. Mhemed, R.; Phillips, W.; Comeau, F.; Aslam, N. Void Avoiding Opportunistic Routing Protocols for Underwater Wireless Sensor Networks: A Survey. *Sensors* **2022**, *22*, 9525. [CrossRef] [PubMed]
7. Esmaiel, H.; Xie, D.; Qasem, Z.A.H.; Sun, H.; Qi, J.; Wang, J. Multi-Stage Feature Extraction and Classifification for Ship-Radiated Noise. *Sensors* **2022**, *22*, 112. [CrossRef] [PubMed]
8. You, H.; Byun, S.-H.; Choo, Y. Underwater Acoustic Signal Detection Using Calibrated Hidden Markov Model with Multiple Measurements. *Sensors* **2022**, *22*, 5088. [CrossRef] [PubMed]
9. Xiao, X.; Gao, X.; Hui, Y.; Jin, Z.; Zhao, H. INAM-Based Image-Adaptive 3D LUTs for Underwater Image Enhancement. *Sensors* **2023**, *23*, 2169. [CrossRef] [PubMed]

10. Kim, Y.; Kim, J.; Hong, J.; Seok, J. The Tacotron-Based Signal Synthesis Method for Active Sonar. *Sensors* **2023**, *23*, 28. [CrossRef] [PubMed]
11. Yue, W.; Xu, F.; Xiao, X.; Yang, J. Track-before-Detect Algorithm for Underwater Diver Based on Knowledge-Aided Particle Filter. *Sensors* **2022**, *22*, 9649. [CrossRef] [PubMed]
12. Alimi, R.; Fisher, E.; Nahir, K. In Situ Underwater Localization of Magnetic Sensors Using Natural Computing Algorithms. *Sensors* **2023**, *23*, 1797. [CrossRef] [PubMed]

Article

INAM-Based Image-Adaptive 3D LUTs for Underwater Image Enhancement

Xiao Xiao [1,2,3], Xingzhi Gao [2,*], Yilong Hui [2], Zhiling Jin [2] and Hongyu Zhao [1]

1   State Key Laboratory of CEMEE, Luoyang 471000, China
2   School of Telecommunications Engineering, Xidian University, Xi'an 710071, China
3   Science and Technology on Complex System Control and Intelligent Agent Cooperation Laboratory Beijing Electro-Mechanical Engineering Institute, Beijing 100074, China
*   Correspondence: gxzhz1988@163.com

Abstract: To the best of our knowledge, applying adaptive three-dimensional lookup tables (3D LUTs) to underwater image enhancement is an unprecedented attempt. It can achieve excellent enhancement results compared to some other methods. However, in the image weight prediction process, the model uses the normalization method of Instance Normalization, which will significantly reduce the standard deviation of the features, thus degrading the performance of the network. To address this issue, we propose an Instance Normalization Adaptive Modulator (INAM) that amplifies the pixel bias by adaptively predicting modulation factors and introduce the INAM into the learning image-adaptive 3D LUTs for underwater image enhancement. The bias amplification strategy in INAM makes the edge information in the features more distinguishable. Therefore, the adaptive 3D LUTs with INAM can substantially improve the performance on underwater image enhancement. Extensive experiments are undertaken to demonstrate the effectiveness of the proposed method.

Keywords: image enhancement; underwater images; instance normalization

Citation: Xiao, X.; Gao, X.; Hui, Y.; Jin, Z.; Zhao, H. INAM-Based Image-Adaptive 3D LUTs for Underwater Image Enhancement. *Sensors* **2023**, *23*, 2169. https://doi.org/10.3390/s23042169

Academic Editors: Haixin Sun and Xuebo Zhang

Received: 24 December 2022
Revised: 24 January 2023
Accepted: 31 January 2023
Published: 15 February 2023

1. Introduction

1.1. Background

In recent years, with the excavation and exploration of marine resources and the ocean world, high-quality underwater images have become increasingly important. The processing and enhancement of underwater signals and images have also attracted a lot of attention. However, the complex underwater environment and lighting conditions significantly pose great challenges for underwater image enhancement, which aims to improve the image visibility and contrast and reduce chromatic aberration. The reasons are as follows. Firstly, underwater images will be affected by noise from marine snow, which increases the scattering effect and dramatically reduces the contrast and visibility of the image. Secondly, underwater images are degraded by wavelength-dependent absorption and scattering including forward and back scattering [1–5], which all limit the practical application of underwater images and videos in marine biology, archaeology, marine ecology, and ocean exploration. In terms of performance and efficiency, underwater image enhancement remains a significant challenge due to the diversity of captured scenes, the complexity of underwater environments, and the fluctuation of underwater lighting conditions.

1.2. Related Works

To solve the above problems, a number of methods have been proposed. (1) Supplementary information-based methods. The method proposed by Narasimhan et al. [6] is to use the supplementary information of multiple images to increase the visibility of the image. Some methods [7–10] use special hardware such as polarization filtering to improve the image visibility. ERH [11] uses three procedures for color compensation, image alignment, and homogenization using a multiscale synthesis strategy. The ERH method cascades the three procedures for underwater image enhancement. (2) Non-physical model-based methods. The method proposed by Iqbal et al. [12] modifies the image pixel values in RGB (Red, Green, Blue) color space and HSV (Hue, Saturation, Value) color space through Non-Physical Model-Based Methods to improve the contrast and saturation of underwater images. (3) Physical model-based methods. Underwater image enhancement is carried out through Physical Model-Based Methods [13]. This method sets a cost function and increases the image contrast by reducing the cost function, thereby obtaining excellent underwater images. GUDCP [14] is a new method for backward scattered light estimation that integrates several prior knowledge and introduces a new scoring formula. This method also develops a white balance method to further modify the appearance of the synthetic image. There are also excellent methods such as SVM-RBF [15] in remote sensing that provide inspiration for underwater image processing.

These years, many deep learning-based image enhancement methods [16–22] have been proposed in the field of computational imaging. MFFN [23] is a multi-scale feature fusion network that can enhance the adaptability and visualization of the scene. SGUIE-Net [24] is a semantic region-based enhancement module that can better learn local enhancement features of semantic regions with multi-scale perception. The fused features are semantically consistent and visually have better enhancement effects. The AGA-based Swin Transformer module [25] is designed to be an end-to-end underwater image enhancement network. It can dynamically select visually complementary channels based on dependencies, reducing the number of further attention parameters. R. Liu et al. developed a bilaterally constrained closed-loop adversarial enhancement module [26] that alleviates the requirements of the unsupervised approach for pairwise data by coupling twin inverse mappings and preserves more informative features. J. Yuan et al. proposed a multi-scale fusion enhancement algorithm [27] to improve sharpness by contrast-based a priori de-fogging of the dark channel in the red-green-blue (RGB) model. However, some of them have complex network structures or expensive computational cost. Zeng, H et al. proposed the learning image-adaptive 3D LUTs (three-dimensional lookup tables) hybrid method [28], combining multi-layer features in deep learning-based methods and image priors in traditional methods. This image enhancement method has high performance, high image quality, high computational efficiency, and low memory consumption. However, the normalization method of instance normalization adopted by the CNN (Convolutional Neural Network)-based weight predictor in this method will lead to the degradation on the network performance. Instance Normalization (IN) [29] is a milestone technique in deep learning that normalizes the distribution of intermediate layers, leading to faster training and better generalization accuracy. However, the residual features' standard deviation is greatly reduced after normalization. The standard deviation reflects the variation of pixel values. When the variation is reduced, the ability of the network distinguishing edges will decrease, further degrading the network performance, which significantly impacts on the performance of underwater image enhancement methods.

1.3. Contributions

In this paper, we propose an Instance Normalization Adaptive Modulator (INAM) that amplifies pixel bias by adaptively predicting modulation factors to solve this issue. The Instance Normalization operation applies normalization to a specific image instance and normalizes it in the *HW* (Height and width) dimension of the tensor, and is initially used for image style transfer. The mean and variance of each channel of the feature map affect the style of the final generated image. We introduce INAM into the learning image-adaptive 3D LUTs for underwater image enhancement. The bias amplification strategy in INAM makes the edge information in the features more distinguishable.

The environmental and lighting conditions underwater lead to poor visibility, low contrast, and large color differences from natural scenes in underwater images. 3D LUT itself is an artificially produced method for image enhancement. Adaptive 3D LUT, on the other hand, can be learned by a deep learning method driven by data, which compares the input poor visibility image with clear ground truth. In this process, the weight prediction of the image is critical. The weights not only determine the weight of the LUT's influence on the image but also affect the parameter learning of the 3D LUT itself under adaptive conditions. Our proposed INAM can effectively improve the problem caused by IN in CNN by amplifying the pixel bias and making the network more capable of discriminating the edges of objects in images. The introduction of INAM enables CNN to predict the weights of images more accurately and further affects the parameters learned by 3D LUTs. It gives the adaptive 3D LUTs a more vital ability to enhance the images. In contrast to other methods, we combine traditional lookup tables and deep learning methods. We exploit the powerful learning ability of deep learning models, which makes manual lookup tables unnecessary, dramatically reduces the workload, and has high accuracy and precision. Our proposed INAM compensates for the IN layer drawback of the weight prediction module, which enables a more efficient and accurate combination of weight prediction and image enhancement. Our model has low complexity, a small number of parameters, and uses only a few computational resources to achieve good enhancement results.

Our approach breaks the limits of deep learning and combines it with traditional methods. It also uses deep learning to simplify the complex work of traditional methods greatly. Significantly, our proposed INAM also reduces the drawbacks of IN, which allows CNNs to ignore the effects of residual feature reduction in weight assignment, substantially improving the edge information recognition and the accuracy of weight assignment. Our method provides a new idea for underwater image enhancement.

The main contributions of the work are summarized as follows:

(1) We propose INAM, which enhances CNN recognition of image edge information by adaptive modulation factors to amplify pixel deviations. Compared with current methods, this module has few parameters, and substantial performance improvement on CNN with almost no increase in computational resources.

(2) We introduce INAM into adaptive 3D LUTs to form adaptive 3D LUTs and train them for data-driven deep learning. We improve the prediction accuracy of the CNN network for image weights in this way. The predicted weights also affect the learning of subsequent 3D LUTs parameters. Our method improves image enhancement by enhancing the weight prediction accuracy and the parameters of 3D LUTs.

(3) We conduct extensive experiments to compare our approach with some existing methods on one public dataset. The results demonstrate the superiority of our method in terms of performance, both data and visual effects.

2. Materials and Methods

2.1. Three-Dimensional Lookup Table

3D LUT is a classic and efficient image enhancement technique that has been widely and effectively used. As shown in Figure 1, a 3D LUT has three dimensional channels, representing the RGB color index. Each color channel is divided into M units so that a 3D LUT consists of M^3 elements $\left\{ \mathbf{V}_{(i,j,k)} \right\}_{i,j,k=0,...,M-1}$, where M is the number of units in any color channel. Each element $\mathbf{V}_{(i,j,k)}$ defines an input $\left\{ r^I_{(i,j,k)}, g^I_{(i,j,k)}, b^I_{(i,j,k)} \right\}$ of RGB value and the corresponding transformed output $\left\{ r^O_{(i,j,k)}, g^O_{(i,i,k)}, b^O_{(i,j,k)} \right\}$. For a given M, the indexed color value $\left\{ r^I_{(i,j,k)}, g^I_{(i,j,k)}, b^I_{(i,j,k)} \right\}_{i,j,k=0,...,M-1}$ can be obtained in a uniform color space. Different 3D LUTs will have different outputs so that different color outputs $\left\{ r^O_{(i,j,k)}, g^O_{(i,j,k)}, b^O_{(i,j,k)} \right\}_{i,j,k=0,...,M-1}$ will be obtained after a given color index. The scale of M determines the accuracy of the color conversion. The larger the M is, the higher the accuracy of the color conversion will be. In our experiments, we set $M = 33$. In this case, a 3D LUT contains $108\,\mathrm{K}\left(3M^3\right)$ parameters.

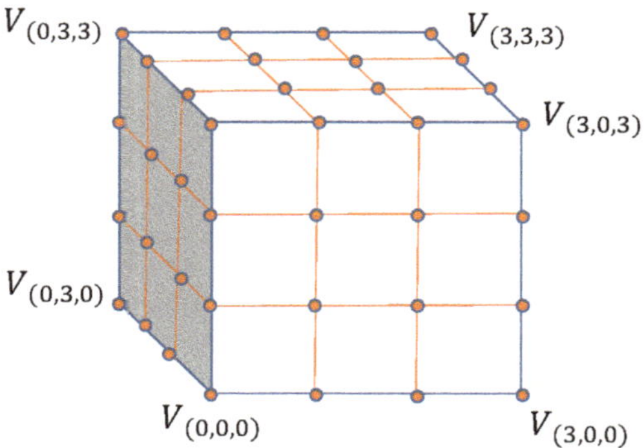

Figure 1. Illustration of a 3D LUT containing 4^3 elements.

The 3D LUT transforms and transmits the color in two steps: lookup and trilinear interpolation. Firstly, given the input RGB color value $\left\{r^I_{(x,y,z)}, g^I_{(x,y,z)}, b^I_{(x,y,z)}\right\}$, the 3D LUT lattice is calculated to find its corresponding position (x, y, z). Here, x, y, z can be expressed as $x = \frac{r^I_{(x,y,z)}}{s}, y = \frac{g^I_{(x,y,z)}}{s}, z = \frac{b^I_{(x,y,z)}}{s}$, respectively, where $s = \frac{C_{max}}{M}$, C_{max} is the maximum value on each color channel. After the position of the input color value is determined, trilinear interpolation can be performed using the eight elements closest to it to obtain the output RGB value. Let $i = \lfloor x \rfloor, j = \lfloor y \rfloor, k = \lfloor z \rfloor$, where $\lfloor . \rfloor$ is the floor function, and let $d_x = \frac{r^I_{(x,y,z)} - r^I_{(i,j,k)}}{s}, d_y = \frac{g^I_{(x,y,z)} - g^I_{(i,j,k)}}{s}, d_z = \frac{b^I_{(x,y,z)} - b^I_{(i,j,k)}}{s}$. The output RGB $\left\{r^O_{(x,y,z)}, g^O_{(x,y,z)}, b^O_{(x,y,z)}\right\}$ after the color transformation will be obtained by trilinear interpolation and can be given by the following expression:

$$
\begin{aligned}
c^O_{(x,y,z)} =\ & (1 - d_x)(1 - d_y)(1 - d_z)c^O_{(i,j,k)} + d_x(1 - d_y)(1 - d_z)c^O_{(i+1,j,k)} \\
& + (1 - d_x)d_y(1 - d_z)c^O_{(i,j+1,k)} + (1 - d_x)(1 - d_y)d_z c^O_{(i,j,k+1)} \\
& + d_x d_y(1 - d_z)c^O_{(i+1,j+1,k)} + (1 - d_x)d_y d_z c^O_{(i,j+1,k+1)} \\
& + d_x(1 - d_y)d_z c^O_{(i+1,j,k+1)} + d_x d_y d_z c^O_{(i+1,j+1,k+1)},
\end{aligned}
\tag{1}
$$

where $c \in \{r, g, b\}$. The above trilinear interpolation is sub-differentiable, and it is easy to derive the gradient of $c^O_{(i,j,k)}$. Since the trilinear interpolation of each input is independent of other pixels, this transformation is convenient for parallel computation.

2.2. Instance Normalization Adaptive Modulator (INAM)

Normalization is a method applied in the data preparation process when features in the data have different ranges to change the values of numeric columns in the data set using an identical scale. The advantages of normalization are as follows. (1) Normalization of each feature maintains the contribution of each feature when some features have higher values than others. This ensures that the network is unbiased. (2) The network activation distribution changes due to changes in network parameters during training. To improve training performance, we use normalization to reduce internal covariance. (3) Normalization makes the loss plane smoother because normalization constrains the size of the gradient more strictly. (4) Normalization makes the optimization faster because it disallows the weights to explode everywhere and limits them to a specific range. (5) Normalization helps the network to apply regularization.

The IN operation is to apply normalization to a specific image instance and normalize it in the HW dimension, which is initially used for image style transfer. The generated result mainly depends on an image instance, and the mean and variance of each feature map channel will affect the final image's style. Therefore, the normalization of the entire batch and entire sample are unsuitable for image stylization, since only H or W dimension is normalized. Model convergence can be accelerated, and the independence between each image instance and channel (various features) is maintained. However, in the instance normalization process, the features' standard deviation will be compressed, and the network's ability to distinguish edge information will be reduced, which results in a decrease in the network's performance. Inspired by Ref. [30], we propose the INAM that amplifies pixel bias by adaptively predicting modulation factors to solve this issue.

As shown in Figure 2, we constructed three simple models to demonstrate our approach. Figure 2a shows the model containing only one Conv layer without instance normalization. The input x is convolved to obtain the output y. Figure 2b shows the model constructed by inserting an IN layer before the Conv layer. The model shown in Figure 2c is constructed by inserting an IN layer before the Conv layer, i.e., the input x is first subjected to an instance normalization operation, and then the result of the operation is convolved to obtain the output y. The model shown in Figure 2c adds the IN layer after the input x, and then the output y' is obtained by convolution. y' is then multiplied by the conditioning factor to obtain the final output $\hat{y}'$. For the sake of our exposition, we simplify this conditioning factor. We will describe the equations for these three models and then discuss the effects of feature normalization in terms of pixel standard deviation.

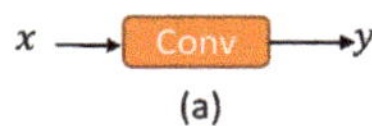

(a)

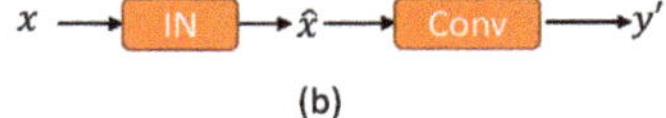

(b)

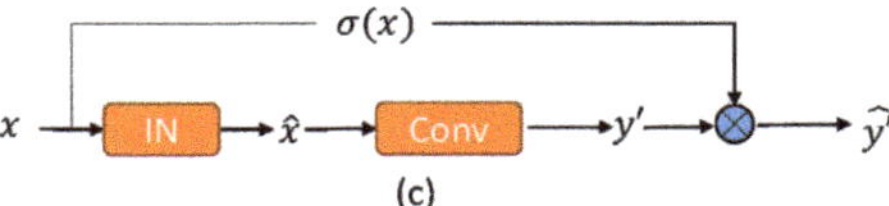

(c)

Figure 2. Three simple models for demonstration. (**a**) shows the model containing only one Conv layer without instance normalization. (**b**) shows the model containing a Conv layer and an instance normalization layer before Conv layer. (**c**) shows the model with an instance normalization layer, a Conv layer and a conditioning factor $\sigma(x)$.

Denote the function of the Conv layer as f_{Conv}, the input of the Conv layer is x, and the output of the Conv layer is y. In Figure 2a, it is evident that we can describe the output y by the following formula:

$$y = f_{\text{Conv}}(x) \tag{2}$$

In Figure 2b, x is a single input sample with four axes (C, H, W, N), and $\hat{x}$ is the transformed feature with IN. For IN, the mean and standard deviation are computed along the (H, W) axes so that we can compute $\hat{x}$ as:

$$\hat{x} = \frac{x - \mu}{\sigma} \tag{3}$$

where μ and σ are scalars shared by all pixels in x, μ is the calculated mean of each instance, and σ is the calculated variance of all instances. Then the output y' can be computed by:

$$y' = f_{\text{Conv}}(\hat{x}) = f_{\text{Conv}}\left(\frac{x - \mu}{\sigma}\right) \tag{4}$$

Rewrite Equation (4) as:

$$y' = \frac{1}{\sigma} f_{\text{Conv}}(x - \mu) \tag{5}$$

Equation (5) can be further expanded using the distributivity:

$$y' = \frac{1}{\sigma} f_{\text{Conv}}(x) - \frac{1}{\sigma} f_{\text{Conv}}(\mu I) \tag{6}$$

where I is the all-ones matrix with the same dimensions as x.

By comparing (2) and (6), it can be found that IN reshapes the pixel distribution of y. We compute the standard deviation (std) of y':

$$\begin{aligned}
\text{std}(y') &= \text{std}\left(\frac{1}{\sigma} f_{\text{Conv}}(x) - \frac{1}{\sigma} f_{\text{Conv}}(\mu I)\right) \\
&= \frac{1}{\sigma} \text{std}(f_{\text{Conv}}(x)) \\
&= \frac{1}{\sigma} \text{std}(y).
\end{aligned} \tag{7}$$

It can be seen that with instance normalization, the pixel bias is reduced to $\frac{1}{\sigma}$. The standard deviation decreases because σ is usually bigger than 1. To compensate for the loss of pixel bias, we multiply y' by σ in the third model, $\hat{y}'$ can be obtained by:

$$\hat{y}' = y' \cdot \sigma \tag{8}$$

2.3. INAM-Based Image-Adaptive 3D LUTs

Learning image-adaptive 3D LUTs is an effective color-mapping operation. Its process is divided into three steps, the first of which is using a weight predictor to predict the weight of the down-sampled low-resolution image and then using a look-up table generated from this weight for looking up and interpolation. For simplicity of description, we do not describe the interpolation operation in the 3D LUT but simplify it to look up in this subsection. Equation (9) represents a mapping function. In the RGB color domain, a classic 3D LUT is defined as a 3D cube containing N^3 elements, where N is the number of bins in each color channel. Each element defines a pixel-to-pixel mapping $\mu(x)$, where x is the input image, and q_o is the output image .

$$q_o = \mu(x) \tag{9}$$

However, the weight predictor used in learning image-adaptive 3D LUTs method to predict weights uses instance normalization (IN), which makes the standard deviation of the features compressed so that the network's performance is reduced. That is, the model's ability to distinguish edge information is reduced, resulting in a reduction in the accuracy of weight assignment. The INAM adaptively adjusts the feature standard deviation based on instance normalization, which improves the model's ability to distinguish edge information and makes the weight prediction more accurate. The overall frame is shown in Figure 3.

As shown in Figure 3, the input of the model is a 480×640 high-resolution image, and the input of the weight predictor is a down-sampled low-resolution image with a resolution of 256×256. The weight predictor consists of five convolution blocks, a dropout layer and a fully connected layer. The output is the weight N which is set to 3 in our experiment. Each convolution block consists of a convolution layer, an IN layer, a leaky Relu layer, and an INAM. The first convolutional layer uses a three-channel image with an input of 256×256, a convolutional kernel of size 3×3, and an output of 16 channels. The second convolutional layer has an input size of 128×128 and an output of 32 channels. The third convolutional layer has an input size of 64×64 and outputs 64 channels. The fourth convolutional layer has an input size of 32×32 and outputs 128 channels. The fifth convolutional layer has an input size of 16×16 and outputs 128 channels. All the convolutional layers use a convolutional kernel size of 3×3, with a stride of 2 and padding of 1. The input size of the fully connected layer is 8×8, the convolutional kernel size is 8×8, and the number of output channels is N = 3. Based on the above data, the complexity of this CNN model can be calculated. The number of its parameters is about 191 k, and the number of FLOPs is 155 M. In Section 2.2, we introduced the basic principles of INAM through three simple models. However, in practical applications, the model will be more complex. For example, bias terms exist in each convolution, and there may be non-linear layers between convolutional layers. Equation (10) describes the INAM in the actual model:

$$\hat{y}' = y' \cdot e^{\phi(\log(\sigma(x)))} \tag{10}$$

where x is the input, σ is the calculated standard deviation of x, y is the feature to be adjusted, y' is the output after modulating the feature, and $\phi(v) := w \cdot v + b$ is a learnable linear model consisting of a weight w and a bias b. During training, w and b can be updated via the backpropagation algorithm. The ϕ function predicts an appropriate modulation factor based on the input value v. In Equation (10), we learn ϕ in logarithmic space for better stability. Finally, the modulation factor is obtained by exponential operation. The model is shown in Figure 4.

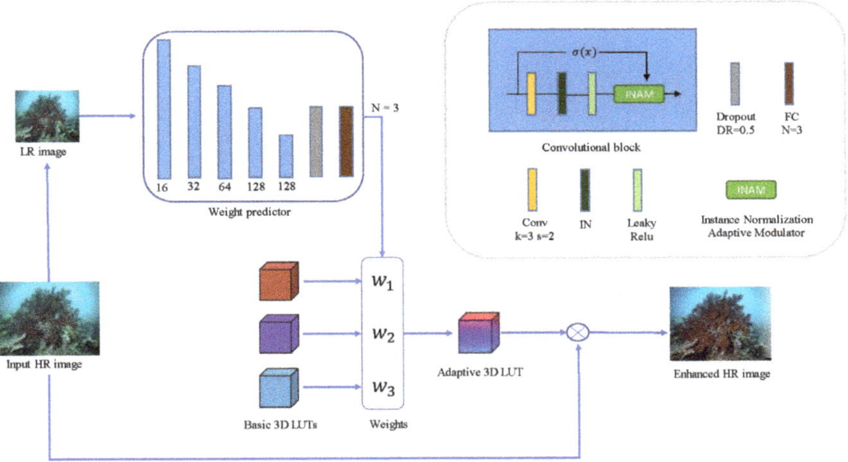

Figure 3. Overview of our proposed framework. The weight predictor outputs weights, and basic 3D LUTs form adaptive 3D LUTs based on the weights. The adaptive 3D LUT then enhances the input image. In the weight predictor, our proposed INAM is used.

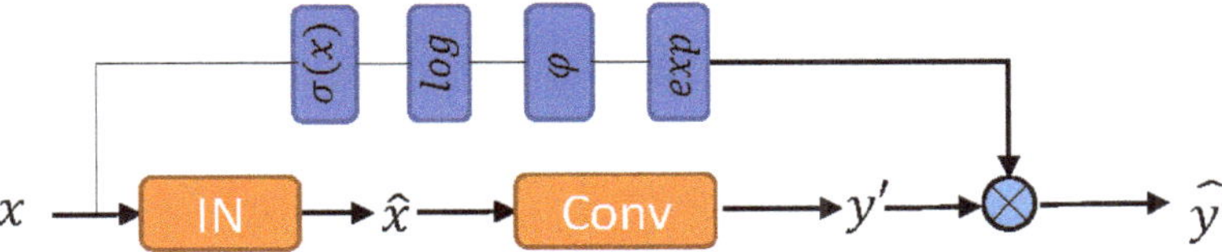

Figure 4. Illustration of INAM in the actual model.

We learn several basic 3D LUTs and INAM-based CNN weight predictors. 3D LUTs are used to process the images and INAM-based CNNs are used to predict the weights of the images. Assuming that the weights obtained from the prediction are $\{w_n\}_{n=1,\dots,N} = f(x)$

and the corresponding 3D LUT processing is $\{\mu_n\}_{n=1,\ldots,N}$, the final enhanced image obtained is:

$$q = \sum_{n=1}^{N} w_n \mu_n(x) \tag{11}$$

where x indicates the input image. The objective function of our learning scheme can be written as follows:

$$\min_{f,\mu_n} \mathcal{L}(q,y) \tag{12}$$

where f and μ_n are the CNN model and the basic 3D LUTs to be learned, $\mathcal{L}(q,y)$ indicates some loss functions and regularization terms.

2.4. Loss Function

We use supervised learning methods to learn image enhancement models. Suppose that there are a number of T training pairs $\{x_t, y_t\}_{t=1,2,\ldots,T}$, where x_t and y_t denote a pair of input and target images, respectively. We employ the Mean Square Error (MSE) loss to train the model:

$$\mathcal{L}_{mse} = \frac{1}{T} \sum_{t=1}^{T} \|q_t - y_t\|^2 \tag{13}$$

Using $\mathcal{L}_{mse}$ loss, we train 3D LUTs with momentum or Adam optimizer and CNN weight predictors using the gradient descent algorithm. However, the optimized 3D LUTs may have unsmooth surfaces. The color mutations in the neighboring lattices of the 3D LUTs may amplify the chromatic aberration after color conversion, resulting in some banding artifacts in the smooth regions of the enhanced images. In order to make the learned 3D LUTs more stable and robust, we introduce two regularization terms in the optimization process.

Smoothing regularization: We introduce a 3D smoothing regularization term in 3D LUTs learning, which converts the input RGB values more stably into the desired color space, thus making the output of the 3D LUTs locally smooth. We choose the L_2 distance in the above term to achieve smoother regularization. To improve the smoothness of the adaptive 3D LUT, we introduce L_2-norm regularization for the prediction weights w_n. The overall smooth regularization term is as follows:

$$\mathcal{R}_s = \sum_{c \in \{r,g,b\}} \sum_{i,j,k} \left(\|c_{(i+1,j,k)}^O - c_{(i,j,k)}^O\|^2 + \|c_{(i,j+1,k)}^O \right.$$
$$\left. - c_{(i,j,k)}^O\|^2 + \|c_{(i,j,k+1)}^O - c_{(i,j,k)}^O\|^2 \right) + \sum_n \|w_n\|^2. \tag{14}$$

Monotonicity regularization: In addition to smoothness, 3D LUTs should also be monotonic. This is because monotonic transformations maintain the relative brightness and saturation of the input RGB values, ensuring natural enhancement results. Monotonicity helps update parameters that may not be activated by the input RGB values, improving the generalization ability of the learned 3D LUTs. Therefore, we adopt a monotonic regularization as follows:

$$\mathcal{R}_m = \sum_{c \in \{r,g,b\}} \sum_{i,j,k} \left[g\left(c_{(i,j,k)}^O - c_{(i+1,j,k)}^O\right) + g\left(c_{(i,j,k)}^O \right. \right.$$
$$\left. \left. - c_{(i,j+1,k)}^O\right) + g\left(c_{(i,j,k)}^O - c_{(i,j,k+1)}^O\right) \right] \tag{15}$$

where $g(\cdot)$ is defined as the standard ReLU operation, i.e., $g(a) = \max(0,a)$. The monotonicity regularization ensures that the output RGB values $c_{(i,j,k)}^O$ increase with the index i,j,k and larger i,j,k indices correspond to larger input RGB values in the 3D LUT lattice.

By incorporating the two regularization terms, the final loss function used in learning is as follows:

$$\mathcal{L} = \mathcal{L}_{\mathrm{mse}} + 0.0001 * \mathcal{R}_s + 10 * \mathcal{R}_m \tag{16}$$

2.5. Experimental Setup

2.5.1. Dataset

We conduct experiments on the EUVP dataset. The dataset contains a large number of paired and unpaired underwater images with good or poor perceptual quality. These images are collected at different locations with different visibility conditions, and most of them are taken during ocean exploration and human-robot cooperation experiments. These images are carefully selected to accommodate a wide range of natural variations in the data. We use paired data of underwater-scenes from EUVP dataset, which contains 2185 raw underwater images covering different underwater scenes, underwater creatures, etc. We randomly select 2000 images as the training set and the remaining 185 images as the test set. We uniformly resize the image to 640 × 480 pixels for the experiment of 480 pixels.

2.5.2. Baselines

We compare the perceptual image enhancement performance of the INAM-based image-adaptive 3D LUTs with the following models: (1) relative global histogram, unsupervised color correction (UCM) [12]; (2) contrast limited adaptive histogram equalization (CLAHE) [31]; (3) underwater dark channel prior (UDCP) [32]; (4) Water-net [33]; and (5) learning image-adaptive 3D LUTs [28]. The first three are physics-based models, and the Water-net is a learning-based model.

2.5.3. Evaluation Metrics

In order to evaluate the image enhancement methods in a multifaceted way, four metrics are used for evaluation. They can be classified as full-reference evaluation metrics and no-reference evaluation metrics.

1. Full-Reference Evaluation:

PSNR is a Full-Reference image evaluation metric. It is one of the most common and widely used objective image evaluation metrics, which is based on the error between the corresponding pixel points, i.e., on error-sensitive image quality evaluation. A larger value indicates less distortion.

SSIM is a Full-Reference image evaluation index, which measures image similarity in brightness, contrast, and structure. Its value range is [0, 1]. A larger value indicates a smaller image distortion.

2. Non-Reference Evaluation:

UCIQE [34] is a linear combination of color intensity, saturation, and contrast, which is used to quantitatively evaluate the non-uniform color shift, blurring, and low contrast of underwater images. It is a Non-Reference (ground-truth) image quality evaluation index, and higher values indicate better image quality.

UIQM [35] is a Non-Reference underwater image quality evaluation index based on the stimulation of the human eye visual system, which adopts color, sharpness, and contrast measurements evaluation basis for the degradation mechanism and imaging characteristics of underwater images. A larger value means better color balance, sharpness, and contrast.

2.5.4. Experiment Settings

Using the EUVP dataset, we experiment with color enhancement pipelines in camera imaging. In such applications, the target image is in the sRGB color space, has 8 bit of dynamic range, and is compressed to JPG format. In the photo retouching application, the input image has the same format as the target image. We learn the color enhancement part of the imaging pipeline instead of learning the entire pipeline from the raw data to the final RGB output. We use 2000 underwater images as the training set for image enhancement. One image per batch is selected with a size of 640 × 480. The model is trained using the ADAM optimizer, and the learning rate is set to 10^{-4} and reduces by half every 200 epochs. The model is trained for a total of 400 epochs. We implement our network with PyTorch and train all modules on an NVIDIA GTX1060ti GPU.

3. Results

3.1. Comparison Experiments

We first perform image enhancement using each method and compare the enhanced image with the ground-truth image using full-reference metrics to evaluate our method against the other methods. As shown in Table 1, our method outperforms UCM by 10.36 points or 71.39% in the PSNR metric. Our method is 6.54 points higher than CLAHE, which is 35.68% higher. Our method outperforms UDCP by 8.12 points or 48.48%. Our method outperforms Water-net by 4.83 points, or 24.10%. Compared to the original learning image-adaptive 3D LUTs method without adding the INAM module, our method improves by 3.85 points. This indicates that the pixel-to-pixel error between the image enhanced by our method and the ground-truth image is minimal. In the SSIM metric, our method is 0.390 points higher than UCM, which is 74.7% higher. It is 0.260 points or 39.87% higher than CLAHE, 0.358 points or 64.62% higher than UDCP, 0.209 points or 29.73% higher than the Water-net method, and 0.056 points or 6.5% higher than the original learning image-adaptive 3D LUTs method. This indicates that the enhanced image of our method is much better than other methods in all three aspects of brightness, contrast, and structure, which proves that our proposed INAM improves the 3D LUT method performance. An intuitive qualitative performance comparison is shown in Figure 5.

Table 1. Quantitative comparison for average PSNR and SSIM values on test images of the EUVP dataset.

Model	PSNR	SSIM
UCM	14.51	0.522
CLAHE	18.33	0.652
UDCP	16.75	0.554
Water-net	20.04	0.703
Learning image-adaptive 3D LUTs	21.02	0.856
INAM-based image-adaptive 3D LUTs	**24.87**	**0.912**

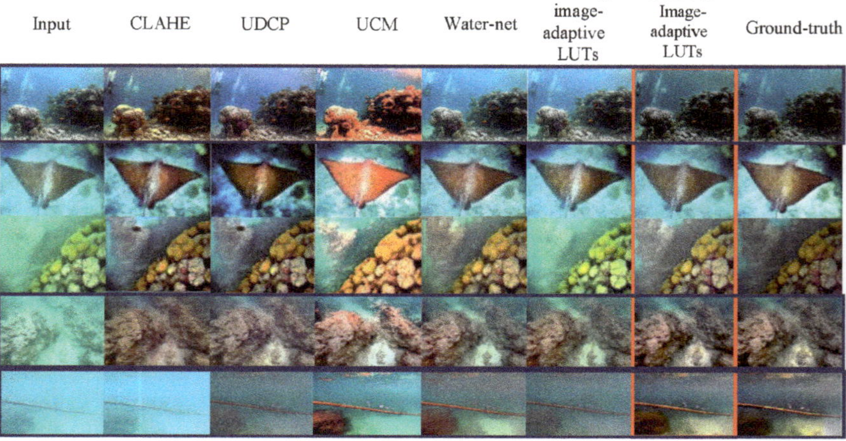

Figure 5. Qualitative performance comparison of UCM [12], CLAHE [31], UDCP [32], Water-net [33], learning image-adaptive 3D LUTs [28], and INAM-based image-adaptive 3D LUTs.

We perform image enhancement using each method and evaluate the enhanced images using the no-reference metrics UCIQE and UIQM. The obtained data are shown in Table 2. In the UCIQE metric, our method is 0.078 points higher than the UCM, which is 13.56% higher. Our method is 0.054 points or 9.01% higher than CLAHE. Our method is 0.068 points or 11.62% higher than UDCP. Our method outperforms Water-net by 0.047 points, or 7.76%. Compared with the learning image-adaptive 3D LUTs method without adding the INAM module, our method improves by 0.030 points, which is 4.8% higher. This indicates that the image quality enhanced by our method is the highest compared with other methods. In the UIQM metric, our method is 0.155 points or 11.27% higher than the UCM. It is 0.135 points higher than CLAHE, which is 9.64% higher. It is 0.120 points or 8.47% higher than UDCP, 0.181 points or 13.36% higher than the Water-net method, and 0.102 points or 7.11% higher than the original learning image-adaptive 3D LUTs method. This demonstrates that the enhanced image of our method is far better than other methods in three aspects: color balance, contrast, and sharpness structure, and it is more in line with human eye perception.

Table 2. Quantitative comparison for average UCIQE and UIQM values on test images of the EUVP dataset.

Model	UCIQE	UIQM
UCM	0.575	1.375
CLAHE	0.599	1.401
UDCP	0.585	1.416
Water-net	0.606	1.355
Learning image-adaptive 3D LUTs	0.623	1.434
INAM-based image-adaptive 3D LUTs	**0.653**	**1.536**

The INAM proposed by us has recovered the residual standard deviation reduced by IN through an operation. In this process, the network's ability to recognize edge information is enhanced, and the prediction accuracy of weight is improved. This enables the model to learn excellent parameters in the learning process quickly. During the test, the image enhanced by the model will have better saturation, brightness contrast, and the highest structural similarity with the ground truth image.

3.2. Ablation Study

The number N of 3D LUTs: To verify the effect of the number of 3D LUTs on the image enhancement, we set the number of LUTs as 1, 2, 3, 4, 5. We still train the model on the EUVP dataset and evaluate the model to observe the effect of the number of LUTs on the image enhancement. We first conducted ablation experiments controlling the number of LUTs and obtained the data as shown in Table 3. From the data in the table, we can see that the values of PSNR, SSIM, UCIQE, and UIQM significantly increase as N increases from 1 to 3. While with N increasing from 3 to 5, the increase of these indicators is relatively small. Therefore, we use the number of LUTs as 3 in the comparison test.

Table 3. Ablation studies on the number (N) of LUTs affecting the INAM-based image-adaptive 3D LUTs.

N	1	2	3	4	5
PSNR	20.43	22.83	24.87	25.81	25.84
SSIM	0.806	0.868	0.912	0.919	0.924
UCIQE	0.596	0.628	0.653	0.665	0.671
UIQM	1.422	1.489	1.536	1.544	1.551

INAM: To verify the effectiveness of our proposed INAM, we set the number of LUTs in the learning image-adaptive 3D LUTs method without adding the INAM module, and train the model on the EUVP dataset and evaluate it. The data are compared with the INAM-based image-adaptive 3D LUTs method to verify the effectiveness of our proposed INAM module for image enhancement. We change the number of LUTs for the learning image-adaptive 3D LUTs method without the INAM and obtain the data in Table 4. We compared it with the INAM-based image-adaptive 3D LUTs method at the same number of LUTs. The comparison results are shown in Figure 6. Our method outperforms the learning image-adaptive 3D LUTs method without INAM for different N in PSNR, SSIM, UCIQE, and UIQM metrics. This comparison verifies the effectiveness of our proposed INAM.

Table 4. Ablation studies on the number (N) of LUTs affecting the image-adaptive-3D LUTs without INAM.

N	1	2	3	4	5
PSNR	18.36	19.84	21.02	21.55	21.86
SSIM	0.784	0.824	0.856	0.863	0.867
UCIQE	0.576	0.608	0.623	0.629	0.633
UIQM	1.314	1.388	1.434	1.446	1.453

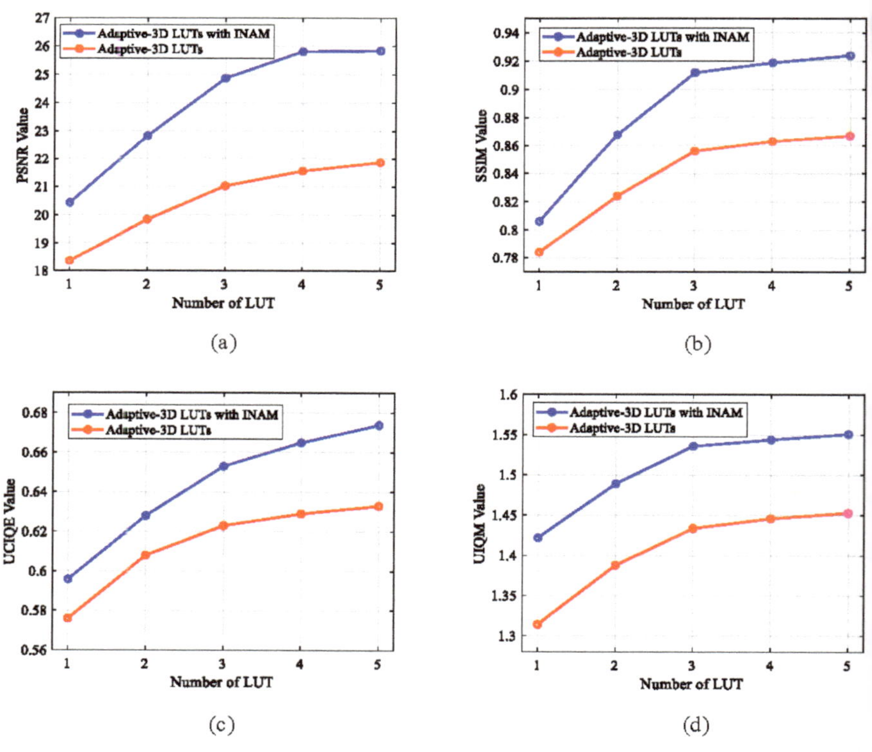

Figure 6. The performance of the learning image-adaptive 3D LUTs method and our proposed INAM-based image-adaptive 3D LUTs are compared in a line graph for each metric. The performance comparison of the two methods on PSNR metrics is shown in (**a**). (**b**) The comparison of the two methods in terms of SSIM performance. (**c**) Comparison of the performance of the two methods on the UCIQE metric. A comparison of the two methods evaluated by the UIQM metrics is shown in (**d**).

4. Conclusions

We introduced the INAM into learning image-adaptive 3D LUTs for underwater image enhancement. After instance normalization, the standard deviation of features will be reduced, which reduces the ability of the network to distinguish edge information and thus, the accuracy of weight assignment will be reduced. Our proposed INAM can compensate for the compressed standard deviation and thus improve the accuracy of the weight predictor. We train and test on the EUVP dataset and evaluate its effectiveness by comparing with other traditional and learning-based methods. The experimental results show that the method outperforms other methods in PSNR, SSIM, UCIQE, and UIQM metrics. We conducted ablation experiments to verify the effect of the number of LUTs on our proposed method. We also verified the performance improvement of our proposed INAM on the image-adaptive 3D LUTs method for different numbers of LUTs.

5. Discussion

In this study, we propose INAM, which can compensate for the change of standard deviation caused by instance normalization through learning. We apply it in the weight prediction module of the adaptive 3D LUT, which significantly improves the underwater image enhancement effect in all indicators without affecting the training speed and generalization accuracy.

We consider the CNN module to be necessary in the image enhancement process. This is because the fraction of LUT weights it predicts directly determines the size of this LUT's role in the subsequent image enhancement process. In past CNN modules that used IN, the standard deviation of the residual features would be greatly reduced. The standard interpolation of the residual features would reflect the variation of the pixel values. We propose the INAM module, which allows CNN to improve the standard deviation of the residual features while maintaining the training speed and generalization accuracy. This operation enables the CNN module to improve the recognition of edges and allows the prediction accuracy of weights to be improved when the CNN and 3D LUTs are trained together. This allows for better coordination between several 3D LUTs in the processing of images and also impacts the parameters being learned by the 3D LUTs. Therefore, our proposed method has an excellent performance in all metrics.

Our proposed INAM can improve the edge recognition ability of CNN by amplifying the pixel bias through adaptive prediction of modulation factors. It may be helpful for other networks, such as edge detection, semantic segmentation, and other tasks.

Author Contributions: X.X. supervised the study, gave suggestions and revised the manuscript; X.G. proposed the original idea, completed the programming and wrote the manuscript. Y.H. revised the manuscript; Z.J. revised the manuscript. H.Z. revised the manuscript. All authors have read and agreed to the published version of the manuscript.

Funding: This work was supported in part by the China Postdoctoral Science Foundation under Grant 2021TQ0260 and Grant 2021M700105; and in part by the Guangzhou Science and Technology Program under Grant 202201010870.

Data Availability Statement: The data used in this work is the EUVP dataset. It can be download from https://irvlab.cs.umn.edu/resources/euvp-dataset (accessed on 15 December 2022).

Conflicts of Interest: The authors declare that there is no conflict of interest.

References

1. McGlamery, B. A computer model for underwater camera systems. *Ocean Opt. VI Int. Soc. Opt. Photonics* **1980**, *208*, 221–231.
2. Jaffe, J.S. Computer modeling and the design of optimal underwater imaging systems. *IEEE J. Ocean. Eng.* **1990**, *15*, 101–111. [CrossRef]
3. Hou, W.; Woods, S.; Jarosz, E.; Goode, W.; Weidemann, A. Optical turbulence on underwater image degradation in natural environments. *Appl. Opt.* **2012**, *51*, 2678–2686. [CrossRef] [PubMed]
4. Akkaynak, D.; Treibitz, T. A revised underwater image formation model. In Proceedings of the IEEE Conference on Computer Vision and Pattern Recognition, Salt Lake City, UT, USA, 18–23 June 2018; pp. 6723–6732.

5. Akkaynak, D.; Treibitz, T.; Shlesinger, T.; Loya, Y.; Tamir, R.; Iluz, D. What is the space of attenuation coefficients in underwater computer vision? In Proceedings of the IEEE Conference on Computer Vision and Pattern Recognition, Honolulu, HI, USA, 21–26 July 2017; pp. 4931–4940.
6. Narasimhan, S.G.; Nayar, S.K. Contrast restoration of weather degraded images. *IEEE Trans. Pattern Anal. Mach. Intell.* **2003**, *25*, 713–724. [CrossRef]
7. Shashar, N.; Hanlon, R.T.; Petz, A.D. Polarization vision helps detect transparent prey. *Nature* **1998**, *393*, 222–223. [CrossRef]
8. Schechner, Y.Y.; Karpel, N. Clear underwater vision. In Proceedings of the 2004 IEEE Computer Society Conference on Computer Vision and Pattern Recognition, Washington, DC, USA, 27 June–2 July 2004; Volume 1, p. I.
9. Schechner, Y.Y.; Karpel, N. Recovery of underwater visibility and structure by polarization analysis. *IEEE J. Ocean. Eng.* **2005**, *30*, 570–587. [CrossRef]
10. Treibitz, T.; Schechner, Y.Y. Active polarization descattering. *IEEE Trans. Pattern Anal. Mach. Intell.* **2008**, *31*, 385–399. [CrossRef]
11. Song, H.; Chang, L.; Chen, Z.; Ren, P. Enhancement-Registration-Homogenization (ERH): A Comprehensive Underwater Visual Reconstruction Paradigm. *IEEE Trans. Pattern Anal. Mach. Intell.* **2022**, *44*, 6953–6967. [CrossRef]
12. Iqbal, K.; Odetayo, M.; James, A.; Salam, R.A.; Talib, A.Z.H. Enhancing the low quality images using unsupervised colour correction method. In Proceedings of the 2010 IEEE International Conference on Systems, Man and Cybernetics, Istanbul, Turkey, 10–13 October 2010; pp. 1703–1709.
13. Liu, H.; Chau, L.P. Underwater image restoration based on contrast enhancement. In Proceedings of the 2016 IEEE International Conference on Digital Signal Processing (DSP), Beijing, China, 16–18 October 2016; pp. 584–588.
14. Liang, Z.; Ding, X.; Wang, Y.; Yan, X.; Fu, X. Gudcp: Generalization of underwater dark channel prior for underwater image restoration. *IEEE Trans. Circuits Syst. Video Technol.* **2022**, *32*, 4879–4884. [CrossRef]
15. Razaque, A.; Ben Haj Frej, M.; Almi'ani, M.; Alotaibi, M.; Alotaibi, B. Improved support vector machine enabled radial basis function and linear variants for remote sensing image classification. *Sensors* **2021**, *21*, 4431. [CrossRef]
16. Hashisho, Y.; Albadawi, M.; Krause, T.; von Lukas, U.F. Underwater color restoration using u-net denoising autoencoder. In Proceedings of the 2019 11th International Symposium on Image and Signal Processing and Analysis (ISPA), Dubrovnik, Croatia, 23–25 September 2019; pp. 117–122.
17. Hu, Y.; Wang, K.; Zhao, X.; Wang, H.; Li, Y. Underwater image restoration based on convolutional neural network. In Proceedings of the Asian Conference on Machine Learning, Beijing, China, 14–16 November 2018; pp. 296–311.
18. Wang, Y.; Zhang, J.; Cao, Y.; Wang, Z. A deep CNN method for underwater image enhancement. In Proceedings of the 2017 IEEE International Conference on Image Processing (ICIP), Beijing, China, 17–20 September 2017; pp. 1382–1386.
19. Li, C.; Anwar, S.; Porikli, F. Underwater scene prior inspired deep underwater image and video enhancement. *Pattern Recognit.* **2020**, *98*, 107038. [CrossRef]
20. Perez, J.; Attanasio, A.C.; Nechyporenko, N.; Sanz, P.J. A deep learning approach for underwater image enhancement. In Proceedings of the International Work-Conference on the Interplay between Natural and Artificial Computation, Corunna, Spain, 19–23 June 2017; Springer: Berlin/Heidelberg, Germany, 2017; pp. 183–192.
21. Islam, M.J.; Xia, Y.; Sattar, J. Fast underwater image enhancement for improved visual perception. *IEEE Robot. Autom. Lett.* **2020**, *5*, 3227–3234. [CrossRef]
22. Fabbri, C.; Islam, M.J.; Sattar, J. Enhancing underwater imagery using generative adversarial networks. In Proceedings of the 2018 IEEE International Conference on Robotics and Automation (ICRA), Brisbane, Australia, 21–25 May 2018; pp. 7159–7165.
23. Chen, R.; Cai, Z.; Cao, W. MFFN: An Underwater Sensing Scene Image Enhancement Method Based on Multiscale Feature Fusion Network. *IEEE Trans. Geosci. Remote Sens.* **2021**, *60*, 1–12. [CrossRef]
24. Qi, Q.; Li, K.; Zheng, H.; Gao, X.; Hou, G.; Sun, K. SGUIE-Net: Semantic Attention Guided Underwater Image Enhancement with Multi-Scale Perception. *arXiv* **2022**, arXiv:2201.02832.
25. Huang, Z.; Li, J.; Hua, Z.; Fan, L. Underwater image enhancement via adaptive group attention-based multiscale cascade transformer. *IEEE Trans. Instrum. Meas.* **2022**, *71*, 1–18. [CrossRef]
26. Liu, R.; Jiang, Z.; Yang, S.; Fan, X. Twin adversarial contrastive learning for underwater image enhancement and beyond. *IEEE Trans. Image Process.* **2022**, *31*, 4922–4936. [CrossRef]
27. Yuan, J.; Cai, Z.; Cao, W. TEBCF: Real-World Underwater Image Texture Enhancement Model Based on Blurriness and Color Fusion. *IEEE Trans. Geosci. Remote Sens.* **2021**, *60*, 1–15. [CrossRef]
28. Zeng, H.; Cai, J.; Li, L.; Cao, Z.; Zhang, L. Learning image-adaptive 3d lookup tables for high performance photo enhancement in real-time. *IEEE Trans. Pattern Anal. Mach. Intell.* **2020**, *44*, 2058–2073. [CrossRef]
29. Ulyanov, D.; Vedaldi, A.; Lempitsky, V. Instance normalization: The missing ingredient for fast stylization. *arXiv* **2016**, arXiv:1607.08022.
30. Liu, J.; Tang, J.; Wu, G. AdaDM: Enabling Normalization for Image Super-Resolution. *arXiv* **2021**, arXiv:2111.13905.
31. Reza, A.M. Realization of the contrast limited adaptive histogram equalization (CLAHE) for real-time image enhancement. *J. VLSI Signal Process. Syst. Signal Image Video Technol.* **2004**, *38*, 35–44. [CrossRef]
32. Drews, P.L.; Nascimento, E.R.; Botelho, S.S.; Campos, M.F.M. Underwater depth estimation and image restoration based on single images. *IEEE Comput. Graph. Appl.* **2016**, *36*, 24–35. [CrossRef]
33. Li, C.; Guo, C.; Ren, W.; Cong, R.; Hou, J.; Kwong, S.; Tao, D. An underwater image enhancement benchmark dataset and beyond. *IEEE Trans. Image Process.* **2019**, *29*, 4376–4389. [CrossRef]

34. Yang, M.; Sowmya, A. An underwater color image quality evaluation metric. *IEEE Trans. Image Process.* **2015**, *24*, 6062–6071. [CrossRef]
35. Panetta, K.; Gao, C.; Agaian, S. Human-visual-system-inspired underwater image quality measures. *IEEE J. Ocean. Eng.* **2015**, *41*, 541–551. [CrossRef]

Article

In Situ Underwater Localization of Magnetic Sensors Using Natural Computing Algorithms

Roger Alimi [1,*], Elad Fisher [1,2] and Kanna Nahir [1]

[1]   Technology Division, Soreq NRC, Yavne 81800, Israel
[2]   Hatter Department of Marine Technologies, University of Haifa, Haifa 3498838, Israel
*   Correspondence: roger@soreq.gov.il

Abstract: In the shallow water regime, several positioning methods for locating underwater magnetometers have been investigated. These studies are based on either computer simulations or downscaled laboratory experiments. The magnetic fields created at the sensors' locations define an inverse problem in which the sensors' precise coordinates are the unknown variables. This work addresses the issue through (1) a full-scale experimental setup that provides a thorough scientific perspective as well as real-world system validation and (2) a passive ferromagnetic source with (3) an unknown magnetic vector. The latter increases the numeric solution's complexity. Eight magnetometers are arranged according to a 2.5 × 2.5 m grid. Six meters above, a ferromagnetic object moves according to a well-defined path and velocity. The magnetic field recorded by the network is then analyzed by two natural computing algorithms: the genetic algorithm (GA) and particle swarm optimizer (PSO). Single- and multi-objective versions are run and compared. All the methods performed very well and were able to determine the location of the sensors within a relative error of 1 to 3%. The absolute error lies between 20 and 35 cm for the close and far sensors, respectively. The multi-objective versions performed better.

Keywords: magnetometers; underwater sensing; genetic algorithm; particle swarm optimization

Citation: Alimi, R.; Fisher, E.; Nahir, K. In Situ Underwater Localization of Magnetic Sensors Using Natural Computing Algorithms. *Sensors* **2023**, *23*, 1797. https://doi.org/10.3390/s23041797

Academic Editors: Sylvain Girard and Galina V. Kurlyandskaya

Received: 7 December 2022
Revised: 29 January 2023
Accepted: 1 February 2023
Published: 5 February 2023

1. Introduction

Underwater sensors arrays have been widely used for various applications such as scientific exploration, sea disaster investigation, and military purposes [1–4]. Figure 1 shows a schematic view of this kind of deployment.

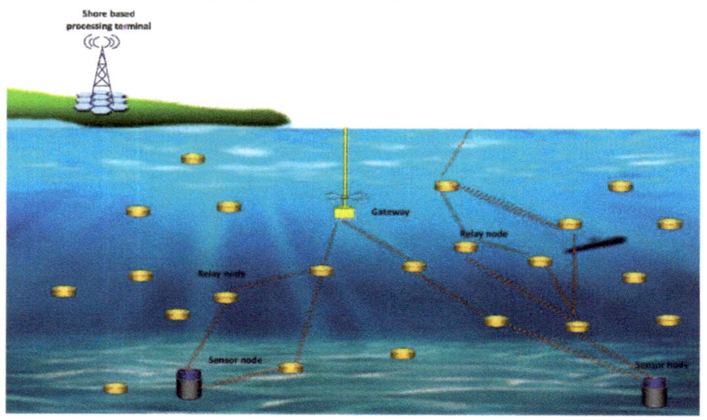

Figure 1. Schematic view of underwater sensor deployment [3].

Acoustic, optical, or RF sensing technologies are commonly used. The challenging task is the precise location estimation for the deployed network, due to the special underwater medium, as well as the lack of GPS data. This is critical to enable the system to provide an accurate description of any investigated phenomenon. Unfortunately, common localization solutions do not provide a satisfactory response to this crucial issue.

In the shallow water regime, solutions for the location of the underwater sensor are often divided into two main categories [5]. Range-based schemes estimate the locations of sensors by using inter-sensor measurements and the prior knowledge of the locations of a few reference sensors [6,7]. Range-free schemes are much simpler, but the location output they supply cannot be very precise [8,9]. Several positioning methods using magnetic technologies have been investigated. The magnetic fields generated at the sensors location, given a known object trajectory, define an inverse problem, in which the precise positions of the sensors are the variables of the equation system.

Callmer et al. [5] utilize triaxial magnetometers and a friendly vessel with a known magnetic dipole to silently localize the sensors. In Yu et al. [10], a solenoid coil carrying a direct current is employed as a magnetic source carried by a boat. This moves along a predetermined trajectory above the sensor field. Using the magnetic measurements, the localization problem is translated into a multi-objective optimization problem. A non-dominated sorting genetic algorithm is used to compute the sensor positions.

Other studies are based on either computer simulations or downscaled laboratory experiments [11–15].

For instance, Bian et al. [11] proposes an oscillating magnetic field-based indoor and underwater positioning system. The magnetic approach generates a bubble-formed magnetic field that is unaffected by environmental variation, unlike radio-wave-based positioning modalities. The proposed system achieves 13.3 cm for the 2D underwater positioning mean accuracy and 19.0 cm for the 3D underwater positioning mean accuracy. In [12], the authors use a magnetic field gradient optimization to localize. The magnetic source in the localization system is a direct current solenoid coil. An objective function is established by measuring the magnetic source's magnetic field in different positions. A multi-swarm particle swarm optimization with a dynamic learning strategy determines the sensor position vector. Recently, John et al. [13] designed an extensible modular multi-sensor platform prototype that can detect and localize objects with different properties in all environments. Their prototype detects and localizes objects using magnetic, acoustic, and electrical sensors. Although the system is ready for underwater measurements, data fusion algorithms are still being developed.

In [14], the authors derive an approximate equation to calculate the external magnetic field of a three-core armored underwater cable based on seafloor environments and the cable structure. Underwater cable localization uses a dual three-axis magnetic sensor array and beetle swarm optimization (BSO) algorithm. An optimization algorithm replaces analytical geometry, and a magnetic flux-density amplitude fitness function improves underwater cable localization. Although quite promising, the results are only based on simulations. Zhang et al. recently conducted real-world experiments (1.5 m depth in a water tank and 9 m depth in a shallow sea) [15]. An electric field induced by a standard current source is used for real-time underwater equipment location. Stationary or moving underwater equipment is tracked and located in shallow and deep seas (9 m) under noisy conditions. A real-time position is estimated using an extended Kalman filter. Although not based on magnetometers, this system is an interesting competitor despite the fact that it exhibits errors larger than ours (0.5–0.7 m vs. 0.25–0.35 m).

As far as resolution methods are concerned, Philippeaux et al. [16] uses a genetic algorithm. The field tests appear promising after extensive magnetic field modeling and an algorithm simulation. The simulation matches the field tests. Only preliminary in-water testing of the system has been conducted. In [17], Hu et al. present a practical localization algorithm that uses magnetic field vector and gradient tensor data to determine the center coordinates and magnetic moments of multiple underwater magnetic objects.

The Levenberg–Marquardt algorithm is used. Hou et al. perform an interesting machine learning approach [18]. They present an Auto Mobile Base Simultaneous localization and mapping (AMB-SLAM) online navigation algorithm based on an artificial neural network (ANN) and measurements from randomly distributed beacons of low-frequency magnetic fields. Their simulations show that a setup of unsophisticated low-cost magnetic beacons can produce geometrically consistent feature maps and an accurate trajectory for AUVs' underwater navigation.

The present work addresses several issues that have not been addressed previously. First, we construct a full-scale experimental setup that provides comprehensive scientific insight as well as a real-life system validation. In addition, instead of the active coil used in previous studies, we use a passive ferromagnetic source. Finally, the magnetic vector moment of the source is a part of the unknown that the algorithms solved.

The algorithms we have developed provide results that closely match reality in this type of scenario. In order to reproduce more faithfully realistic conditions, we have voluntarily disturbed one of the testing events. Even in this case, the results are still reliable. We also present a comparison between two classes of meta-heuristics, GA and PSO, for which single- and multi-objective implementations are considered.

2. Experimental Setup and Methodology

Our setup consists of eight vector magnetometers arranged on the ground according to the grid shown in Figure 2. The geometry is chosen to imitate a shallow water deployment in the vicinity of a harbor entrance. Six meters above, a plastic tube is attached on the laboratory ceil. Within the tube, a non-ferromagnetic winch-drive activated by a pulley allows for the motion of the ferromagnetic object. The velocity is motor-controllable from the ground. The only unknown variable remains the moment of the moving object, which we let the algorithms calculate together with the coordinates of the sensors.

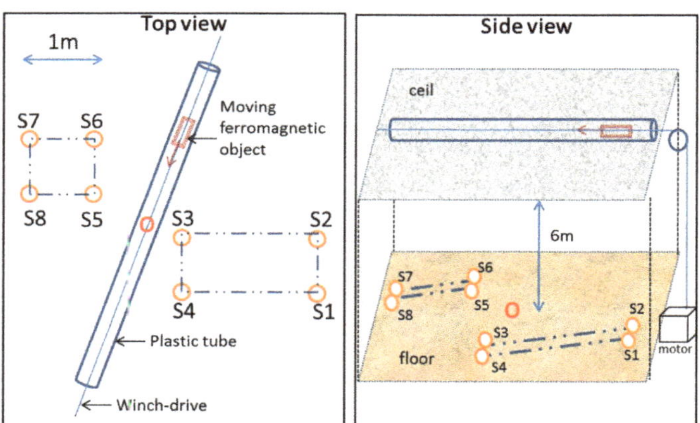

Figure 2. Top and side views of the experimental setup used in the experiment. Note the point O in red, which is the center of the searching box used in both GA and PSO.

Natural computing or nature-inspired algorithms refer to a wide class of computational paradigms in which selected features of natural phenomena or behaviors are studied and imitated to solve complex mathematical systems [19]. Two important subclasses are evolutionary algorithms and swarm intelligence [20]. The former is inspired by the Darwin theory of evolution; the latter is based on emergent properties of the collective behavior of a large number of small, simple agents. Among evolutionary algorithms, the genetic algorithm (GA) is one of the more well-studied ones that has been implemented in a wide variety of applications [21]. Among the swarm intelligence schemes, we choose particle swarm optimization (PSO), which is also a well-known algorithm that has proven itself in a large range of scientific problems [22].

2.1. Physical Problem

We assume the dipole approximation of a ferromagnetic object moving on a straight line at constant velocity [23]. This motion deforms the ambient magnetic field recorded by the sensor. The signal shape depends on the trajectory parameters, the magnetic moment of the moving object, and the position of the sensor. The components of the field, B_x, B_y, and B_z, are given by Equation (1):

$$\begin{pmatrix} B_x \\ B_y \\ B_z \end{pmatrix} = \frac{\mu_0}{4\pi R^5} \begin{vmatrix} 3x^2 - R^2 & 3xy & 3xz \\ 3yx & 3y^2 - R^2 & 3yz \\ 3zx & 3zy & 3z^2 - R^2 \end{vmatrix} \begin{pmatrix} M_x \\ M_y \\ M_z \end{pmatrix},$$

where $x = x_{sensor} - x_{object}$ (likewise for y and z); R is the distance between the sensor and the object; M_x, M_y, and M_z are the object moment components; and μ_0 is the vacuum magnetic permeability.

In our situation, the geometric parameters of the trajectory are known, including velocity and direction. The sensors' positions are the main unknowns. Since the moving object is not a controllable magnet or beacon, we cannot assume given values for the moment vector that we must consider as additional variables in the optimization problem. The latter then includes six parameters: three for the sensor position and three for the moment. Since the trajectory is known, there is no other coupling term between the fields measured by the sensors. Hence, each sensor is treated separately.

The algorithm looks for the set of variables that minimizes an error functional expressing the difference between the calculated and measured fields. The proximity of the graphs of two 2D functions is estimated by considering how well the graphs superpose, e.g., evaluate, the cross correlation of the functions. This is achieved using the normalized dot product of the two functions. If P and Q are the curves we compare, the cross-correlation fitness, f_{cc}, can be written using Equation (2):

$$f_{cc} = \frac{P \cdot Q}{max[P \cdot P, Q \cdot Q]},$$

where

$$P \cdot Q = \sum_{i=1}^{length(P)} P_i \cdot Q_i.$$

For each sensor, when the errors coming from the three axis curves are summed into one single expression, we have a single-objective scheme. When each sensor axis error is treated separately, we refer to it as multi-objective optimization. The error to minimize is a 3D function of the three axis errors. Although more computationally complex, it has the advantage of benefitting more from the spatial information provided by the vector magnetometer. For both GA and PSO, we have tested both single- and multi-objective versions.

In its most general form, a multi-objective optimization (MO) problem consists of finding, in a set of admissible solutions, a subset of solutions minimizing (or maximizing) its objectives. The main issue in MO is the definition of order (sorting) of two vectors in a space that has the dimension of the number of objectives to optimize. Such a relation is called dominance, and it serves to define a Pareto front. Consider a vector, **f(u)**, of the decision vector, *u*, having n variables or objectives. The MO problem can be expressed as minimizing the value of **f** for every variable:

$$\min f(u) = (f_1(u), f_2(u), \ldots, f_n(u))), \quad u \in \Omega$$

Then, **f** can be interpreted as a mapping of the decision space to the objective space. Next, the dominance relation is defined in the decision space (for a minimization problem). Given two vectors, *u* and *v*, *u* dominates v if and only if $f(u)$ is not larger than $f(v)$ for any objective, and it is less for at least one objective. *u* and *v* are equivalent if and only if $f(u)$ and $f(v)$ are the same for all objectives; *u* and *v* cannot be compared if and only if neither dominates the other nor are they not equivalent. The dominance relation is noted $\prec$ and is formally defined by

$$f(u) \prec f(v) \ if$$

$$\forall i \in \{ 1, 2, \ldots, n \}, \ f_i(u) \leq f_i(v), \ \wedge \exists i, \ f_i(u) < f_i(v) . \tag{5}$$

Given the dominance relation, the Pareto optimal set, ρ^*, is defined as the set of all Pareto optimal vectors, where a vector is called optimal if and only if it is not dominated by any other vector of the decision space. Formally speaking:

$$\rho^* = \{ u \in \Omega \ \neg \exists v \in \Omega, \ f(u) \prec f(v) \} . \tag{6}$$

The image set of ρ^* in the objective space is called a Pareto front:

$$\rho f^* = \{ f(u) \mid u \in \rho^* \} . \tag{7}$$

Solving the MO problem means finding solutions that are closed to the Pareto front and are uniformly distributed. Figure 3a–c show the Pareto front formation and evolution for one of our Multi Objective Genetic Algorithms (MOGA) runs.

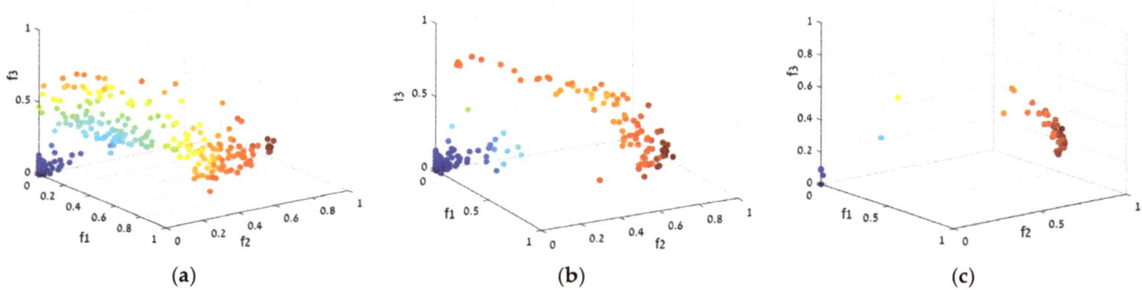

(a) (b) (c)

Figure 3. Pareto front dynamics in multi-objective optimization of the genetic algorithm. (**a–c**) refer to 5, 15, and 50 generations of the GA, respectively.

2.2. Genetic Algorithm

Genetic algorithms encode potential solutions on a simple chromosome-like structure and apply recombination operators to preserve critical information. They apply the principal concept of "survival of the fittest" on genes achieving best results. An implementation of a genetic algorithm begins with a population of random chromosomes. Each chromosome is a possible solution, which carries six genes, one for each variable of the magnetic equation: three for the sensor position and three for the moment. These structures are evaluated and reproductive opportunities are allocated in such a way that those chromosomes that represent a better solution are given more chances to reproduce than others. Mutation operators are applied to guarantee good exploration of the solution space.

For multi-objective GA, we used a modified version of the fast, elitist Non Selective Genetic Algorithm (NSGAII) algorithm [24]. This algorithm is based on a classification of individuals into several levels. Since, as we shall see later, the NSGAII produces the best solution, we provide a more detailed description of the algorithm here.

The NSGA algorithm does not fundamentally differ from classical genetic algorithms. No choice is imposed on the genetics operators of mutation, crossing, selection, or insertion.

The only difference is at the level of the implementation of the assignment operator of the adaptation value of individuals. We have upgraded the classical GA operators by using an adaptive and dynamic mutation mechanism. Moreover, the crossover was modified by adding a simulated annealing moderating process. Finally, we update the size the population according to the convergence of the best fitness values. These three additions greatly increase the convergence rate of the algorithm without deteriorating the quality of the final results.

This assignment operator is based on the rank of the Pareto set to which the solution belongs. In this procedure, all the solutions of the first set are given with the same adaptation value. For all other sets, the adaptation value is equal to the smallest value of the edge solution that precedes it, minus ε, which is a small positive number. This mechanism prevents the situation in which two solutions belonging to two different sets have the same adaptation value. The main disadvantage of this algorithm is the lack of elitism. The NSGA-II algorithm fills this gap.

The presence of elites increases the chances of creating better children, leading to a much faster convergence. In the case of single-criterion optimization problems, the elites are easily identified. They have the best value with respect to the objective function. For multi-criteria optimization problems, this statement is, of course, no longer valid.

In order to manage elites, the NSGA-II algorithm uses the rank of the Pareto set to which the solution belongs. The elites are identified as the solutions that are part of the first Pareto set. To manage diversity at the level of Pareto sets, the algorithm uses a particular implementation of the selection operator called "crowded tournament". In the crowded distance assignment procedure, the density of the solutions surrounding a given solution are evaluated. This value has the effect of reducing the chances of survival of a solution in a region where several other solutions are concentrated.

In our scheme, the density is measured in the objective space and not in the decision space. The method directly includes a normalization, which is essential when calculating distances.

To summarize, the NSGA-II algorithm uses two populations: a parent population P of size N and an offspring population Q, which consists of the set of individuals that have been created by the application of the GA operators. At each iteration, the two populations are combined in an intermediate population, R, and sorted to obtain the Pareto sets. Finally, a new population P′ consists of the best Pareto sets of population R. To achieve this, the solutions of the Pareto sets are included in the population, until the size becomes greater than or equal to the size, N, of the initial population. If, following the addition of the last possible set of Pareto, the population size is greater than N, then the last set is sorted according to the crowded distance, and the solutions with the smallest distances are eliminated until size N is achieved. The selection operator, crossing, and mutation are then applied to P′ to create the new population, Q′, and so on. A basic flowchart of NSGA-II is shown in Figure 4.

GA has already been used as an optimization technique for underwater sensor positioning [25]. However, in that study, GA was employed to calculate the best deployment of the sensor nodes not the exact position of these nodes *after* deployment, as we show in our work. In another important study, already cited above [10], GA was used to compute the position of the sensor, given the known trajectory of a ferromagnetic object like in our study. There are, however, two important differences. First, Yu's paper reports a laboratory reduced-scale setup, which is more similar to a computer simulation than to a real-world full-scale experiment. Second, the authors use, as the source, a solenoid coil with a given, controllable magnetic moment. This ideal setup reduces, by a factor of 2, the number of unknown parameters (genes in GA terminology) and simplifies the search procedure. We release this constraint by allowing any ferromagnetic object that is big enough to be used, leaving the algorithm to compute, by itself, the source of the magnetic field. Later, we show that this choice does not affect the ability of our GA to successfully determine the exact position of the sensors.

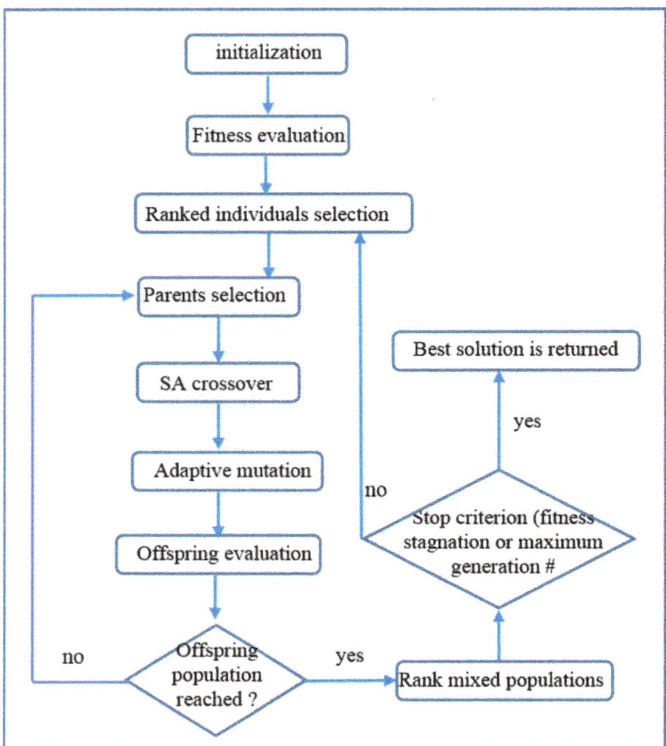

Figure 4. Basic flowchart of the NSGA-II algorithm.

2.3. Particle Swarm Optimization

In the PSO paradigm [22], the chromosomes of the GA are replaced by particles moving in the space spanned by the 6 directions defined by the problem's unknowns (the genes in the GA). From a given position in the space, the new particle direction and amplitude is computed as a weighted combination of its current inertia, its own best past position, and the best past position of the swarm. This is usually understood as a compromise between pure individual and social behaviors.

A particle, i, is defined by its position vector, x_i, and its velocity vector, v_i. In every iteration, each particle changes its position according to the new velocity, see Equations (8) and (9):

$$v_i^{t+1} = \omega v_i^t + c_1 r_1 \left(xbest_i^t - x_i^t \right) + c_2 r_2 \left(gbest_i^t - x_i^t \right) , \tag{8}$$

$$x_i^{t+1} = x_i^t + v_i^{t+1} \cdot t \tag{9}$$

where x_{best} and g_{best} denote the best particle position and best group position, respectively; and the parameters ω, c_1, c_2, r_1, and r_2 are inertia weight, two positive constants, and two random parameters within $[0, 1]$, respectively. In the baseline particle swarm optimization algorithm, ω is selected as unit, but an improvement of the algorithm is found in its inertial implementation using $\omega \in [0.5\ 0.9]$. Usually, maximum and minimum velocity values are also defined, and initially the particles are distributed randomly to encourage the search in all possible locations.

A position is "better" defined according to its fitness value. The fitness function is the same functional error that is defined above. The multi-objective version is inspired by [26].

3. Results and Discussion

Among the several events (the motion of the target above the grid) performed during the experiments, we selected six of them for testing the algorithms. Events #1, #3, and #6 are the reverse directions of events #2, #4, and #5, respectively. The velocities varied between 0.5 to 1.5 m/s. During event #6, a distractor was moved within the laboratory, creating a degradation of the signal. We see the influence of it in the results. All four algorithms were run: single- and multi-objective GA (NSGA-II) and single- and multi-objective PSO.

Some comments regarding the raw data are required before the experiment results and analysis.

The distance between the sensors and the moving object is at least 6 m. The object dimensions are not longer than 0.5 m in any dimension. In these conditions, we can perfectly assume a dipole approximation for the magnetic field model. In this condition, we know that the magnetic field should decay with a power of 3 of the distance between the sensor and the source.

The total moment is around 10 Am^2, as shown later. The magnetic field that the sensor records at a distance of 6 m should be roughly 7 nT [23]. We employ three-axial fluxgate sensors with an internal noise of about 0.25 nT rms (Mag634 by Bartington Instruments Ltd., Oxon OX28 4GG, UK), which is an order of magnitude less than the anticipated signal strength. We calculated an Signal to Noise Ratio (SNR) value of approximately 20, for most events. This is more than sufficient to quantify the magnetic anomaly produced by the moving object.

The sensors' differences can be seen in two ways: first, in a slight but nevertheless well-discernible difference in the field magnitude; second, in the changes in the phase as a function of time (we use vector magnetometers). These variations enable the localization of each sensor on its own, and our SNR enables a high absolute value precision.

Figure 5 displays the sensor-measured raw data. Keep in mind that the maximum amplitude values for the near and distant sensors are different: [7.16 6.91 6.71 6.68] nT for sensors #3, #5, #6, and #4, respectively, and [5.42 5.56 6.44 6.63] nT for sensors #1, #2, #7, and #8, respectively. With just an amplitude variation, the sensors on the same side of the trajectory display identical phases throughout each axis. For the sensors located on a separate side of the trajectory, the x phase is the opposite.

Both PSO and GA require a bounded search space. For each sensor, a random point was chosen inside a 5 × 5 m box centered on the point O, as shown in Figure 2. This box has a width of 2 m and is centered on the floor level. The searching space is then defined by a square of 10 m side around this initial point. The algorithm has almost no idea where each sensor is localized when it starts searching.

Both the GA and PSO parameters were optimized using event #1 as the test case (an arbitrary choice). The NSGAII algorithm was slightly modified during the ranking process, and one population from the two generated from the optimal Pareto front was selected. The algorithms were first run 10 times for each of the six events. From these runs, we chose the best result according to the best fitness value. This procedure was repeated 10 times in order to obtain some statistics. We then calculated the weighted error, relative to the confidence of each run, and computed the median error for each coordinate.

Figure 6 shows the median error of all the sensors (deviation from their exact position), and Table 1 looks at each sensor averaged over all events. Taken together, Figure 6 and Table 1 compare the four algorithms.

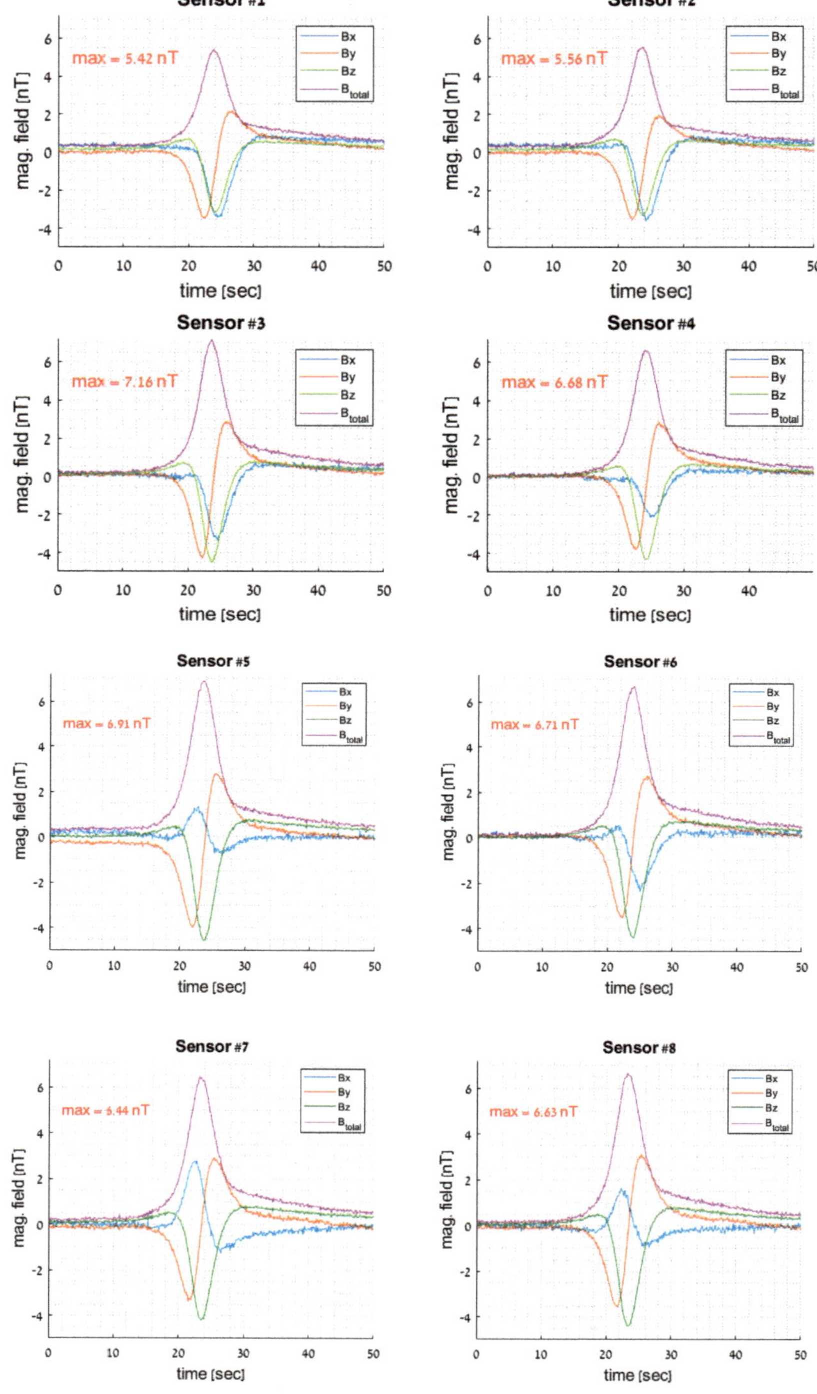

Figure 5. Typical raw data recorded during one of the events.

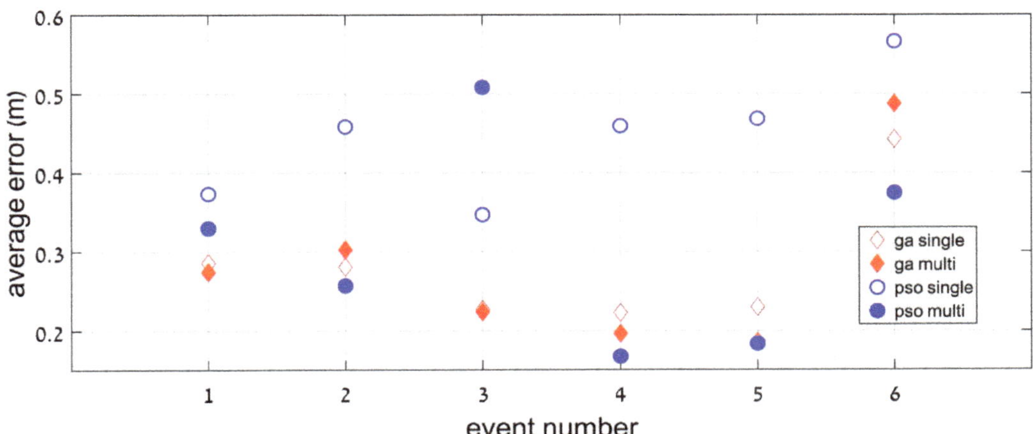

Figure 6. Deviation from exact location (in m) for each event (averaged over the eight sensors).

Table 1. Deviation from exact location (in m) for each sensor (averaged over the six events).

	GA		PSO	
	Single	**Multi**	**Single**	**Multi**
Sensor #1	0.26	0.28	0.44	0.25
Sensor #2	0.32	0.29	0.57	0.40
Sensor #3	0.29	0.26	0.42	0.32
Sensor #4	0.28	0.31	0.44	0.22
Sensor #5	0.22	0.21	0.44	0.33
Sensor #6	0.25	0.22	0.42	0.17
Sensor #7	0.23	0.22	0.47	0.25
Sensor #8	0.24	0.21	0.43	0.24

An overall look at Table 1 and Figure 5 shows that the results are excellent. This is particularly encouraging regarding the very large size of the search space. Figure 5 shows that the algorithms were insensitive to the direction, velocity, and moment of the crossing event. Event #6 shows a larger error due to the distractor presence, although its best median location error was only 40 cm. This means that even in more realistic conditions, our method can provide a precise and reliable location of the grid. Overall, the multi-objective optimization is the most appropriate method, either for the GA or PSO algorithm. A remarkably small averaged error (less than 30 cm) was found in both cases.

Figure 7 shows a more detailed view of the results, where each sensor's performance is shown for each event separately. Figure 7a shows the overall best results (obtained with MOGA), while Figure 7b shows the overall worst results (obtained with Single Objective Particle Swarm Optimization (SOPSO)).

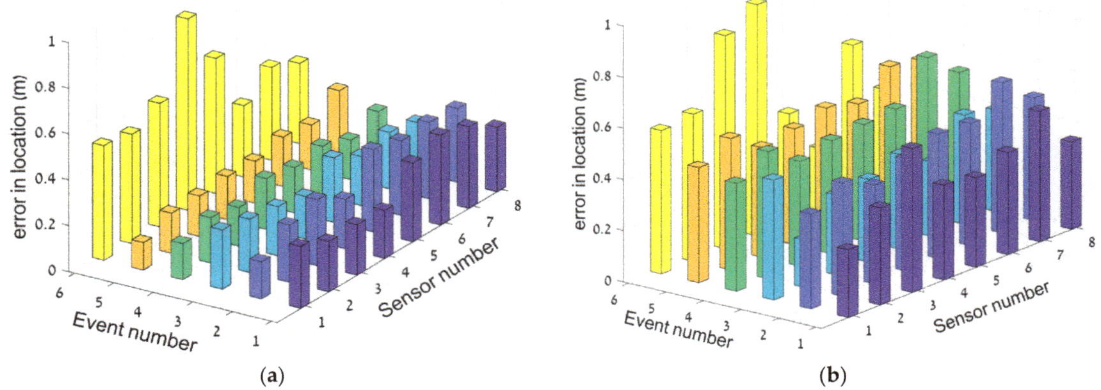

Figure 7. Deviation from exact location (in m) for each event and for each sensor. (**a**) and (**b**) are the best (MOGA) and the worst (SOPSO) results, respectively.

The MOGA algorithm performs slightly better that the Multiple Objective Particle Swarm Optimization (MOPSO) algorithm, although the best local solution was obtained using MOFSO; this is encouraging, since it has a smaller number of free parameters. Still, GA looks more stable, probably because of its ability through the mutation operator to exit local minima and find a better optimum. Adding this kind of operator to MOPSO should improve the performance toward errors of less than 20 cm.

As already mentioned, our methods do not suppose a known moment but rather calculate it as a part of the global solution of the inverse problem. The results are shown in Table 2. The values are consistent from sensor to sensor and from event to event, which is a fact that gives us confidence in our calculation schemes. The results from events #1, #3, and #6 show consistently higher values than those from events #2, #4, and #5. This can be expected, since the same moment can have a different projection on the Earth's magnetic field according to its direction of motion.

Table 2. Calculated total moments (A m^2) using the MOGA method.

		Sensor #							
		1	**2**	**3**	**4**	**5**	**6**	**7**	**8**
	1	9.05	9.35	8.42	8.69	9.55	9.61	9.05	**9.04**
	2	8.64	7.9	8.22	8.18	8.02	8.15	8.25	**7.73**
Event #	**3**	9.23	9.02	9.18	8.59	8.17	8.52	8.68	**8.89**
	4	7.89	8.17	8.25	8.21	7.75	7.87	7.96	**7.88**
	5	8.4	8.08	7.86	8.01	7.96	8.18	8.12	**7.82**
	6	9.21	8.84	9.25	8.76	8.48	9.01	9.16	**9.4**

4. Conclusions

A full-scale experiment was performed that mimics the underwater positioning of magnetic sensors. The position of eight magnetometers was successfully calculated, where the signals generated by a known trajectory of a ferromagnetic object were solved by two kinds of natural computing algorithms. Single- and multi-objective versions of a GA and a PSO were able to compute the position of the sensors, with an error of 25 cm around their exact locations. This is a remarkable result regarding the realistic scale of the setup and the almost blind initial guess locations.

Multi-objective optimization seems to be the most appropriate heuristic technique, despite its higher computation complexity. This is probably because, when looking for the three coordinates of the sensors, the 3D information available in the signals must be taken advantage of in the best way possible. MOGA generally performs better than MOPSO, although the latter sometimes shows better local results.

Finally, we believe our methods can offer an answer to several open challenges listed in [2], such as reliability, node mobility, and efficiency. Future work includes testing the system in a true maritime environment and using a shallow water deployment of the magnetometers. Several configurations can be considered, from relatively small-scale applications involving securing harbor facilities to larger-scale deployments for monitoring ship motion.

Author Contributions: Conceptualization, R.A. and E.F.; methodology, R.A. and E.F.; software, R.A. and K.N.; validation, K.N.; writing—R.A.; writing—review and editing, R.A. and E.F.; supervision, R.A. All authors have read and agreed to the published version of the manuscript.

Funding: This research received no external funding.

Institutional Review Board Statement: Not applicable.

Informed Consent Statement: Not applicable.

Data Availability Statement: The data presented in this study are available on request from the corresponding author. The data are not publicly available due to legal restrictions.

Acknowledgments: The helpful discussions with Eyal Weiss are gratefully acknowledged.

Conflicts of Interest: The authors declare no conflict of interest.

References

1. Felemban, E.; Shaikh, F.K.; Qureshi, U.; Sheikh, A.; Qaisar, S.B. Underwater Sensor Network Applications: A Comprehensive Survey. *Int. J. Distrib. Sens. Netw.* **2015**, *11*. [CrossRef]
2. Su, X.; Ullah, I.; Liu, X.; Choi, D. A Review of Underwater Localization Techniques, Algorithms, and Challenges. *J. Sens.* **2020**, *2020*, 6403161. [CrossRef]
3. Wang, H.; Wang, S.; Bu, R.; Zhang, E. A Novel Cross-Layer Routing Protocol Based on Network Coding for Underwater Sensor Networks. *Sensors* **2017**, *17*, 1821. [CrossRef] [PubMed]
4. Cayirci, E.; Tezcan, H.; Dogan, Y.; Coskun, V. Wireless sensor networks for underwater surveillance systems. *Ad Hoc Netw.* **2006**, *4*, 431–446. [CrossRef]
5. Callmer, J.; Skoglund, M.; Gustafsson, F. Silent Localization of Underwater Sensors Using Magnetometers. *EURASIP J. Adv. Signal Process.* **2010**, *2010*, 709318. [CrossRef]
6. Erol, M.; Vieira, L.F.M.; Caruso, A.; Paparella, F.; Gerla, M.; Oktug, S. Multi stage underwater sensor localization using mobile beacons. In Proceedings of the 2nd International Conference on Sensor Technologies and Application, SENSORCOMM 2008, Cap Esterel, France, 25–31 August 2008; pp. 710–714.
7. Cheng, X.; Shu, H.; Liang, Q.; Du, D.H.-C. Silent positioning in underwater acoustic sensor networks. *IEEE Trans. Veh. Technol.* **2008**, *57*, 1756–1766. [CrossRef]
8. Wong, S.Y.; Lim, J.G.; Rao, S.; Seah, W. Multihop localization with density and path length awareness in non-uniform wireless sensor networks. In Proceedings of the 2005 IEEE 61st Vehicular Technology Conference, Stockholm, Sweden, 30 May–1 June 2005.
9. Chandrasekhar, V.; Seah, W. An area localization scheme for underwater sensor networks. In Proceedings of the OCEANS 2006-Asia Pacific, Singapore, 16–19 May 2006.
10. Yu, Z.; Xiao, C.; Zhou, G. Multi-Objectivization-Based Localization of Underwater Sensors Using Magnetometers. *IEEE Sens. J.* **2014**, *14*, 1099–1106. [CrossRef]
11. Bian, S.; Hevesi, P.; Christensen, L.; Lukowicz, P. Lukowicz. Induced Magnetic Field-Based Indoor Positioning System for Underwater Environments. *Sensors* **2021**, *21*, 2218. [CrossRef] [PubMed]
12. Zhou, G.; Wang, Y.; Wu, K.; Wang, H. Localization Approach for Underwater Sensors in the Magnetic Silencing Facility Based on Magnetic Field Gradients. *Sensors* **2022**, *22*, 6017. [CrossRef] [PubMed]
13. John, F.; Schmidt, S.O.; Hellbruck, H. Multisensor Platform for Underwater Object Detection and Localization Applications. In Proceedings of the OCEANS 2022, Hampton Roads, Hampton Roads, VA, USA, 17–20 October 2022.
14. Huang, W.; Pan, Z.; Xu, Z. Underwater cable localization method based on beetle swarm optimization algorithm. *IEEE/CAA J. Autom. Sin.* **2022**, 1–3. [CrossRef]

15. Zhang, J.-W.; Yu, P.; Jiang, R.-X.; Xie, T.-T. Real-time localization for underwater equipment using an extremely low frequency electric field. *Def. Technol.* 2022; in press. [CrossRef]
16. Philippeaux, H.; Dhanak, M. Real-Time Localization of a Magnetic Anomaly: A Study of the Effectiveness of a Genetic Algorithm for Implementation on an Autonomous Underwater Vehicle. In Proceedings of the OCEANS 2018 MTS/IEEE Charleston, Charleston, SC, USA, 22–25 October 2018.
17. Hu, S.; Tang, J.; Ren, Z.; Chen, C.; Zhou, C.; Xiao, X.; Zhao, T. Multiple Underwater Objects Localization With Magnetic Gradiometry. *IEEE Geosci. Remote Sens. Lett.* **2019**, *16*, 296–300. [CrossRef]
18. Hou, G.; Shao, Q.; Zou, B.; Dai, L.; Zhang, Z.; Mu, Z.; Zhang, Y.; Zhai, J. A Novel Underwater Simultaneous Localization and Mapping Online Algorithm Based on Neural Network. *ISPRS Int. J. Geo-Inf.* **2019**, *9*, 5. [CrossRef]
19. Brabazon, A.; O'Neill, M.; McGarraghy, S. *Natural Computing Algorithms*; Springer Publishing Company, Inc.: New York, NY, USA, 2015.
20. Bansal, J.; Singh, P.; Pal, N. *Evolutionary and Swarm Intelligence Algorithms*; Springer Nature: Cham, Switzerland, 2018.
21. Affenzeller, M.; Winkler, S.; Wagner, S.; Beham, A. *Genetic Algorithms and Genetic Programming: Modern Concepts and Practical Applications*; Chapman & Hall/CRC: New York, NY, USA, 2009.
22. Clerc, M. From Theory to Practice in Particle Swarm Optimization. In *Handbook of Swarm Intelligence: Adaptation, Learning, and Optimization*; Panigrahi, B.K., Shi, Y., Lim, M.H., Eds.; Springer: Berlin/Heidelberg, Germany, 2011; Volume 8.
23. Weiss, E.; Alimi, R. *Low-Power and High-Sensitivity Magnetic Sensors and Systems*; Artech House: Boston, MA, USA; London, UK, 2018.
24. Deb, K.; Pratap, A.; Agarwal, S.; Meyarivan, T. A fast and elitist multi-objective genetic algorithm: NSGA-II. *IEEE Trans. Evol. Comput.* **2002**, *6*, 182–197. [CrossRef]
25. Iyer, S.; Rao, D.V. Genetic algorithm based optimization technique for underwater sensor network positioning and deployment. In Proceedings of the IEEE Underwater Technology (UT), Chennai, India, 23–25 February 2015; pp. 1–6.
26. Coello, C.A.C.; Toscano-Pulido, G.T.; Lechuga, M.S. Handling multiple objectives with particle swarm optimization. *IEEE Trans. Evol. Comput.* **2004**, *8*, 256–279. [CrossRef]

Article

The Tacotron-Based Signal Synthesis Method for Active Sonar

Yunsu Kim [1], Juho Kim [2], Jungpyo Hong [1,*] and Jongwon Seok [1,*]

1   Department of Information and Communication Engineering, Changwon National University, Changwon 51140, Republic of Korea
2   Sonar System PMO, Agency for Defense Development, Changwon 51618, Republic of Korea
*   Correspondence: hansin@changwon.ac.kr (J.H.); jwseok@changwon.ac.kr (J.S.)

Abstract: The importance of active sonar is increasing due to the quieting of submarines and the increase in maritime traffic. However, the multipath propagation of sound waves and the low signal-to-noise ratio due to multiple clutter make it difficult to detect, track, and identify underwater targets using active sonar. To solve this problem, machine learning and deep learning techniques that have recently been in the spotlight are being applied, but these techniques require a large amount of data. In order to supplement insufficient active sonar data, methods based on mathematical modeling are primarily utilized. However, mathematical modeling-based methods have limitations in accurately simulating complicated underwater phenomena. Therefore, an artificial intelligence-based sonar signal synthesis technique is proposed in this paper. The proposed method modified the major modules of the Tacotron model, which is widely used in the field of speech synthesis, in order to apply the Tacotron model to the field of sonar signal synthesis. To prove the validity of the proposed method, spectrograms of synthesized sonar signals are analyzed and the mean opinion score was measured. Through the evaluation, we confirmed that the proposed method can synthesize active sonar data similar to the trained one.

Keywords: active sonar; deep learning; signal synthesis; Tacotron

Citation: Kim, Y.; Kim, J.; Hong, J.; Seok, J. The Tacotron-Based Signal Synthesis Method for Active Sonar. *Sensors* **2023**, *23*, 28. https://doi.org/10.3390/s23010028

Academic Editors: Haixin Sun and Xuebo Zhang

Received: 25 November 2022
Revised: 16 December 2022
Accepted: 17 December 2022
Published: 20 December 2022

1. Introduction

Sonar stands for sound navigation and ranging, and refers to equipment or methodology that identifies the existence, location, and characteristics of an underwater target object. As water is used as a medium in which propagation proceeds, detection is performed using sound waves [1]. Passive sonar is a receiver-only system that detects vibrations originating from objects, such as the vessel's engines and propellers themselves. Relatively, it is simple to design and inexpensive to build. However, it requires a vast amount of data to distinguish only the desired signal by receiving all signals from animals and other ships. On the other hand, active sonar detects echo signals which are radiated from the transmitter, reflected by targets, and returned to the receiver. Since the radiated signal has a preset frequency characteristic and matched filtering can be applied to improve the signal-to-noise ratio (SNR) with the knowledge of the transmitted signal, active sonar is promising for underwater target detection in spite of the reverberation [2,3].

Active sonar modeling refers to estimating a returned echo signal reflected by an underwater target. In general, an active sonar modeling system consists of a transmitter, a receiver, and a target, and the transmitter and the receiver are located in different places to perform radiation and reception [4]. Various studies have presented methods for the simulated generation of sonar data [5–8], one of which is a simulation module provided by the North Atlantic Treaty Organization (NATO) submarine research center. In the simulation module, the signal emitted by the transmitter is simulated with the target through statistical calculation [5]. It also produces a more realistic signal by providing a target fading effect between sensors as seen in real-world sea environment datasets. However, in simplifying the sonar equation, the modeled signal inevitably differs from the data collected in the real ocean. In [6], La Cour et al. developed a multi-everything sonar simulation (MESS), reflecting the reverberation and simplified sea environment. However, the MESS also failed to closely realize the real ocean data because simplified sea environment parameters were added to the existing sonar equation. In addition, a simulator of non-acoustic and acoustics (SIMONA) simulators generate signals that reflect contact states and reverberations, as well as target shapes, multipath fading, and waveform types [7]. For the full simulation, beamforming and reverberation calculations, which are required to be input to the matching filter module, play a major role in the realistic data generation. Therefore, studies on generating reverberation in real time only for bidirectional active sonar have also been conducted [8]. However, these mathematical modeling-based methods are limited in accurately simulating by vast and complicated underwater environments.

Meanwhile, with the rapid development of deep neural networks (DNN), a lot of interest and research has been conducted on the technology of generating complicated time series signals of variable length from simple text information in the speech synthesis area [9–13]. Representatively, WaveNet [9] is a voice signal synthesis model that presented a remarkable performance in audio sample generation. However, there is a limitation in that it is only used as a kind of vocoder that uses a mel spectrogram, which contains linguistic features of the desired voice, not natural language text, as an input. In addition, DeepVoice [10] is a method that replaced conventional text-to-speech (TTS) pipelines with DNN. However, the method is limited because the learning process is not an end-to-end system. Subsequently, a model of the encoder–decoder structure is proposed to improve synthesis performance, and the importance is calculated using a pre-trained hidden Markov model (HMM) to predict the vocoder parameters [11]. Furthermore, a Char2Wav [12] model is designed to enable end-to-end learning, but additional preprocessing is still required in that it predicts the Vocoder parameter. Finally, Tacotron [13] is an end-to-end TTS model that succeeds in training a linear spectrum of speech data in natural language text at once. It consists of an encoder, a decoder, and attention modules, showing high generation performance enough to be used in commercial applications, and is widely used as a basic structure of the TTS models.

Therefore, in this paper, we propose a signal synthesis method for active sonar using the Tacotron model. To achieve our goal, we modified several main blocks of the Tacotron model to be operated for sonar signal synthesis. Starting with the introduction of the related works in Section 2, we explain the proposed method in detail in Section 3. Through experiments, we verify the effectiveness of the proposed method in Section 4 and conclude in Section 5.

2. Related Works

2.1. Active Sonar Target Signal Generation Based on the Highlight Model

Active sonar modeling means simulating a reflected signal against an underwater target. When a pulse signal is emitted to an underwater object in a steady state, various types of reflective signals are generated due to factors such as hull, medium, structural characteristics, frequency of incident waves, and pulse width. The echo signal of an active sonar using high frequency is produced by the reflection of the object's representation, along with several equivalent scattering inside, characterized by the spatial distribution of the object's highlights. Simulating a sonar signal is to consider everything that may occur in this process of reflection. Entering each point that hits the target into the reflection tracking algorithm has an infinite number of cases, thus the concept of highlights that simulate the target as a series of points is introduced [14].

At long range, an underwater target is represented by a single point generated from a single highlight. However, at short range, the distribution of highlights needs to be properly expressed because the target can have a distribution characteristic that varies with time and angle. Assuming that target is a submarine, the concept of a spheroid-placed highlight is used. The concept of a spheroid-placed highlight discontinuously recognizes the surface of the target that varies with the angle of incidence of the highlights attached at a specific position. Figure 1 shows the concept of the corresponding highlight:

$$p_b(r,\mathbf{x}) = \sum_{g=0}^{N} h_g(r_g,\mathbf{x}) * p_i(x) \tag{1}$$

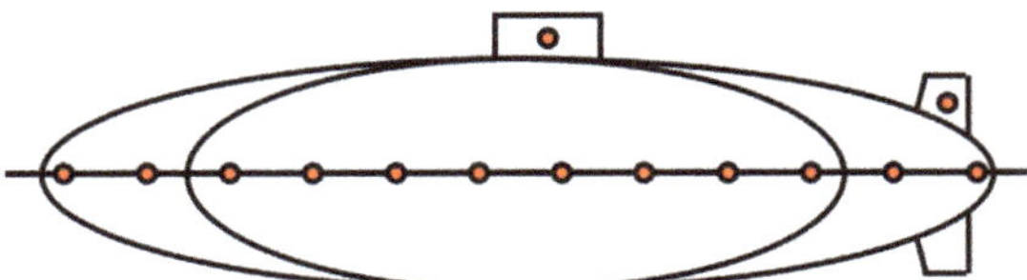

Figure 1. Spheroid-placed highlight modeling.

Given the time delay of each highlight, the signal p_b reflected on all the highlights of the target can be expressed as a sum, as shown in (1). The receiver is r, the target is x, and there are a total of N highlight points in a multi-highlight system, including short-range underwater targets. At this time, the object transfer function of each highlight is called h_g and the incident signal is p_i. This highlight modeling is simple but widely used due to high environmental approximation.

2.2. The Tacotron for TTS Modeling

At long range, an underwater target is represented by a single point generated from a single highlight. However, at short range, the distribution of highlights needs to be properly expressed because the target can have a distribution characteristic that varies with time and angle. Assuming that target is a submarine, the concept of a spheroid-placed highlight is used. The concept of a spheroid-placed highlight discontinuously recognizes the surface of the target that varies with the angle of incidence of the highlights attached at a specific position. Figure 2 shows the concept of the corresponding highlight.

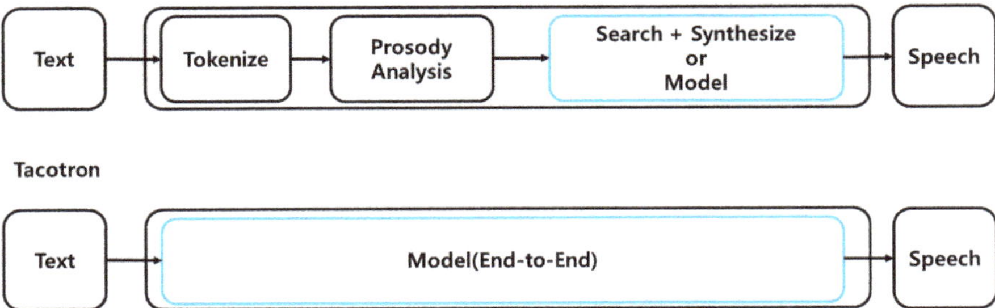

Figure 2. The Tacotron compared to conventional methods.

The encoder–decoder structure and attention mechanism are the core building blocks of the Tacotron. The <Text, Speech> pair consists of the input and output of the model, respectively. The input uses natural language raw text and, as an output, linear and mel spectrograms are generated, respectively. Finally, the spectrograms are reconstructed as a WAV audio file through post-processing. The encoder receives text data and outputs a kind of text embedding, a vector that best represents the meaning of the input text sequence. The embedded text vector is used as information for reference when the decoder sequentially generates audio samples.

In addition, attention techniques determine the importance of text embedding vectors used by decoders to generate audio sequences at each time step. In the recurrent neural network (RNN) based sequence-to-sequence (seq2seq) model, the vanishing gradient problem in which the information itself slowly disappears when it is located at the beginning of the sentence exists. However, the attention technique successfully alleviates the problem. With these advantages, the Tacotron became the cornerstone of the end-to-end TTS model.

3. Proposed Method

3.1. System Structure

Figure 3 presents the overall structure of the proposed signal synthesis method for active sonar. The entire system is largely divided into four stages: dataset configuration, preprocessing, signal synthesis, and post-processing. In this paper, the dataset configuration part used a highlight-based active sonar simulator for data generation because the amount of real ocean datasets is insufficient to train the proposed system model. However, this data generation part has to be replaced by real ocean data ultimately. The dataset generated in this way is converted into data to be an input of the model through the preprocessor, and the input is synthesized by the DNN model and outputs a corresponding linear spectrogram. By estimating the phases corresponding to the synthesized spectrograms using the Griffin–Rim algorithm, the synthesized waveform signal is reconstructed through post-processing.

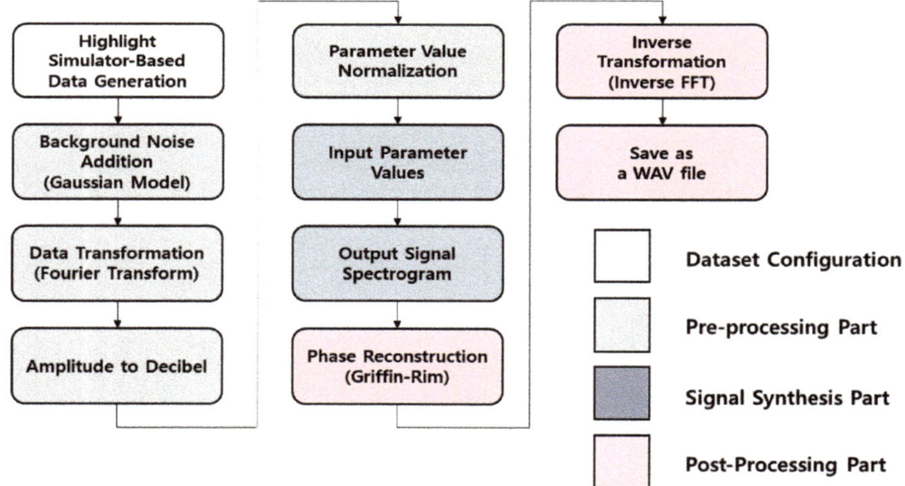

Figure 3. System structure.

Figure 4 compares the inputs and outputs of Tacotron models utilized in speech and sonar signal synthesis areas. The two Tacotron models are in common with yielding linear spectrograms corresponding to the provided inputs, whereas they are different in relevance with time-order dependency. In other words, the input of the TTS model for Korean synthesis combines 80 symbols in a time-ordered sequence, but the input of the proposed model for sonar signal synthesis consists of 14 parameters regardless of time order. Therefore, in order to achieve our goal of reflecting the difference to the Tacotron model, we modified several main blocks and the modifications will be explained in detail in the following section.

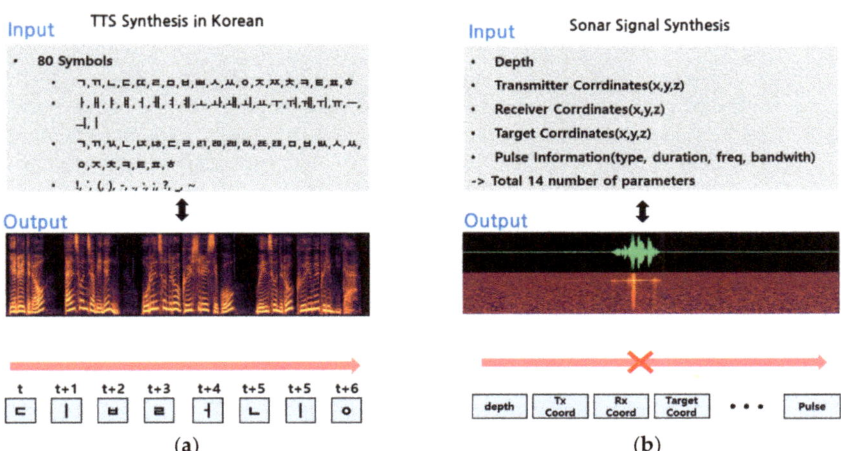

Figure 4. Input and output configuration. (**a**) The input/output structure of the TTS model. A total of 80 symbols are arranged in time order. (**b**) The input/output structure of the proposed sonar signal synthesis model. Fourteen marine environmental parameters are arranged regardless of time order.

3.2. The Tacotron-Based Sonar Signal Synthesis Model

Figure 5 shows the structure of the proposed Tacotron-based sonar signal synthesis model. As the input of the model, the parameter values used in the configuration of the dataset are normalized to real numbers in the range of [0, 1] in the order of depth, transmitter, receiver, target coordinates, and pulse information. After that, it goes through an encoder network to extract and convert from parameter information necessary for signal simulation to information necessary for synthesizing a linear spectrogram. The information vector output of the encoder is input into the decoder and goes through a process of synthesizing an active sonar echo signal corresponding to the input marine environment parameters. The active sonar echo signal is sequentially synthesized through multiple steps using decoder RNN in the form of a spectrogram. In the decoding step, the attention RNN refers to the necessary information from the parameters input to the model when synthesizing the frequency coefficient of the corresponding time step.

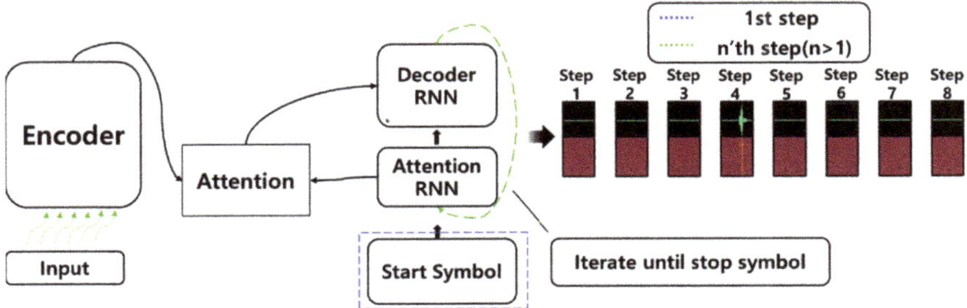

Figure 5. Model structure of the proposed sonar signal synthesis.

3.2.1. The Sonar Environment Parameter Embedding Layer

In the conventional TTS model, tokenization is performed in the process of converting natural language text into vectors. After converting the order in which the word appears within a preset word dictionary into a one-hot vector, the neural network can judge the meaning of the word within the sentence by itself through a text embedding layer. It is a more effective approach in that it estimates meaning specific to each task than conventional word embedding algorithms, such as count vectorization [15], bag-of-words [16], and term frequency–inverse document frequency (TF–IDF) [17,18]. Figure 6 shows the operation process of this text embedding layer. Although it appears to compute the context vector h_c of the word as the matrix product of the weights W_{in}^c and the one-hot vector c, the h_c can be easily obtained by selecting the corresponding row of the weight W_{in}^c.

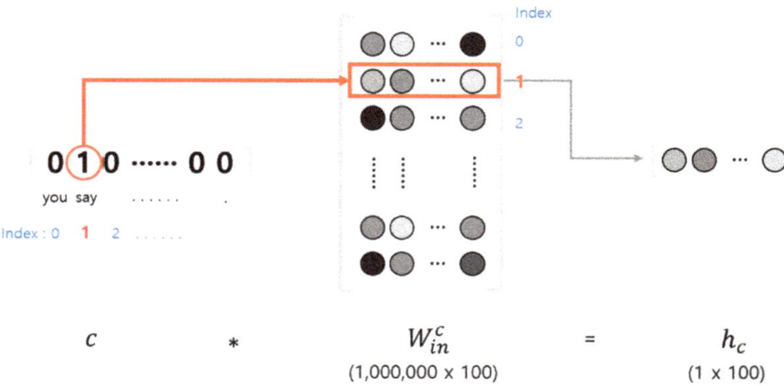

Figure 6. Word embedding layer.

In this paper, however, the input of DNN used to synthesize sonar signals represents a series of numerical vectors of environmental parameter values such as depth, pulse information, transmitter, receiver, and target coordinates. Unlike text embedding layers that estimate only the meaning of words that exist within a dictionary, sonar environment parameter embedding layers are continuous numbers and the number of cases can be infinite. In addition, due to the nature of the one-hot vector, meaningless zero values occupying only space are filled as elements, but sonar environment parameter vectors are denser and have unique meanings for each element. Therefore, it is necessary to design a weight vector so that the meaning can be inferred individually according to each parameter.

The operation process of the sonar environment parameter embedding layer is depicted in Figure 7. Unlike the sparse text embedding vector c, the sea environment parameter vector requires processing as a dense structure. The *i*th element s_i of the vector S is assigned a weight W and a bias b_i to output h_i^s, which transforms the meaning of the element into the information needed for signal synthesis purposes with a offset dimension *n_embed*. By performing an operation, such as a fully connected layer for each element of the input, it becomes possible to convert a single number into a nonlinear context vector. This allows the proposed model to synthesize more realistic active sonar data.

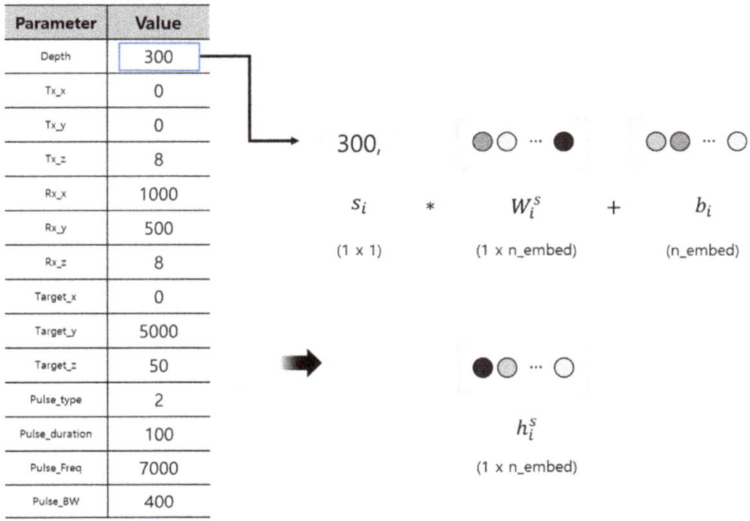

Figure 7. Sonar parameter embedding layer.

3.2.2. Attention Layer

The proposed model has an autoregressive structure that synthesizes variable-length signals in units of a specific number of frames and again uses them as input to the decoder cell to output frames of the next time step. The RNN-based seq2seq model [19] inputs an entire sequence, referencing one information vector output from the encoder equally across all steps, and iterating the process until an end-of-sequence (EOS) token appears. However, after the Transformer model [20] came out, an attention layer that acts as an intermediary between the encoder and the decoder was introduced. Although the seq2seq structure is used as it is in the decoder output of the model, the attention layer determines its important input features at that time step, helping to process and generate more flexible performance. In this paper, we also use a structure that introduces an attention mechanism to enable the extraction and processing of encoded information necessary to form a signal spectrum output at frame time t. A Bahdanau attention mechanism [21] was used as the method of calculating attention in the same manner as the Tacotron, and its configuration is as follows:

$$Q = S_{t-1}, \quad K = H, \quad V = H \tag{2}$$

$$e^t = W_a tanh\left(W_q Q + W_k K\right) \tag{3}$$

$$a^t = softmax(e_t) \tag{4}$$

$$c^t = V a^t \tag{5}$$

$$S_t = D\left(concat\left(c^t, I_t\right)\right) \tag{6}$$

We performed the importance calculation process by repeating up to the last frame generation point T with a total of three vectors: $Q, K,$ and V, which mean queries, keys, and values. S_{t-1} refers to the decoder cell's hidden state at the point just before the point t and H refers to the encoder cell's hidden states at all points in time. Similarly, three types of weights: $W_a, W_q,$ and W_k correspond to attention values, queries, and keys and are calculated to obtain attention score values e^t. Furthermore, e_t becomes the attention value a^t via the softmax function, and computes a context vector c^t that utilizes only important information from the encoded information vector via a dot product operation with V. Finally, the calculated c^t is concatenated to the input I_t of the current decoder $D(x)$, resulting in S_t. In this way, determining how important information is in synthesizing signals plays a crucial role in improving synthesis performance.

3.2.3. Positional Decoding

The biggest difference between speech synthesis and the proposed sonar signal synthesis is in time information. The text sequence, which is the input of the speech Tacotron model, is representative time series data, and the list of each word in the sentence affects each other a lot in order, which also directly affects the output speech spectrogram. However, the sonar Tacotron model simultaneously receives a number of sea environment parameters as input. The corresponding values have a profound effect on the output of each element, but the arrangement order of the parameters does not affect the output. This temporal mismatch causes confusion as the decoder of the model does not correspond to the input in yielding the output sequence sequentially. Therefore, in this paper, we add a term to the input of the cell under decoding to indicate at what point in the entire file the frame corresponds to so that the decoder can track the context of the output point. The added temporal term is expressed in the form of a normalized floating point of [0, 1], and each time point t is expressed in the order of frames rather than information in seconds; thus, t divides the total number of frames in the generation file by T and uses it as location information. This alleviates the problem of perception confusion between inputs and outputs of models that do not correspond to each other in time, as described above.

3.2.4. Target Masked L1 Loss

The design of appropriate cost functions is essential for the optimization of DNN models. To design a cost function comparing the linear spectrogram of the model's output speech signal with the actual one, the conventional method used mean absolute error (MAE), as shown in (7):

$$L_{total} = \frac{1}{T}\frac{1}{N}\sum_{t=0}^{T}\sum_{i=0}^{N}\left|o_t^i - \hat{o}_t^i\right| \tag{7}$$

Time information, i.e., the total number of frames, is T, the ith frequency spectrum coefficient of the tth frame output by the model is $\hat{o}_t^i$, and the frequency spectrum coefficient of the reference signal is set to o_t^i. The L1 distance of o_t^i and $\hat{o}_t^i$ was averaged over the entire frame and coefficients as a loss value, which presented better performance than using mean squared error (MSE) [13].

However, as described above, the sonar signal synthesis model does not effectively pass time information in the decoding step. Using positional decoding to provide temporal information to decoders is only an auxiliary role and is not a fundamental solution. In addition, due to the nature of the sonar signal, the background noise or clutter occupies most of the time except for the target portion at a specific point in time, so it is necessary to design the cost function to focus more on reducing the difference from the original. To solve this problem, we propose a target-masked MAE. A frequency coefficient of the target signal is mainly larger than the magnitude of the background noise. We calculate a binary mask M that is 1 where the target is estimated, as shown in (8–11), and 0 where there is no target. We add $L_{maskedLinear}$ to the overall cost function L_{total}, which allows the energy to be compared only to the target locations through the element-specific product of the output value of the neural network model and the frequency spectrum of the original signal.

$$\mu = \frac{1}{N}\sum_{i=0}^{N} s_t^i \tag{8}$$

$$M = \begin{cases} 0, & s_t^i \leq 2\mu \\ 1, & s_t^i > 2\mu \end{cases} \tag{9}$$

$$L_{maskedLinear} = \frac{1}{T}\frac{1}{N}\sum_{t=0}^{T}\sum_{i=0}^{N}\left| M \odot s_t^i - M \odot \hat{s}_t^i \right| \tag{10}$$

$$L_{total} = L_{linear} + L_{maskedLinear} \tag{11}$$

4. Experiments

4.1. Dataset Configuration

As described in Section 1, the proposed method aims to synthesize more realistic echo signals but requires more than a certain amount of data due to the nature of the data-driven approach. Because it is difficult to collect large amounts of sonar data in practice, data generated by an active sonar simulator are used for training the proposed Tacotron model. Ultimately, this generated data should be replaced by real ocean data when the real data is sufficiently collected.

In order to generate highlight-based active sonar data introduced in Section 2, the active sonar simulator receives parameters, including the coordinates of the transmitter, target, and receiver, calculates the signal reflected on the target, and outputs it in the form of a waveform. The input parameters of the highlight-based active sonar signal generator are summarized in Table 1.

Table 1. Input parameters of the highlight-based active sonar signal generator.

Parameter	Description	Setting Range
depth	depth information (unit: m)	[0, 5000]
tx	transmitter coordinates (x,y,z) (unit: m)	[0, range]/[0, range]/[0, range]
rx	receiver coordinates (x,y,z) (unit: m)	[0, range]/[0, range]/[0, depth]
target	target coordinates (x,y,z) (unit: m)	[0, range]/[0, range]/[0, depth]
svp	sound velocity profile (unit: m/s)	$[0,\infty]/[0,\infty]$
highlight	highlight information (number of points, relative position)	$[0,\infty]/[0,\infty]$
range	calculation limit range (min, max, and step) (unit: m)	$[0,\infty]$
pulse	pulse information (type, duration, center frequency, and bandwidth) (unit: linear frequency modulation, and continuous wave/sec/Hz/Hz)	$[0,\infty]$

When the entire range and the step are set to 15,000 m and 10 m, respectively, the sound ray is tracked until the entire distance is reached by the interval by the set parameter. The tracking altitude is calculated by dividing the [−20,20] degree range set as the default value by 400, the number of indexes, as shown in Figure 8.

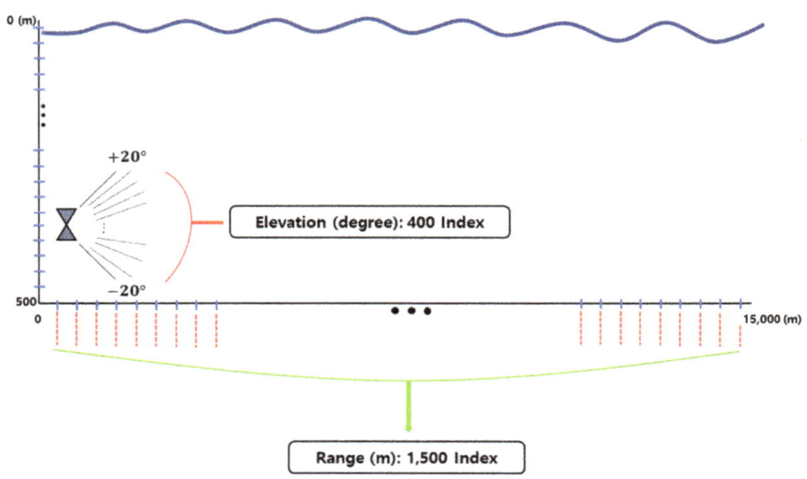

Figure 8. The ray tracing process.

In the experiment of this paper, a dataset was constructed by changing a total of three parameters that have a noticeable influence on the characteristics of sonar signals: depth, pulse length, and pulse center frequency. The sound velocity profile used for data generation is presented in Table 2. Highlight points were set to 10 and Gaussian noise is added to generated sonar signals corresponding to 10 dB SNR. The total number of cases considered in this experiment for training the model is summarized in Table 3.

Table 2. The sound velocity profile.

Depth (m)	Sound Speed (m/s)
0	1502
31	1504
60	1480
90	1477
120	1482
150	1481
180	1482
210	1476
240	1477
360	1479
⋮	⋮
5000	1480

Table 3. Dataset configuration.

Parameter	Step	Number of Files
depth	100–1000 m, 10 m per step	100
pulse duration	100–1000 m, 10 m per step	100
pulse center frequency	1000–7000 Hz, 60 Hz per step	100

4.2. Experimental Settings

Instead of loading a single long signal file and entering the entire file, it divides into frames and goes through all processes such as processing, input, and training. This section describes all parameters used in the experimental process. It is divided into two categories, audio processing and DNN training, and consists of parameter names, numerical values, and parameter descriptions.

(a)   Audio processing parameters.

- num_mels: 80, the number of mel filters to obtain the mel spectrogram.
- num_freq: 1025, the number of frequency coefficients.
- frame_length_ms: 50 ms, the length of the frame.
- frame_shift_ms: 12.5 ms, the length of the shift between frames.

(b)   Model training parameters.

- Parameter embedding dimension (same as encoder input dimension): 256.
- Encoder output dimension: 128.
- Attention type: the Bahdanau attention.
- Attention dimension: 256.
- Decoder input dimension: 256.
- Decoder output dimension (meaning the final output dimension of the model): num_freq and num_mels.

4.3. Experimental Results

To evaluate the synthesis performance of the proposed model, we examined three aspects: comparing spectrograms, checking attention alignment, and measuring mean opinion scores (MOS). The evaluation was conducted using an untrained test file, and 10 were separated for each parameter.

(a)   Spectrogram comparison.

Spectrograms of the original sonar data generated by an active sonar simulator and spectrograms of the synthesized signal according to changes in depth, pulse length, and pulse center frequency are presented in Figures 9–11. As can be seen in the figures, a target echo signal is successfully synthesized in each parameter condition. The time of the signal, which means the distance of the target and its shape are, similarly synthesized to the original signal. However, attenuation of the background noise level, which is synthesized by simple repetition, is observed.

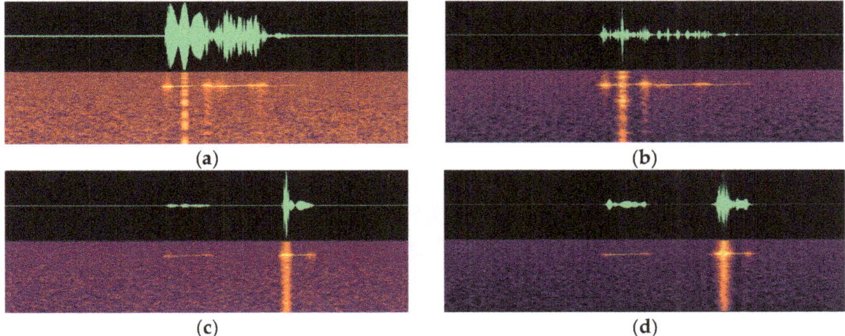

Figure 9. Spectrograms of active sonar signals with various depth parameters. (**a**) Original signal with a depth of 180 m. (**b**) A synthesized signal with a depth of 180 m. (**c**) An original signal with a depth of 660 m. (**d**) A synthesized signal with a depth of 660 m.

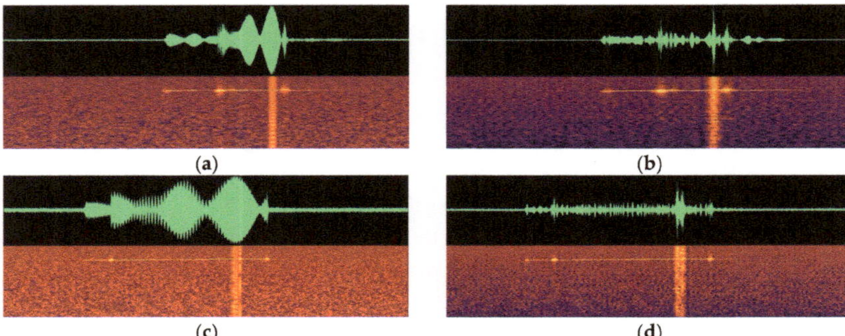

Figure 10. Spectrograms of active sonar signals with various pulse duration parameters. (**a**) An original signal with a pulse duration of 160 ms. (**b**) A synthesized signal with a pulse duration of 160 ms. (**c**) An original signal with a pulse duration of 770 ms. (**d**) A synthesized signal with a pulse duration of 770 ms.

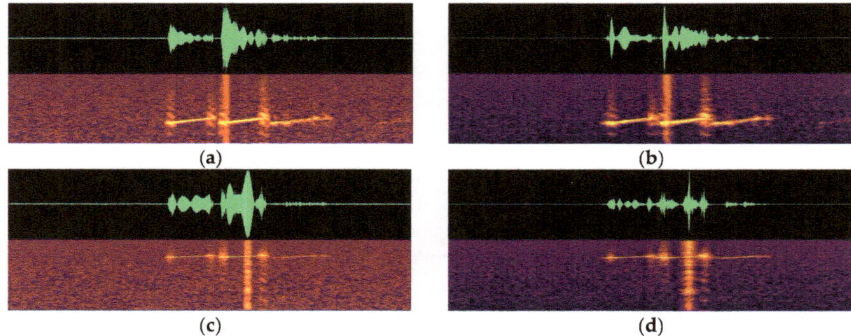

Figure 11. Spectrograms of active sonar signals with various pulse center frequency parameters. (**a**) An original signal with a pulse center frequency of 2080 Hz. (**b**) A synthesized signal with a pulse center frequency of 2080 Hz. (**c**) An original signal with a pulse center frequency of 5920 Hz. (**d**) A synthesized signal with a pulse center frequency of 5920 Hz.

(b)   Attention alignment.

In order to check the attention mechanism, we visualized the importance of parameters for synthesizing sonar signals. As shown in Figure 12, the high parameter importance resulted in the corresponding training cases. Thus, we confirmed that the attention mechanism for model training was normally operated.

(c)   The MOS score.

In order to measure the subjective quality between the generated and synthesized data, we conducted an MOS test [22]. A total of five persons participated in this experiment, and each type of data, i.e., generated and synthesized data, are evaluated. The average score of the participants for each sea environment parameter is shown in Table 4.

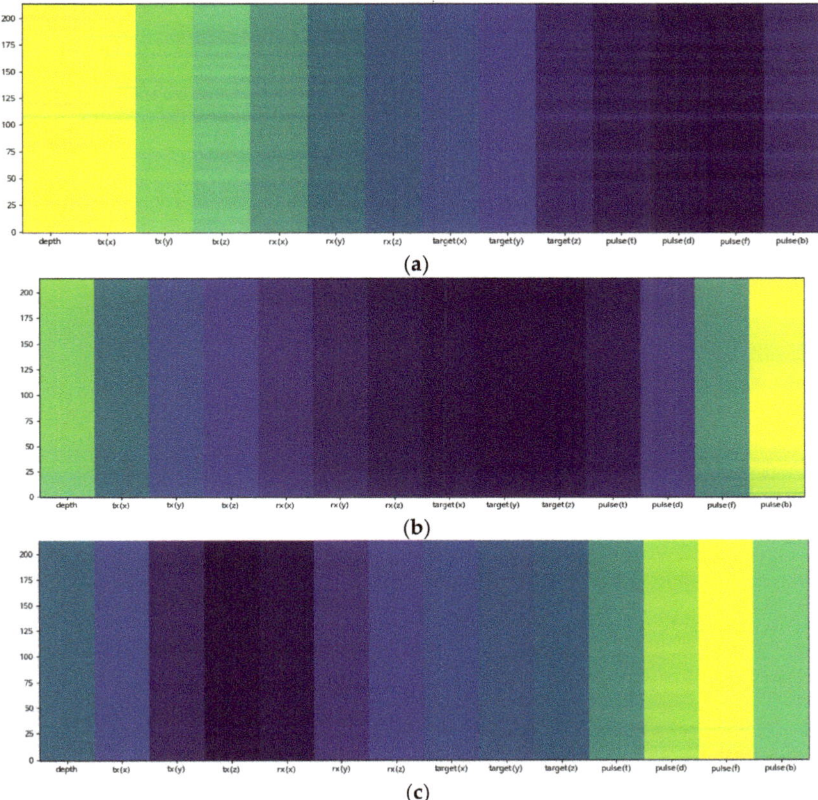

Figure 12. Attention alignments in the test. (**a**) Training only depth parameters. (**b**) Training only pulse center frequency parameters. (**c**) Training only pulse duration parameters.

Table 4. The MOS score.

	Depth	Pulse Duration	Pulse Center Frequency
Generated	4.03	4.10	4.51
Synthesized	3.68	3.78	4.32

As shown in Table 4, the MOS score of the signal synthesized by the proposed model is similar to the generated original sonar signal. From these results, it can be seen the sonar signal synthesized by the proposed model generates a signal similar to the trained signals.

5. Conclusions

In this paper, we proposed a Tacotron model based on DNN for active sonar signal synthesis. The proposed Tacotron-based sonar signal synthesis method is suitably modified for active sonar. It consists of three submodels: an encoder that turns the input vector into an information vector needed to simulate the environment, a decoder that sequentially generates output based on the received information vector, and an attention module that extracts and processes only the information needed at each point in time when decoding. To evaluate the proposed method, we performed spectrogram comparison, attention results checking, and MOS tests. Through the evaluation, we confirmed that the proposed Tacotron model successfully synthesized almost similar data used for training. Furthermore, the proposed Tacotron model can be improved using variable signal generation models, such as Tacotron2 [23], combined with the WaveNet Vocoder and Flowtron [24] from NVIDIA, but it remains to be tested in a future work.

Author Contributions: Writing—original draft, Y.K.; Writing—review & editing, Y.K.; Supervision, J.H. and J.S.; Project administration, J.K. All authors have read and agreed to the published version of the manuscript.

Funding: This work was supported by the Agency for Defense Development (ADD), Republic of Korea, under grant UD210005DD.

Institutional Review Board Statement: Not applicable.

Informed Consent Statement: Informed consent was obtained from all subjects involved in the study.

Data Availability Statement: Not applicable.

Conflicts of Interest: The authors declare no conflict of interest.

References

1. Knight, W.C.; Pridham, R.G.; Kay, S.M. Digital signal processing for sonar. *Proc. IEEE* **1981**, *69*, 1451–1506. [CrossRef]
2. Winder, A.A. Sonar System Technology. *ITSU IEEE Trans. Sonics Ultrason.* **1975**, *22*, 291–332. [CrossRef]
3. Haesang, Y.; Sung-Hoon, B.; Keunhwa, L.; Youngmin, C.; Kookhyun, K. Underwater Acoustic Research Trends with Machine Learning: Active SONAR Applications. *J. Ocean. Eng. Technol.* **2020**, *34*, 277–284. [CrossRef]
4. Coraluppi, S.; Grimmett, D.; de Theije, P. Benchmark Evaluation of Multistatic Sonar Trackers. In Proceedings of the 9th International Conference on Information Fusion, Florence, Italy, 10–13 July 2006.
5. Grimmett, D.; Coraluppi, S. Contact-Level Multistatic Sonar Data Simulator for Tracker Performance Assessment. In Proceedings of the 9th International Conference on Information Fusion, Florence, Italy, 10–13 July 2006.
6. La Cour, B.; Collins, C.; Landry, J. Multi-Everything Sonar Simulation (MESS). In Proceedings of the 9th International Conference on Information Fusion, Florence, Italy, 10–13 July 2006.
7. De Theije, P.A.M.; Groen, H. Multistatic Sonar Simulations with SIMONA. In Proceedings of the 9th International Conference on Information Fusion, Florence, Italy, 10–13 July 2006; pp. 1–6. [CrossRef]
8. LePage, K.D.; Harrison, C.H.; Strode, C.; Oddone, M. Real-time reverberation time series simulation for embedded simulation of bistatic active sonar. In Proceedings of the OCEANS 2021: San Diego—Porto, San Diego, CA, USA, 20–23 September 2021; pp. 1–4. [CrossRef]
9. Oord, A.; Dieleman, S.; Zen, H.; Simonyan, K.; Vinyals, O.; Graves, A.; Kalchbrenner, N.; Senior, A.; Kavukcuoglu, K. WaveNet: A generative model for raw audio. *arXiv* **2016**, arXiv:1609.03499.
10. Arik, S.O.; Chrzanowski, M.; Coates, A.; Diamos, G.; Gibiansky, A.; Kang, Y.; Li, X.; Miller, J.; Ng, A.; Raiman, J.; et al. Deep Voice: Real-time Neural Text-to-Speech. *arXiv* **2017**, arXiv:1702.07825.
11. Mehta, S.; Szekely, E.; Beskow, J.; Henter, G.E. Neural HMMs are all you need (for high-quality attention-free TTS). *arXiv* **2022**, arXiv:2108.13320.
12. Sotelo, J.; Mehri, S.; Kumar, K.; Santos, J.F.; Kastner, K.; Courville, A.; Bengio, Y. Char2Wav: End-to-end speech synthesis. In Proceedings of the ICLR 2017, Toulon, France, 24–26 April 2017.
13. Wang, Y.; Skerry-Ryan, R.J.; Stanton, D.; Wu, Y.; Weiss, R.J.; Jaitly, N.; Yang, Z.; Xiao, Y.; Chen, Z.; Bengio, S.; et al. Tacotron:Towards end-to-end speech synthesis. In Proceedings of the Interspeech 2017, Stockholm, Sweden, 20–24 August 2017; pp. 4006–4010.
14. Boo Il, K.; Hyeong Uk, L.; Myung Ho, P. A study on highlight distribution for underwater simulated target, ISIE 2001. In Proceedings of the 2001 IEEE International Symposium on Industrial Electronics Proceedings (Cat. No.01TH8570), Pusan, Republic of Korea, 12–16 June 2001; Volume 3, pp. 1988–1992. [CrossRef]
15. Kozhevnikov, V.A.; Pankratova, E.S. Research of the text data vectorization and classification algorithms of machine learning. *ISJ Theor. Appl. Sci.* **2020**, *5*, 574–585. [CrossRef]
16. Qader, W.A.; Ameen, M.M.; Ahmed, B.I. An Overview of Bag of Words; Importance, Implementation, Applications, and Challenges. In Proceedings of the 2019 International Engineering Conference (IEC), Erbil, Iraq, 23–25 June 2019; pp. 200–204. [CrossRef]
17. Das, B.; Chakraborty, S. An Improved Text Sentiment Classification Model Using TF-IDF and Next Word Negation. *arXiv* **2018**, arXiv:1806.06407.
18. Wang, C.; Nulty, P.; Lillis, D. A Comparative Study on Word Embeddings in Deep Learning for Text Classification. In Proceedings of the NLPIR 2020: 4th International Conference on Natural Language Processing and Information, Seoul, Republic of Korea, 18–20 December 2020; pp. 37–46. [CrossRef]
19. Sutskever, I.; Vinyals, O.; Le, Q.V. Sequence to sequence learning with neural networks. In Proceedings of the 27th International Conference on Neural Information Processing Systems—Volume 2 (NIPS'14), Montreal, QC, Canada, 8–13 December 2014; MIT Press: Cambridge, MA, USA, 2014; pp. 3104–3112.

20. Vaswani, A.; Shazeer, N.; Parmar, N.; Uszkoreit, J.; Jones, L.; Gomez, A.N.; Kaiser, L.; Polosukhin, I. Attention is all you need. In Proceedings of the Advances in Neural Information Processing Systems, Long Beach, CA, USA, 4–9 December 2017; pp. 6000–6010.
21. Bahdanau, D.; Kyunghyun, C.; Bengio, Y. Neural machine translation by jointly learning to align and translate. *arXiv* **2014**, arXiv:1409.0473.
22. Subjective Performance Assessment of Telephone-Band and Wideband Digital Codecs. Document ITU-T P.830. Available online: https://www.itu.int/rec/T-REC-P.830-199602-I/en (accessed on 25 September 1996).
23. Shen, J.; Pang, R.; Weiss, R.J.; Schuster, M.; Jaitly, N.; Yang, Z.; Chen, Z.; Zhang, Y.; Wang, Y.; Skerrv-Ryan, R.; et al. Natural TTS synthesis by conditioning wavenet on mel spectrogram predictions. In Proceedings of the ICASSP 2018, Calgary, AB, Canada, 15–20 April 2018; pp. 4779–4783.
24. Valle, R.; Shih, K.; Prenger, R.; Catanzaro, B. Flowtron: An autoregressive flow-based generative network for text-to-speech synthesis. *arXiv* **2020**, arXiv:2005.05957.

sensors

Article

Track-before-Detect Algorithm for Underwater Diver Based on Knowledge-Aided Particle Filter

Wenrong Yue [1,2], Feng Xu [1], Xiongwei Xiao [1,2] and Juan Yang [1,*]

1 Ocean Acoustic Technology Laboratory, Institute of Acoustics, Chinese Academy of Sciences, Beijing 100190, China
2 University of Chinese Academy of Sciences, Beijing 100049, China
* Correspondence: yangjuan@mail.ioa.ac.cn; Tel.: +86-159-108-363-96

Abstract: This work studies the underwater detection and tracking of diver targets under a low signal-to-reverberation ratio (SRR) in active sonar systems. In particular, a particle filter track-before-detect based on a knowledge-aided (KA-PF-TBD) algorithm is proposed. Specifically, the original echo data is directly used as the input of the algorithm, which avoids the information loss caused by threshold detection. Considering the prior motion knowledge of the underwater diver target, we established a multi-directional motion model as the state transition model. An efficient method for calculating the statistical characteristics of echo data about the extended target is proposed based on the non-parametric kernel density estimation theory. The multi-directional movement model set and the statistical characteristics of the echo data are used as the knowledge-aided information of the particle filter process: this is used to calculate the particle weight with the sub-area instead of the whole area, and then the particles with the highest weight are used to estimate the target state. Finally, the effectiveness of the proposed algorithm is proved by simulation and sea-level experimental data analysis through joint evaluation of detection and tracking performance.

Keywords: active sonar; track-before-detect; knowledge-aided; particle filter; non-parametric kernel density estimation

Citation: Yue, W.; Xu, F.; Xiao, X.; Yang, J. Track-before-Detect Algorithm for Underwater Diver Based on Knowledge-Aided Particle Filter. *Sensors* **2022**, *22*, 9649. https://doi.org/10.3390/s22249649

Academic Editors: Mark Shortis, Haixin Sun and Xuebo Zhang

Received: 16 November 2022
Accepted: 7 December 2022
Published: 9 December 2022

Publisher's Note: MDPI stays neutral with regard to jurisdictional claims in published maps and institutional affiliations.

1. Introduction

The signal-to-reverberation ratio (SRR) decreases with the increasing complexity of targets and marine environments, resulting in reduced detection performance of sonar equipment. Traditional active sonar regards detection and tracking as two separate subsystems. The detection system sets the threshold using constant false alarm technology to detect the echo signal. If the echo intensity exceeds the threshold, it is considered a target; otherwise, it is regarded as clutter and filtered out. The obtained point-trace information is then passed to the tracking system. When the target motion model is known, some common tracking algorithms estimate the target trajectories using the obtained point trace information. In this method, the accuracy of detection depends on tracking performance. However, in low SRR, the lack of target information reduces tracking accuracy since the threshold filters the weak target. If a lower threshold is used to improve the detection rate of weak targets, it will cause many false alarms. This significantly increases both the difficulty of associating and the computational cost.

The track-before-detect (TBD) methods address the target detection and tracking problem in low SRR [1]. This method accumulates the test statistic according to possible target trajectories, and the threshold decision is then made. Finally, target tracking is realized through traceback. The commonly used TBD algorithms include the dynamic programming TBD algorithms (DP-TBD) [2–8], Hough transform TBD algorithms (HF-TBD) [9–11], and particle filter TBD algorithms (PF-TBD) [12–24]. The DP-TBD algorithms generally require state space discretization and are very computationally intensive. The

HF-TBD algorithms are only suitable for a rectilinear motion target. The PF-TBD algorithms are applied to nonlinear non-Gaussian problems and are more flexible than the former two methods.

At present, the PF-TBD method has been applied in infrared [13–15], radar [16–19], sonar [20–24], and other fields. Most researchers evaluate the PF-TBD method by simulation. However, in [16], Guerraou et al. use the PF-TBD method to detect and track the real target on marine radar. To address weak target detection and tracking in multi-spectral infrared images, a PF-TBD algorithm based on a measurement fusion strategy is proposed by [13]. Bao et al. derive a multi-model optimal particle filter track-before-detect (MMPF-TBD) algorithm for maneuvering weak targets [15]. This algorithm can estimate the target's state and the existence of the target separately, and improve the particle utilization rate. To solve the problem of target loss or poor tracking accuracy caused by particle impoverishment, Huang et al. propose an improved TBD method that combines the auxiliary particle filter and the multiple-model filter (AUX-MMPF-TBD) [17]. Tian et al. propose a PF-TBD method based on the spring model firefly algorithm, and this method can guide low-weight particles to move in the direction of the high likelihood region, thereby improving particle quality [12]. Awadhiya presents a weight update method based on previous moment feedback for PF-TBD [18]. For an underwater target, Jing combines a standard particle filter with the track-before-detect method to solve the problem of underwater target detection and tracking [20]. Yi proposed a PF-TBD method for passive array sonar target detection [23]. This method minimizes the Cramer-von (CV) distance to obtain the statistical characteristics of the spectrum measurement data.

The TBD method has two key factors: the test statistic and the target motion path. First, a good test statistic can distinguish the target from the clutter. Common test statistics for point targets are amplitude [13], complex amplitude [9], or the likelihood ratio between the power of the target and that of clutter [7]. However, active sonar has the characteristics of high resolution, and the target appears in multi-resolution cells. Although some point spread functions (PSFs) [25] are used to model the energy diffusion over the resolution cells [26], the limited PSFs are too simple to reflect the variations in the characteristics of the target in the sonar system. Secondly, the exact target motion paths on frames are established to accumulate the test statistics in the correct direction. However, whether in single-mode or multi-mode PF-TBD algorithms, the established motion models are mostly uniform motion or uniform turning motion without considering the target motion characteristics in practical applications.

A track-before-detect based on a knowledge-aided particle filter (KA-PF-TBD) algorithm for an underwater diver was proposed. First, based on the motion characteristics of the diver target, a multi-direction motion model set is developed as the target state transition model, which can guide the particle state transition more accurately. To address the problem that the likelihood ratio calculation is inaccurate due to the difficulty in modeling the weak extended underwater target, we use the non-parametric kernel density estimation method to simulate the statistical characteristics of echo data in the sub-area rather than the whole area, which reduces the calculation time with little loss of detection and tracking performance. Therefore, we solve the problem of the underwater diver detecting and tracking by using the multi-directional movement model set of the diver, and the statistical characteristics of the real echo data, as knowledge-aided information in the filtering process. Finally, joint detection and tracking performance indicators are proposed to evaluate the algorithm performance.

The structure of this paper is as follows: In Section 2, the sonar measurement model and target state transition models are given. In Section 3, we develop a KA-PF-TBD method framework for active sonar systems, which includes constructing the diver multi-directional movement model set, and acquiring the measurement data likelihood function and particle filtering process. In Section 4, the effectiveness and efficiency of the KA-PF-TBD method are confirmed both in simulation and in the sea trial data. In Section 5, we summarize the results.

2. System Models

2.1. Sonar Measurement Model

Assuming that the position of the target x^t located in the far field at time t is x_{p_t} and y_{p_t}, and the corresponding velocities are v_{x_t} and v_{y_t}, respectively, then the target state at time t is $x^t = \left[x_{p_t}, v_{x_t}, y_{p_t}, v_{y_t} \right]$. In an ideal condition without considering noise, the relationship between sonar distance measurement, angle measurement, and target location is:

$$\begin{cases} r_t = \sqrt{x_{p_t}^2 + y_{p_t}^2} \\ \theta_t = \arctan\left(\frac{x_{p_t}}{y_{p_t}}\right) \end{cases},$$ (1)

The sonar system consists of a transmitting array element and a uniform linear array. M is the number of receiving array elements, and d is the array element spacing. If the transmitted signal is $s(t)$, the signal received by M array elements is written as a vector:

$$r(t) = a(\theta_t)As(t - \tau_{tt+tr})e^{jw(t-\tau_{tt+rt})} + v(t),$$ (2)

where A is the reflection coefficient of the target; the carrier angular frequency is w; tt stands for signal from transmission to target; tr represents the signal from the target to the receiver; the propagation time delay between transmission and reception is $\tau_{tt+tr} = \frac{2r_t}{c}$; and c denotes the speed of sound. The propagation delay is denoted by $\tau_m(\theta_t) = (m-1)d\sin\theta_t/c$; $r(t) = \left[r^1(t) \quad \cdots \quad r^m(t) \right]'$ denotes the vector that is composed of the received signal; $a(\theta_t) = \left[1 \quad e^{-jw\tau_1(\theta_t)} \quad \cdots \quad e^{-jw\tau_m(\theta_t)} \right]'$ denotes the steering vector; and the noise vector is $v(t) = \left[v^1(t) \quad \cdots \quad v^m(t) \right]'$.

In general, because the complex carrier does not carry any useful information, we only consider complex baseband signals [27]. Thus, the discrete form of (2) is:

$$r(k) = a(\theta_k)As(k - \tau_{tt+tr}) + v(k),$$ (3)

where k denotes the k-th sample time.

In this paper, after matched filtering and beamforming, the echo signal $r(k)$ is used as the original measurement of the track-before-detect algorithm. Taking the location of the sonar as the origin, the observation area of interest is limited. The distance range is $[R_{\min}, R_{\max}]$, which is divided into N_r distance units, and the azimuth range is $[\theta_{\min}, \theta_{\max}]$, which is divided into N_b azimuth units, according to Equation (4):

$$N_r = \frac{2(R_{\max} - R_{\min})}{c} \times F_s,$$ (4)

where F_s is the sampling frequency in array signal processing; and N_b can be determined by the direction resolution unit, $\Delta\theta$.

Then the measurement $\Delta\theta$ at time k contains $N_r \times N_b$ data, which is defined as:

$$z_k = \begin{cases} g(y(k)) + w_k & \text{The target exists.} \\ w_k & \text{The target does not exist.} \end{cases}$$ (5)

Among them, $g(\cdot)$ is the mapping between signal $y(k)$ and the measurement, and its form affects the specific form of its measurement; and w_k is the measurement of noise and clutter of the system at time, k.

2.2. Target State Transition Model

If the target state transition model matches the actual target motion, the filter has good tracking performance. If the filter fails to track, it is important for the tracking system to establish an appropriate target state transition model. The general expression for target state transition is:

$$x_{k+1} = f(x_k, \tau_k) + v_k,\tag{6}$$

where the target state is denoted by x_k; τ_k represents the target motion model type; v_k is the corresponding process noise; and $f(\cdot)$ is the state transition matrix under different models.

In this paper, $\tau_k = 1$ and $\tau_k = 2$ are the commonly used uniform and cooperative turning motion models, respectively, and w_k is the cooperative turning rate.

$$f(x_k, \tau_k = 1) = \begin{bmatrix} 1 & T & 0 & 0 \\ 0 & 1 & 0 & 0 \\ 0 & 0 & 1 & T \\ 0 & 0 & 0 & 1 \end{bmatrix},\tag{7}$$

$$f(x_k, \tau_k = 2) = \begin{bmatrix} 1 & \frac{\sin(w_k T)}{w_k} & \frac{\cos(w_k T)-1}{w_k} & 0 \\ 0 & \cos(w_k T) & 0 & \sin(w_k T) \\ 0 & \frac{1-\cos(w_k T)}{w_k} & 1 & \frac{\sin(w_k T)}{w_k} \\ 0 & \sin(w_k T) & 0 & \cos(w_k T) \end{bmatrix}.\tag{8}$$

The transition probability between multiple models at time k can be represented by a Markov chain:

$$P\{\tau_k = j | \tau_{k-1} = i\} = p_{ij}, i, j = 1, 2.\tag{9}$$

3. Algorithm Development

We propose a KA-PF-TBD method for underwater diver target tracking in active sonar systems. Figure 1 shows the flowchart of this method, and as shown, the input of the method is the unthresholded measurements, and the output of the method is the target tracking results. The important steps include the construction of the diver multi-directional movement model set, the acquisition of the measurement data likelihood function, and the particle filtering process.

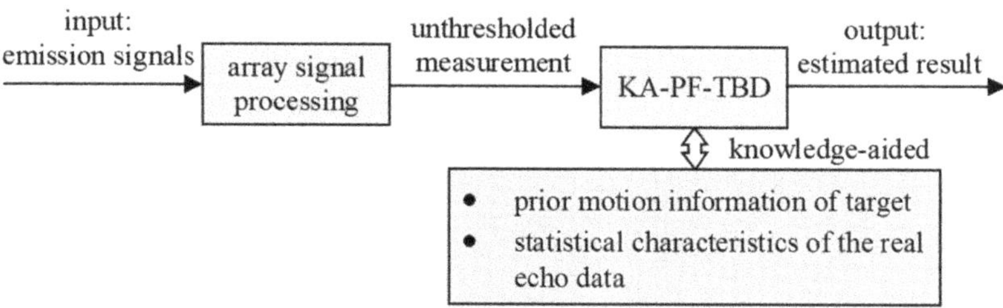

Figure 1. The KA-PF-TBD signal processing procedure flowchart.

3.1. The Construction of the Diver Multi-Directional Movement Model Set

Compared with the maneuvering target, the diver target has its own unique motion characteristics. It is difficult for its motion direction to be predicted, and there is little change in the speed of motion between adjacent moments. To describe the low-speed and high-directional change rate of the diver, as shown in Figure 2, a multi-directional motion model set composed of 8 directions and 16 uniform linear motions is established in this paper.

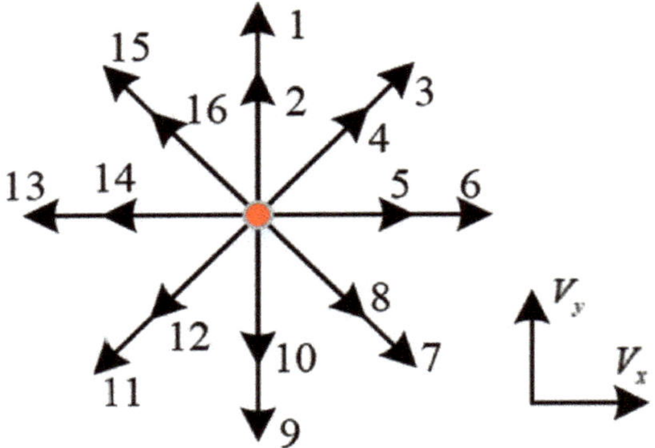

Figure 2. The diver multidirectional movement model set.

In Figure 2, the red circle represents the position of the diver target at the current moment, and the direction indicated by the eight arrows is the possible predicted direction of the target. Each direction includes two uniform linear motion models with speeds of 0.75 and 1.25 times the target speed range at the previous moment. V_x and V_y denote the velocity direction of the target state estimation at the current time.

3.2. The Acquisition of the Measurement Data Likelihood Function

The non-parametric kernel density estimation does not make any assumptions about the distribution of measurement data, and only models the probability density function based on the sample data itself. This method is frequently used in financial risk prediction and estimation [28], industrial machinery residual life prediction and estimation [29], and so on. This paper uses the non-parametric kernel density estimation theory to fit the statistical properties of sonar measurement data.

Assuming that $z_1, z_2, \cdots, z_n$ are the n measurement data samples, and $\hat{g}(z)$ is the kernel density estimation of the sample probability density function, then the expression of $\hat{g}(z)$ is:

$$\hat{g}(z) = \frac{1}{nh} \sum_{i=1}^{n} K\left(\frac{z - z_i}{h}\right),\tag{10}$$

where n is the number of independent identically distributed samples; $K(\cdot)$ represents the kernel function, which determines the role of each sample data point $z_i, i = 1, \cdots, n$ in the density estimation of random variable z; and h is the window width that affects the smoothness of the probability density estimation.

After obtaining the statistical characteristics of real echo data, the likelihood function of the measurement resolution unit (i, j) can be calculated from:

$$l\left(z_k^{(i,j)}|x_k, E_k\right) = \begin{cases} \dfrac{g_1\left(z_k^{(i,j)}|x_k, E_k=1\right)}{g_0\left(z_k^{(i,j)}|E_k=0\right)} & E_k = 1 \\ 1 & E_k = 0 \end{cases},\tag{11}$$

where $g_1(\cdot)$ denotes the statistical property of measurement data when the target exists, and $g_0(\cdot)$ denotes the statistical property of measurement data when the target does not exist. Since the target occupies multiple resolution units in the measurement space, the likelihood function is expressed as,

$$l(z_k|x_k, E_k) = \prod_{i,j \in \Theta} l\left(z_k^{(i,j)}|x_k, E_k\right),\tag{12}$$

where Θ is the range of target influence resolution units, which will be detailed in Section 4.

3.3. Algorithm Steps

If the particle set $\left\{ x_{k-1}^{A,i}, w_{k-1}^{i} \right\}_{i=1}^{N_c}$ at time $k-1$ can be used to describe the posterior probability density $p\left(x_{k-1}^{A}, E_{k-1} | z_{1:k-1} \right)$, then one iteration of the algorithm is shown in Figure 3.

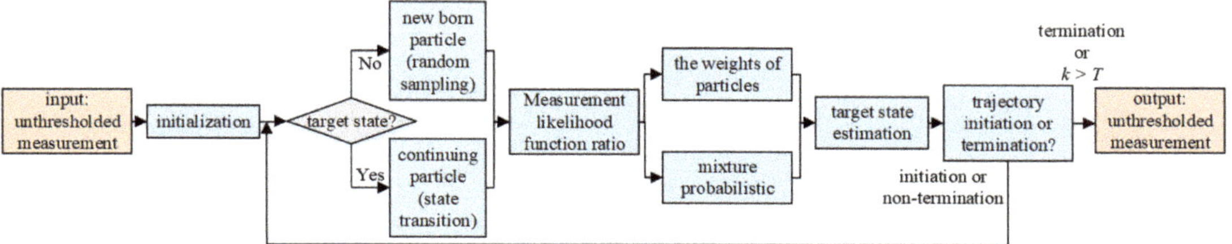

Figure 3. Iteration steps.

Steps 1: At the initial moment, only newly born particles are generated. If the prior distribution of the target is known, the particle is generated according to its distribution; if there is no prior information on the target, the samples are uniformly sampled in the observation area.

Steps 2: According to the prior probability distribution, N_b new-born particles are generated; using the established diver motion model as knowledge-aided information, the state of the N_c continuing particles is estimated as follows:

$$x_k^{(b)i} \sim q(x_k | E_k = 1, E_{k-1} = 0, z_k). \tag{13}$$

$$x_k^{(c)i} \sim q(x_k | x_{k-1}, E_k = 1, E_{k-1} = 1, z_k). \tag{14}$$

Steps 3: The statistical characteristics of measurement in the previous section are used as knowledge-aided information. The weight of new particles and continuing particles are calculated and normalized as follows.

$$\widetilde{w}_k^{(b)i} = \frac{l\left(z_k \middle| x_k^{(b)i}, E_k^{(b)i} = 1 \right) p\left(x_k^{(b)i} \middle| E_k^{(b)i} = 1, E_{k-1}^{(b)i} = 0 \right)}{N_b q\left(x_k^{(b)i} \middle| E_k^{(b)i} = 1, E_{k-1}^{(b)i} = 0, z_k \right)}. \tag{15}$$

$$\widetilde{w}_k^{(c)i} = \frac{l\left(z_k \middle| x_k^{(c)i}, E_k^{(b)i} = 1 \right)}{N_c}. \tag{16}$$

$$w_k^{(b)i} = \frac{\widetilde{w}_k^{(b)i}}{\sum_{i=1}^{N_b} \widetilde{w}_k^{(b)i}}. \tag{17}$$

$$w_k^{(c)i} = \frac{\widetilde{w}_k^{(c)i}}{\sum_{i=1}^{N_c} \widetilde{w}_k^{(c)i}}. \tag{18}$$

Steps 4: Non-normalized weights are used to calculate the mixing probability and then it is normalized.

$$\widetilde{M}_b = P_b \left(1 - \hat{P}_{k-1} \right) \sum_{i=1}^{N_b} \widetilde{w}_k^{(b)i}. \tag{19}$$

$$\widetilde{M}_c = (1 - P_d)\hat{P}_{k-1}\sum_{i=1}^{N_c} \widetilde{w}_k^{(c)i}. \tag{20}$$

$$M_b = \frac{\widetilde{M}_b}{\widetilde{M}_b + \widetilde{M}_c}. \tag{21}$$

$$M_c = \frac{\widetilde{M}_c}{\widetilde{M}_b + \widetilde{M}_c}. \tag{22}$$

Steps 5: The weight of new born and continuing particle is scaled according to the mixing probability.

$$\widehat{w}_k^{(b)i} = M_b w_k^{(b)i}. \tag{23}$$

$$\widehat{w}_k^{(c)i} = M_c w_k^{(c)i}. \tag{24}$$

Steps 6: The new and continuing particles form a complete particle set. They are resampled to obtain $\left\{ x_k^{A,i} \right\}_{i=1}^{N_c}$, and then the target state $\left\{ x_k^{A,i} \right\}_{i=1}^{N_c}$ is estimated. In the next simulation experiments, we use the system resampling method.

$$\left\{ \left(x_k^{(t)i}, \widehat{w}_k^{(t)i} \right) | i = 1, \cdots, N_t, t = b, c \right\}. \tag{25}$$

$$\hat{x}_k = \frac{\sum_{i=1}^{N_c} x_k^i}{N_c}. \tag{26}$$

Steps 7: In this algorithm, target detection and tracking are realized by particles, and the initiation and termination of trajectories are actually related to the initiation and termination of particles. The particles are generated by importance sampling, so the importance density can be designed to determine the trajectory initiation and termination.

According to reference [21], for each target with N_t frames, the Σ_w sum of all particles weight is calculated.

$$\Sigma_w = \sum_{j=1}^{N_t} \sum_{i=1}^{N_b+N_c} \widetilde{w}_j^i. \tag{27}$$

If:

$$\Sigma_w < \eta_d, \tag{28}$$

then the target initiation fails or terminates, and η_d is the likelihood threshold.

Steps 8: The target track results are generated as output.

4. Numerical Results

In this section, the performance of the proposed KA-PF-TBD method is evaluated by analyzing the simulated data and the trial data which are compared with the MMPF-TBD [15] and the AUX-MMPF-TBD [17]. The experiments are performed on our computer using an Intel i5-12500H CPU (2.50 G) and 16 GB memory.

4.1. Evaluation Indicators

Evaluating a track-before-detect algorithm is itself a challenge. A common quantitative evaluation indicator is the position's root mean square error. However, this indicator does not always accurately measure tracking performance. For instance, if an algorithm tracks a target properly in most cases but fails to do so in a few cases, the mean error may be higher than that generated by the algorithm without accurate tracking. In addition, the indicator only evaluates the tracking performance without considering the detection performance. For the above reasons, we also use the following indicators in addition to the position's root mean square error.

The position's root mean square error (RMSE): This is used to evaluate the tracking performance well, and it is defined as

$$RMSE = \sqrt{\sum_{i=1}^{m}\left[\left(\hat{x}_k^i - x_k\right)^2 + \left(\hat{y}_k^i - y_k\right)^2\right]/m},\qquad(29)$$

where m is the Monte Carlo (MC) experiment times.

The accurate detection probability sequence P_d: This consists of the N-frame accuracy detection probability of the target. The single-frame accurate detection probability is the probability that the frame is accurately detected in multiple MC simulations. The steeper the rising edge of the accurate detection probability sequence curve is, the faster is the effective track formed. The accurate detection probability sequence curve tends to be stable, and the smaller the fluctuation is, the stronger the robustness of the algorithm.

The stable detection and tracking probability P_{dt}: In the MC experiments, if the proportion of the number of traces of the effective track in the total number of traces exceeds the predetermined proportion, the experiment is said to have achieved stable detection and tracking of the target. The proportion of the number of experiments to achieve stable detection and tracking relative to the total number of experiments is called the stable detection and tracking probability. Compared with the traditional single-frame detection probability, this indicator no longer considers the plot alone but also the probability of stable detection and tracking of the target based on the effective track. It is a joint evaluation of detection and tracking performance.

The precision plot: This plot shows the percentage of frames for which the estimated object location was within some threshold distance of the actual position. If the algorithm has higher tracking accuracy at a lower threshold, it can achieve more accurate target tracking.

4.2. Simulation Experiments

In this section, to test the performance of the proposed KA-PF-TBD method, we designed a simulation in which the diver target makes a compound movement in the plane. For our purposes, we make the assumption that the target shows up at 10 s and disappears at 120 s. The initial state is set as follows:

$$x_0 = [100, 0.5, 150, 0.5].\qquad(30)$$

The reference trajectory is shown in Figure 4.

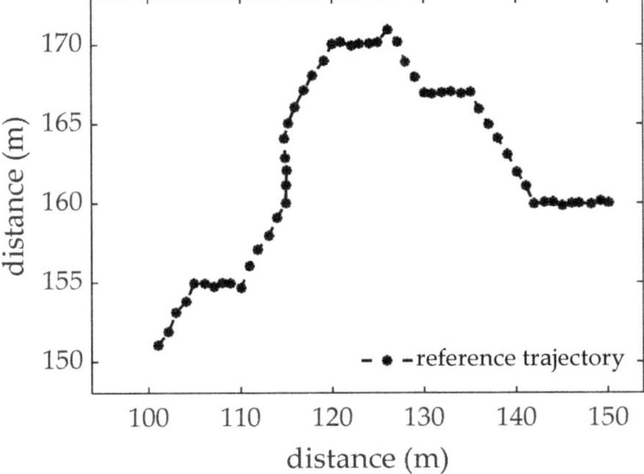

Figure 4. The reference trajectory.

The sonar emits linear frequency modulation (LFM) signals and the sampling interval is $T = 2$ s. The measurement range is $R_{\min} = 150$ m to $R_{\max} = 250$ m; and angle range is $\theta_{\min} = 20°$, $\theta_{\max} = 80°$. Taking SRR = 10 dB and SRR = -5 dB as examples, the non-parametric kernel density estimation method is used to simulate the statistical characteristics of the echo data. The effect of the size of the sub-area on the performance of the algorithm—and how to select the appropriate sub-area through the tradeoff between efficiency and accuracy—are explained.

Figure 5a,b displays the fitting results of the statistical characteristics of the echo data when the SRR is 10 dB and -5 dB, respectively. It can be seen from Figure 5 that the smaller the SRR is, the larger the overlap between the noise and the target statistical characteristics, and the more difficult it is to distinguish the target from the clutter.

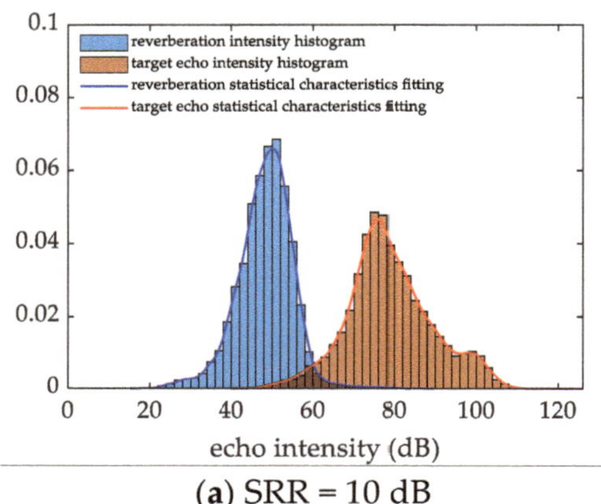

(a) SRR = 10 dB

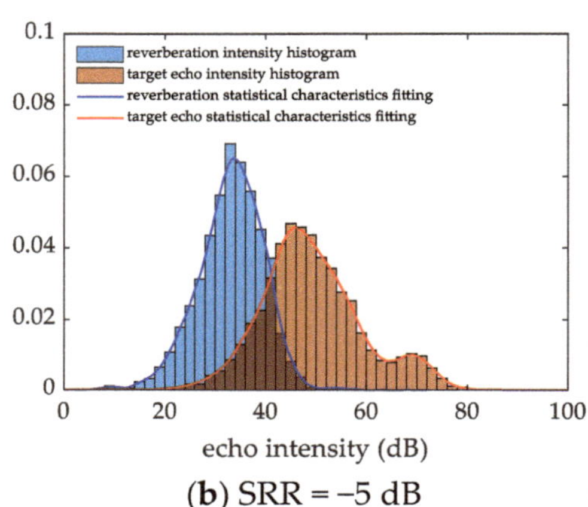

(b) SRR = -5 dB

Figure 5. Statistical characteristics of reverberation and target echo data, showing the fitting results when: (a) the SRR is 10 dB. (b) the SRR is -5 dB.

The six setups represent six different sizes of the sub-area. The stable detection and tracking probability, the position's root mean square error, and the computation time are compared in Table 1.

Table 1. Simulation results of the different setups.

	Setup 1	Setup 2	Setup 3	Setup 4	Setup 5	Setup 6
$N_{wr} \times N_{wa}$	2×2	5×5	10×5	20×5	50×10	*all*
$P_{dt}^{\text{SNR}=10}$	0	0.798	0.79	0.902	0.934	1
$RMSE^{\text{SNR}=10}$		2.075 m	2.11 m	1.696 m	1.331 m	0.19 m
$t^{\text{SNR}=10}$	46.3 s	52.1 s	54.2 s	56.6 s	90.4 s	1805.3 s

It can be seen from Table 1 that using a sub-area instead of the whole area to calculate the likelihood ratio can greatly improve computational efficiency. The smaller the size of the sub-area, the less computation time is required. However, the size of the subarea affects the accuracy of the algorithm. In all settings, setup 1, 2, and 3 have higher computational efficiency, but they have obvious performance degradation. Although the detection and tracking performance of setup 5 and 6 are good, the computational efficiency is low, and cannot meet the real-time requirements. Compared to other setups, setup 4 has higher computational efficiency with little loss of accuracy, so in the subsequent experiments, the likelihood ratio is calculated on the basis of setup 4 instead of the whole area.

To ensure the fairness of the algorithm comparison, the three methods keep the same parameter settings. The number of particles is 1500, the birth probability of particles is

$P_b = 0.85$, and the death probability of particles is $P_d = 0.15$. Birth information is an a priori for target births, similar to [30].

$$
\begin{aligned}
x_{p_0} &\sim U[90, 160] \\
y_{p_0} &\sim U[140, 170] \\
v_{x_0} &\sim U[-1, 1] \\
v_{y_0} &\sim U[-1, 1]
\end{aligned}
$$

The transition probability matrix of the commonly used model sets is:

$$
P = \begin{bmatrix} 0.85 & 0.15 \\ 0.15 & 0.85 \end{bmatrix}.
$$

The transition probability matrix of the multi-directional diver motion model set is:

$$
P = \begin{bmatrix} 0.85 & 0.01 & \cdots & 0.01 \\ 0.01 & 0.85 & \cdots & 0.01 \\ \vdots & \vdots & \ddots & \vdots \\ 0.01 & \cdots & 0.01 & 0.85 \end{bmatrix}_{16 \times 16}.
$$

To ensure the reliability of the results, we carried out 100 MC simulation experiments. Figure 6a,b shows that the stable detection and tracking probability of the three methods decrease with the decrease in SRR. The MMPF-TBD algorithm suffers from poor detection and tracking performance when the SRR is lower than 5 dB. Because it uses two conventional state transition models to limit the direction of particle transfer, the prediction results are inconsistent with the actual diver motion state. The performance of AUX-MMPF-TBD is slightly better than that of the MMPF-TBD algorithm because the former uses auxiliary particles to improve particle utilization. Compared with the former two methods, the KA-PF-TBD algorithm has better detection and tracking performance when the target SRR is low. The reason for the performance improvement is that it uses the original measurement as its input and combines the motion characteristics of the diver target to provide the necessary knowledge assistance for detection and tracking.

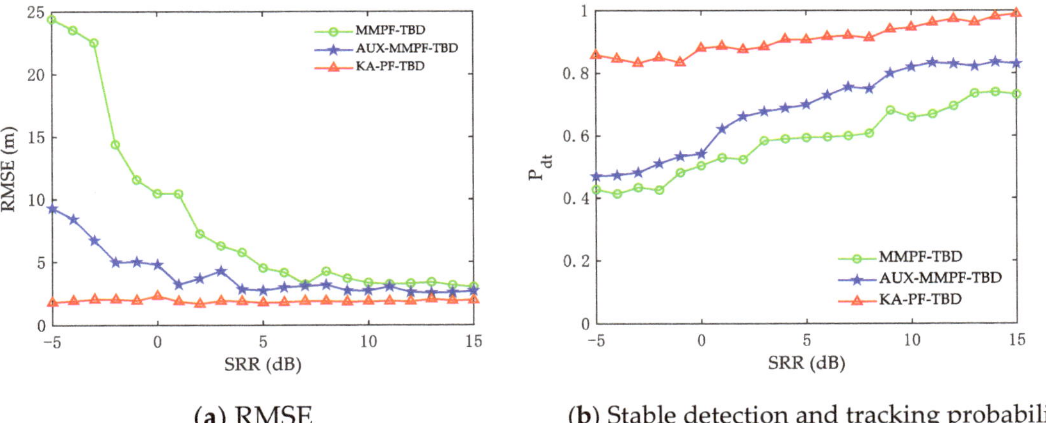

(a) RMSE   (b) Stable detection and tracking probability

Figure 6. (**a**) The RMSE under different SRR. (**b**) The stable detection and tracking probability under different SRR.

Figure 7 shows the reference and tracking trajectories obtained using the three algorithms when SRR = 4 dB. In the low SRR scene with an uncertain target direction, the

AUX-MMPF-TBD and KA-PF-TBD algorithms can accurately estimate the target direction and position. In contrast, the MMPF-TBD algorithm performs poorly and only tracks the target in a few frames. The first two methods make particles move toward the high likelihood ratio region by using auxiliary particles and improving the target state transition model, so their performance is improved. Since both of them are tracking before the detection algorithms come into effect, the detection decision is made after tracking the target, and the tracking will affect the subsequent detection results.

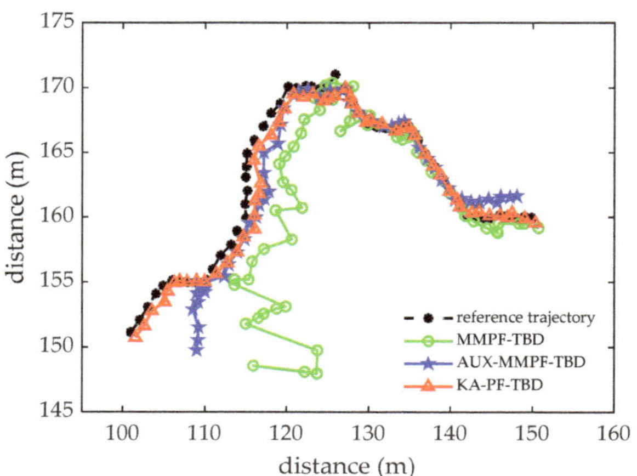

Figure 7. The tracking results.

Figure 8a compares the position RMSE results of the three algorithms under each frame. The position RMSE of the KA-PF-TBD algorithm is much smaller than that of the other two algorithms, and the fluctuation is not large. The tracking accuracy plot of the three methods is shown in Figure 8b. This shows that the KA-PF-TBD method can obtain higher accuracy at a lower threshold, so its tracking is more accurate.

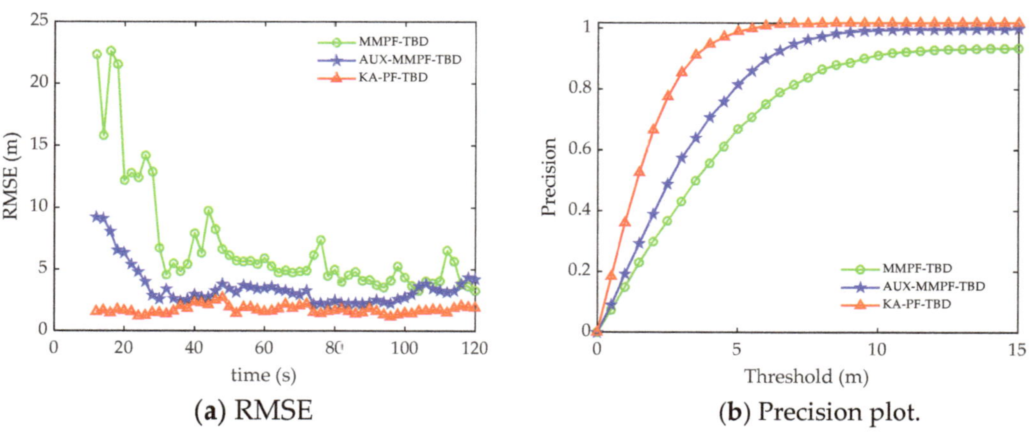

(a) RMSE   **(b)** Precision plot.

Figure 8. Simulation results. (**a**) The RMSE plotted against time. (**b**) Precision plotted against time.

Figure 9 presents the accurate detection probability sequence of the three methods. Compared with the MMPF-TBD and AUX-PF-TBD methods, the KA-PF-TBD method has a higher detection probability, which is consistent with the corresponding tracking results.

Moreover, because the accurate detection probability curve of the KA-PF-TBD algorithm fluctuates less, the algorithm is more robust.

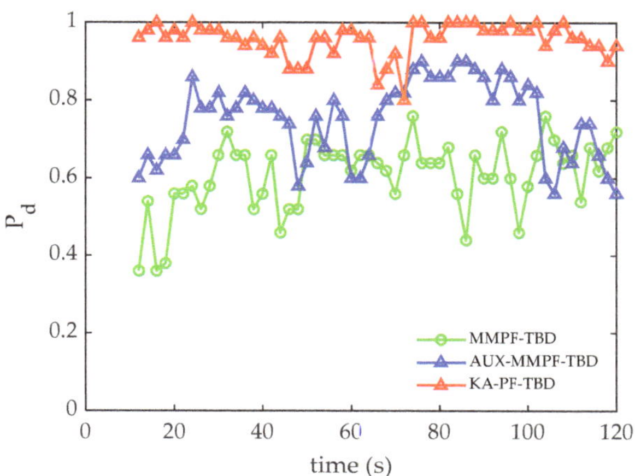

Figure 9. The accurate detection of probability sequences.

4.3. Sea Trial Experimental Data Processing

In this section, the performance of the KA-PF-TBD method is further confirmed using a set of sea trial experimental data. We collected the trail data in a shallow sea, and recorded the actual position of the diver target using GPS devices. The voyage with reverberation around the target is chosen for data processing to test the effectiveness of the proposed algorithm. First, the non-parametric kernel density estimation method is used to fit the statistical characteristics of the measured data. Figure 10 shows the histogram of the reverberation and target echo data and the fitting results. As shown in the figure, the reverberation and target statistical characteristics largely overlap, which proves that the SRR is low.

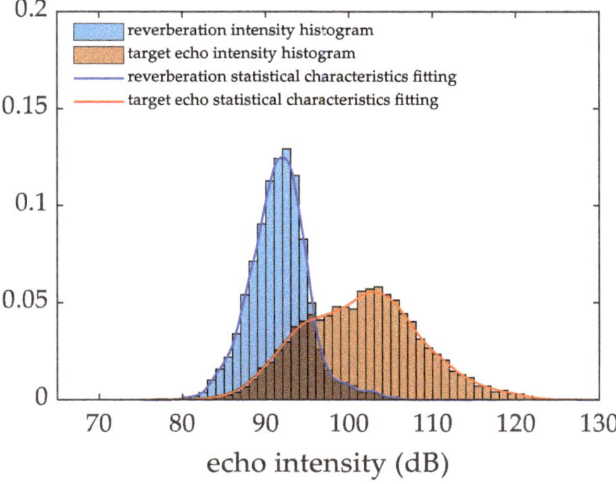

Figure 10. Statistical characteristics of reverberation and target echo data.

We then carried out 100 MC simulation experiments. The number of particles used in both algorithms was 3000, the probability of particle birth was $P_b = 0.85$, and the probability of particle death was $P_d = 0.15$. Figure 11 shows the tracking results of the three methods. Compared with the MMPF-TBD and AUX-PF-TBD methods, the KA-PF-TBD algorithm estimates the target state more accurately. Because the latter establishes a multi-directional motion model based on the motion characteristics of the diver, the surviving particles can transfer to all directions of the target motion; that is, the particles that predict the target state of the next frame will appear in multiple directions. We then calculated the particle likelihood ratio according to the measurement statistical characteristics of the original data. Finally, the particles with the high likelihood ratio from resampling were selected to estimate the target state.

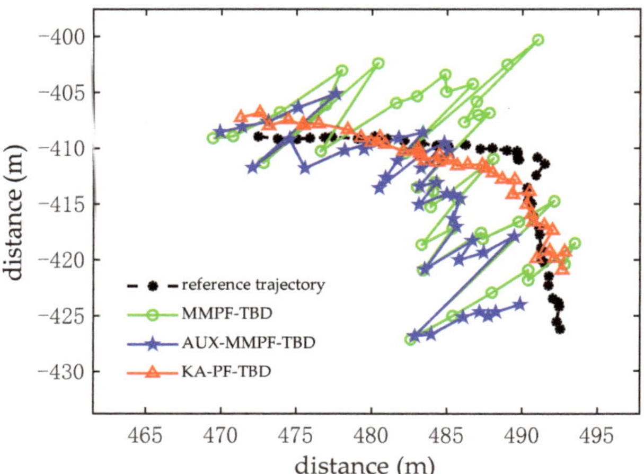

Figure 11. The tracking results.

Figure 12a,b shows the root mean square error and accuracy curves of the three algorithms at each time. Figure 13 shows the accurate detection probability sequence of the three algorithms at each time, indicating that the proposed KA-PF-TBD method can achieve weak target detection and can track more accurately than the other two methods.

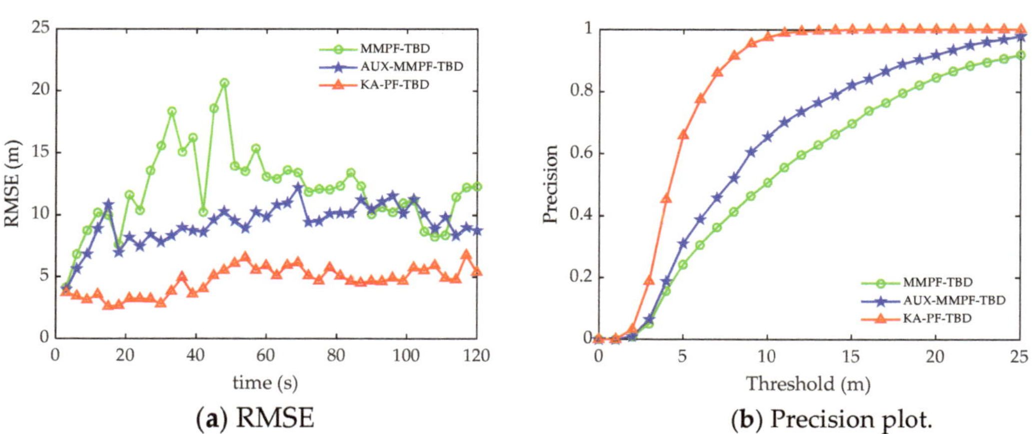

Figure 12. The trial data processing results. (**a**) The RMSE plotted against time. (**b**) Precision plotted against time.

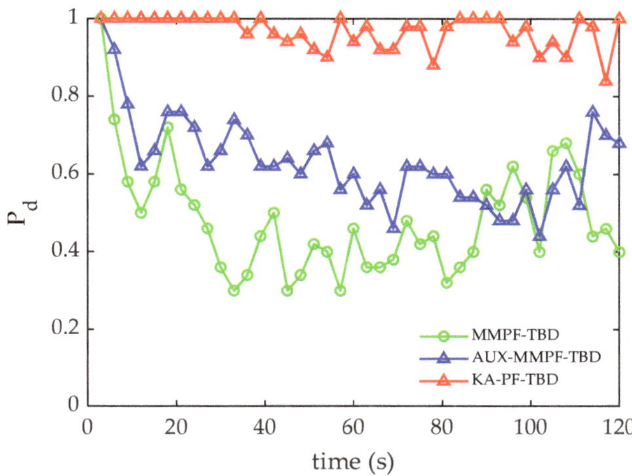

Figure 13. Accurate detection of probability sequences.

5. Conclusions

This work considers underwater diver target tracking in active sonar systems with low signal-to-reverberation (SRR). We proposed a particle filter track-before-detect based on a knowledge-aided (KA-PF-TBD) algorithm to enhance the tracking performance of low SRR diver targets. This method establishes a diver multi-directional motion model set for an underwater diver by making full use of the prior information about the diver target, and solves the problem that the conventional model set does not match the actual target motion. The received original measurement data is directly used as the input of the KA-PF-TBD method to avoid the loss of the target due to threshold processing. We adopt the non-parametric kernel density to simulate the statistical characteristics of echo data, which is used to calculate the particle likelihood ratio with the sub-area instead of the whole area. The effectiveness of the KA-PF-TBD method was verified by simulation and sea trial data processing. Compared with the MMPF-TBD and AUX-MMPF-TBD algorithms, the proposed method detects the diver target with a detection probability higher than 90 % in the sea trial data with a low SRR.

The proposed method in this paper only models the motion state of the diver, but is difficult to apply to all underwater targets due to the diversity of underwater target motion types. The kernelized correlation filter tracker does not depend on any predefined target state transition model [30,31]. Thus, it can obtain the maximum test statistics with a fast and exhaustive search, and track targets with multiple motion types. In addition, extracting the image features or signal features of the target can improve the target recognition ability [32,33]. We will extend the kernelized correlation filter theory-based track-before-detect methods and consider assistance based on feature knowledge in the future.

Author Contributions: W.Y. conceived the main idea, designed the main algorithm, and wrote the manuscript. X.X. designed the main experiments under the supervision of J.Y. and F.X. The experimental results were analyzed by W.Y. and X.X. J.Y. and F.X. provided suggestions for the proposed algorithm. All authors have read and agreed to the published version of the manuscript.

Funding: This research was funded by the National Key Research and Development Program of China, grant number 2018YFC0824103, and the Major Science and Technology Program of Hainan Province, grant number ZDKJ2020010.

Institutional Review Board Statement: Not applicable.

Informed Consent Statement: Not applicable.

Conflicts of Interest: The authors declare no conflict of interest.

References

1. Barniv, Y. Dynamic programming solution for detecting dim moving targets. *IEEE Trans. Aerosp. Electron. Syst.* **1985**, *21*, 144–156. [CrossRef]
2. Davey, S.J.; Rutten, M.G.; Cheung, B. Using phase to improve track-before-detect. *IEEE Trans. Aerosp. Electron. Syst.* **2012**, *48*, 832–849. [CrossRef]
3. Grossi, E.; Lops, M.; Venturino, L. A novel dynamic programming algorithm for track-before-detect in radar systems. *IEEE Trans. Signal Process.* **2013**, *10*, 2608–2619. [CrossRef]
4. Yi, W.; Morelande, M.R.; Kong, L.; Yang, J. An efficient multi-frame track-before-detect algorithm for multi-target tracking. *IEEE J. Sel. Top. Sign. Process.* **2013**, *7*, 421–434. [CrossRef]
5. Yan, B.; Xu, L.; Li, M.; Yan, J.Z. Track-before-detect algorithm based on dynamic programming for multi-extended-targets detection. *IET Signal Process.* **2017**, *11*, 674–686. [CrossRef]
6. Zheng, D.; Wang, S.; Qin, X. A dynamic programming track-before-detect algorithm based on local linearization for non-gaussian clutter background. *Chin. J. Electron.* **2016**, *25*, 583–590. [CrossRef]
7. Yi, W.; Fang, Z.; Li, W.; Hoseinnezhad, R.; Kong, L. Multi-frame track-before-detect algorithm for maneuvering target tracking. *IEEE Trans. Veh. Technol.* **2020**, *69*, 4104–4118. [CrossRef]
8. Zhu, Y.; Li, Y.; Zhang, N.; Zhang, Q. Candidate-plots-based dynamic programming algorithm for track-before-detect. *Digit. Signal Process.* **2022**, *123*, 103458. [CrossRef]
9. Yan, B.; Xu, N.; Zhao, W.B.; Xu, L.P. A three-dimensional Hough transform-based track-before-detect technique for detecting extended targets in strong clutter backgrounds. *Sensors* **2019**, *19*, 881. [CrossRef]
10. Moyer, L.R.; Spark, J.; Lamanna, P. A Multi-dimensional Hough transform-based track-before-detect technique for detecting weak targets in strong clutter backgrounds. *IEEE Trans. Aerosp. Electron. Syst.* **2011**, *47*, 3062–3068. [CrossRef]
11. Li, Y.; Wang, G.; Zhang, X.; Yu, H. A Hough transform TBD algorithm in three-dimensional space for hypersonic weak target under range ambiguity. *J. Astro.* **2017**, *38*, 979–988.
12. Tian, M.; Chen, Z.; Wang, H.; Liu, L. An intelligent particle filter for infrared dim small target detection and tracking. *IEEE Trans. Aerosp. Electron. Syst.* **2022**, *1*, 1–17. [CrossRef]
13. Zhang, G.; Ma, L.; Ge, J.; Zhang, D.; Liu, G.; Zhang, F. Dim Target Track-Before-Detect based on Particle Filtering. In Proceedings of the IEEE Conference on Industrial Electronics and Applications (ICIEA), Kristiansand, Norway, 9–13 November 2020; pp. 1588–1593.
14. Jia, L.; Rao, P.; Zhang, Y.; Su, Y.; Chen, X. Low-SNR infrared point target detection and tracking via saliency-guided double-stage particle filter. *Sensors* **2022**, *22*, 2791. [CrossRef] [PubMed]
15. Bao, Z.; Jiang, Q.; Liu, F. Multiple model efficient particle filter based track-before-detect for maneuvering weak targets. *J. Syst. Eng. Electron.* **2020**, *31*, 647–656.
16. Guerraou, Z.; Khenchaf, A.; Comblet, F.; Leouffre, M.; Lacrouts, O. Particle Filter Track-Before-Detect for Target Detection and Tracking from Marine Radar Data. In Proceedings of the IEEE Conference on Antenna Measurements & Applications (CAMA), Kuta, Indonesia, 23–25 October 2019; pp. 304–307.
17. Huang, C. Simultaneously Track and Detect Using Auxiliary Multiple-Model Particle Filter. In Proceedings of the IEEE 9th Joint International Information Technology and Artificial Intelligence Conference (ITAIC), Chongqing, China, 11–13 December 2020; pp. 14–18.
18. Awadhiya, R. Particle Filter Based Track Before Detect Method for Space Surveillance Radars. In Proceedings of the IEEE Radar Conference (RadarConf22), New York, NY, USA, 21–25 March 2022; pp. 1–6.
19. Kreucher, C.; Bell, K.L. A geodesic flow particle filter for nonthresholded radar tracking. *IEEE Trans. Aerosp. Electron. Syst.* **2018**, *54*, 3169–3175. [CrossRef]
20. Jing, C.; Lin, Z.; Li, J. Detection and tracking of an underwater target using the combination of a particle filter and track-before-detect. In Proceedings of the OCEANS, Shanghai, China, 10–13 April 2016; pp. 1–5.
21. Xu, L.; Liu, C.; Yi, W.; Li, G.; Kong, L. A particle filter based track-before-detect procedure for towed passive array sonar system. In Proceedings of the IEEE Radar Conference (RadarConf), Seattle, DC, USA, 8–12 May 2017; pp. 1460–1465.
22. Northardt, T.; Nardone, S.C. Track-before-detect bearings-only localization performance in complex passive sonar scenarios: A case study. *IEEE J. Ocean. Eng.* **2019**, *44*, 482–491. [CrossRef]
23. Yi, W.; Fu, L.; García-Fernández, Á.F.; Xu, L.; Kong, L. Particle filtering based track-before-detect method for passive array sonar systems. *Signal Process.* **2019**, *65*, 303–314. [CrossRef]
24. Zhang, D.; Gao, L.; Sun, D.; Teng, T. Soft-decision detection of weak tonals for passive sonar using track-before-detect method. *Appl. Acoust.* **2022**, *188*, 108549. [CrossRef]
25. Xiong, Y.; Peng, J.; Ding, M.; Xue, D. An extended track-before-detect algorithm for infrared target detection. *IEEE Trans. Aerosp. Electron. Syst.* **1997**, *33*, 1087–1092. [CrossRef]
26. Jiang, H.; Yi, W.; Kirubarajan, T.; Kong, L.; Yang, X. Multiframe radar detection of fluctuating targets using phase information. *IEEE Trans. Aerosp. Electron. Syst.* **2017**, *53*, 736–749. [CrossRef]

27. Krim, H.; Viberg, M. Two decades of array signal processing research: The parametric approach. *IEEE Signal Process. Mag.* **1996**, *13*, 67–94. [CrossRef]

28. Hesamian, G.; Akbari, M.G. Nonparametric kernel estimation based on fuzzy random variables. *IEEE Trans. Fuzzy Syst.* **2017**, *25*, 84–99. [CrossRef]

29. Patel, R.D.; Nazaripouya, H.; Akhavan-Hejazi, H. Non-parametric probabilistic demand forecasting in distribution grids; kernel density estimation and mixture density networks. In Proceedings of the 52nd North American Power Symposium (NAPS), Tempe, AZ, USA, 11–13 April 2021; pp. 1–5.

30. Zhou, Y.; Wang, T.; Hu, R.; Su, H.; Liu, Y.; Liu, X.; Suo, J.; Snoussi, H. Multiple kernelized correlation filters (MKCF) for extended object tracking using x-band marine radar data. *IEEE Trans. Signal Process.* **2019**, *67*, 3676–3688. [CrossRef]

31. Zhou, Y.; Su, H.; Tian, S.; Liu, X.; Suo, J. Multiple kernelized correlation filters based track-before-detect algorithm for tracking weak and extended target in marine radar systems. *IEEE Trans. Aerosp. Electron. Syst.* **2022**, 3411–3426. [CrossRef]

32. Li, Y.; Gao, P.; Tang, B.; Yi, Y.; Zhang, J. Double Feature Extraction Method of Ship-Radiated Noise Signal Based on Slope Entropy and Permutation Entropy. *Entropy* **2022**, *24*, 22. [CrossRef] [PubMed]

33. Li, Y.; Tang, B.; Yi, Y. A novel complexity-based mode feature representation for feature extraction of ship-radiated noise using VMD and slope entropy. *Appl. Acoust.* **2022**, *196*, 108899. [CrossRef]

sensors

Article

HW/SW Platform for Measurement and Evaluation of Ultrasonic Underwater Communications

Unai Fernández-Plazaola, Jesús López-Fernández, Eduardo Martos-Naya, José F. Paris and Francisco Javier Cañete *

Communications and Signal Processing Lab, Telecommunication Research Institute (TELMA), ETS Ingeniería de Telecomunicación, Universidad de Málaga, 29010 Málaga, Spain; unai@ic.uma.es (U.F.-P.); jlf@ic.uma.es (J.L.-F.); eduardo@ic.uma.es (E.M.-N.); paris@ic.uma.es (J.F.P.)
* Correspondence: francis@ic.uma.es

Abstract: The purpose of this work is to present a flexible system that supports the study of wideband underwater acoustic communications (UAC). It has been developed both to measure channels and to test transmission techniques under realistic conditions in the ultrasonic band. This platform consists of a hardware (HW) part that includes multiple hydrophones, projectors, analog front-ends, acquisition boards, and computers, and a software (SW) part for the generation, reception, and management of acoustic sounding signals and noise. UAC channels are among the most hostile ones and exhibit an important attenuation and distortion, essentially due to both multipath propagation, which results in a very long channel impulse response, and time-varying behavior, which produces a notable Doppler spread. To cope with this challenging medium, sophisticated transmission techniques must be employed. In this sense, adequate signal processing algorithms have been designed aiming not only at the analysis and characterization of underwater communication channels but also at the evaluation of diverse modulation, detection, and coding schemes, from Orthogonal Frequency Division Multiplexing (OFDM) to single-carrier digital modulations with a single-input multiple-output (SIMO) configuration that takes advantage of diversity techniques. Wideband sounding signals, to be injected into the sea from the transmitter side, are created with patterns that allow multiple tests on a batch. With offline processing of the captured data at the receiver side, different trials can be carried out in a very flexible manner. The different aspects of the platform are described in detail: the HW equipment used, the SW interface to control acquisition boards, and the signal processing algorithms to estimate the UAC channel response. The platform allows the analysis and design of new proposals for underwater communications systems that improve the performance of the current ones.

Keywords: underwater acoustic communications; ultrasonic frequencies; fading channels; broadband channel measurements; channel estimation; sounding signals; OFDM; MIMO; filter bank; Doppler spread

Citation: Fernández-Plazaola, U.; López-Fernández, J.; Martos-Naya, E.; Paris, J.F.; Cañete, F.J. HW/SW Platform for Measurement and Evaluation of Ultrasonic Underwater Communications. *Sensors* **2022**, 22, 6514. https://doi.org/10.3390/s22176514

Academic Editors: Haixin Sun, Xuebo Zhang and Sylvain Girard

Received: 20 July 2022
Accepted: 26 August 2022
Published: 29 August 2022

1. Introduction

This paper is focused on underwater acoustic communications (UAC) for shallow water applications and deals with a measurement system designed for UAC channels in the ultrasonic range. In this section, we give context to the work, especially from an experimental perspective.

UAC channels impose relevant restrictions like strong attenuation, mainly due to water absorption; severe time-dispersion, because of multipath propagation; and also, frequency-dispersion, due to the Doppler broadening caused by the medium and terminals motion [1–3]. All these factors impede the use of high data-rate communication systems [4,5], because the high delay spread and the low speed of sound put UAC channels around the limit, even crossing it, of being overspread [6–9]. Hence, the applications are mostly oriented to low-rate communications, like a collection of data from sensor networks, environmental monitoring, surveillance, etc. [10].

Several works have reported measurements of sea trials trying to characterize both the UAC channel response and the received additive noise. The authors of [11,12], provide estimations of the multipath channel impulse response obtained by means of single-carrier sounding signals, with a direct sequence spread spectrum (DSSS) scheme and binary phase-shift keying (PSK) modulation at a carrier frequency of 40 kHz. In [3], a wide set of trials are described, with signals located in bandwidths up to 32 kHz, to analyze the channel behavior with emphasis on its doubly selective nature, in time and frequency. Less attention has been devoted to noise measurements, but we mention [1] that gives a general view of the expected power spectral density (PSD) and [13] that presents some trials to study UAC noise in the audio frequency band in shallow water scenarios and provides estimations of the noise PSD as well.

There are interesting papers that deal with communication system trials for lower frequencies and for ultrasonic bands. Among the firsts, it is worth mentioning the pioneering work in [14], which describes experimental tests in a lake in deep waters, the authors employed a frequency-hopping scheme with PSK signals to reach data transmissions of about 5–6 kb/s. The more recent work in [15] analyzes the experimental behavior of a communication system working at audio frequencies in a trial at the sea with a water depth of 100 m approximately, using a SIMO configuration and comparing diverse techniques based on an OFDM system for low data rates. Likewise, multiple-input multiple-output (MIMO) systems have been deployed in shallow water scenarios, for narrow-band signals in the audible frequencies, and their performance is discussed in [16–18]. Other notable works that show measurements of systems in similar scenarios and operating at the ultrasonic band, but in low frequencies around 30 kHz, can be found in [19,20].

There are few efforts trying to develop systems for higher bit-rate applications, like e.g., video transmission, although of modest quality [21]. Some interesting ideas of how to transmit video signal through UAC channels are provided in [22], proposing a selective frequency mapping of the video components according to their importance, with simulation tests over modeled channels for a band from 4 to 30 kHz. Likely, the more ambitious experimental UAC trial for video transmission of wide bandwidth and at high ultrasonic frequencies (between 60 and 100 kHz) is reported in [23]. Although the link is established at deep waters (at 100 m depth), with more favorable channel conditions, and off-line simulations were performed to apply the receiver algorithms over the recorded data, which included adaptive decision-feedback equalizers whose complexity is not described. There exist also works presenting measurement trials for other applications like in [24] for underwater acoustic localization, by using DSSS signals at 15 kHz, or for military applications ([25] and references therein), but in this latter case the technical details are not usually unveiled. All this literature supports the increasing interest in the development of new systems that improves the current technology for UAC applications.

The main contribution of this paper is to describe the details of a platform that has been designed for UAC measurements in our research group. It is based on the signal processing of sounding signals that are injected into the water from a transmitter and are registered at a remote receiver. Both terminals are autonomous and contain specific hardware (HW) equipment that is controlled by ad-hoc software applications. The presented set-up allows carrying out trials in underwater scenarios over a wide band of ultrasonic frequencies, up to approximately 200 kHz. Nevertheless, there is a trade-off between bandwidth (and, hence, data rate) and signal-to-noise ratio (SNR), because the higher the frequency the higher the UAC channel attenuation due to the water absorption [1]. For that reason, we limit our measurements up to 130 kHz. The presented platform has been successfully used both in narrow-band UAC trials, for channel characterization [26], and in broad-band UAC trials, with a twofold aim: channel characterization [9] and communication systems assessment based on OFDM [27]. To the best of the authors' knowledge, no research group has published any description of this type of platform in such detail. This paper will facilitate the work of those researchers who want to create their own measurement setups. In addition, a signal processing algorithm for estimating UAC channels based on filter banks is presented and its better performance is justified as compared with other classical ones based on a correlator receiver.

We summarize here the main distinctive facts of our system that makes it innovative:

1. High operating frequencies in the ultrasonic band, 32–128 kHz.
2. Very broad bandwidth of 96 kHz, two octaves.
3. A robust and rather automated system that allows registering long signal records.

The first one is a challenge, because the UAC channel conditions at such frequencies are more hostile, the transducers operations are less favorable and it is more difficult to reach long distances. The second one represents a trade-off. On one hand, the SNR is less homogeneous in the whole frequency band and receiver algorithms must face more difficulties to reach good performance. On the other, we can exploit the diversity in frequency, e.g., by means of coded OFDM [27], and employ apparently simpler schemes, with lower spectral efficiencies, but effective and reliable. The third one is an advantage to studying the time-varying nature of UAC channels with higher resolution [9] and testing broadband communication systems that deal with it.

The paper is organized as follows. After this introduction, we will describe the UAC scenarios that are targeted in the trials. In the third section, the equipment employed to carry out the measurements is explained whereas, in the fourth one, it is the software used for the control and management of the involved signals that are shown. The fifth section deals with the algorithms designed for signal processing of the data obtained from the trials and, in particular, for synchronization and channel estimation. Finally, some conclusions are given in the last section.

2. UAC Scenarios

The measurement and emulation platform presented in this paper is designed to investigate certain scenarios in the field of UAC, especially within the 3-octave ultrasonic band from 32 kHz to 128 kHz. This section describes in detail these communication scenarios and their associated UAC channels.

At first, our platform is designed for the measurement/emulation of both horizontal and vertical UAC channels. The *horizontal UAC channel* is characteristic of shallow waters in littoral zones. As shown in Figure 1, possible transmission paths include direct line-of-sight (LoS) (red), surface reflection (blue) and bottom reflection (green). There is an essential difference between the two reflections mentioned. Reflection at the surface usually occurs with a reflection coefficient whose modulus is close to one due to the large difference between the characteristic impedance of seawater and air, this implies that little energy is lost in such reflection. On the other hand, reflection at the bottom depends very much on the geological nature of it. If the bottom is sandy, a significant part of the incident energy is usually absorbed, while if the bottom is rocky, little energy is lost for similar reasons to what occurs at the seawater-air interface. The full configuration for the UAC horizontal channel is also shown in Figure 1. The platform is able to measure/emulate a 1 × 4 single-input multiple-output (SIMO) UAC channel, i.e., transmitting through one projector and receiving through up to four hydrophones. The number of receiver hydrophones can be variable from one to four, including UAC single-input single-output (SISO) channel measurement.

The *vertical UAC channel* is shown in Figure 2a. In this case, the paths followed by the acoustic waves are predominantly orthogonal to the bottom plane and the effect of reflection on the seabed is non-existent or almost negligible. In general, the vertical UAC channel is much less hostile for communication than the horizontal one. The main reason is that multipath components other than LoS usually reach the receiver in a smaller number, with lower energy and with less temporal dispersion. As shown in Figure 2a, measuring vertical channels at the receiver requires an extension cable, which in some scenarios can be large (the green cable). In addition to horizontal and vertical UAC channels, the platform is capable of measuring other types of geometric configurations that are a mixture of the two.

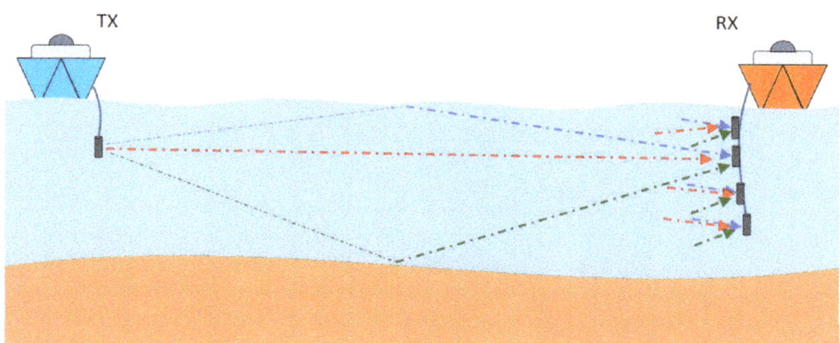

Figure 1. 1×4 SIMO UAC horizontal channel.

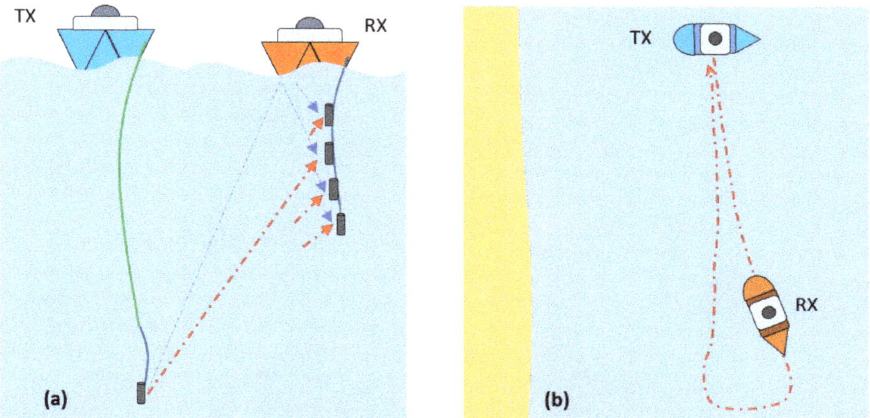

Figure 2. Diagrams for UAC measurement scenarios: (**a**) 1×4 SIMO vertical channel; (**b**) dynamic channel.

Secondly, the platform is capable of measuring both quasi-static and dynamic UAC channels. *Quasi-static channels* are those in which both the transmitter and receiver do not explicitly move relative to each other, and channel variations are due to oscillations of both around a given equilibrium position, as well as variations and movements of the physical environment (waves, currents, tides, marine life, etc.). However, quasi-static UAC channels in the ultrasonic band exhibit strong temporal variations due to the low speed of sound propagation in seawater ($\approx$1500 m/s). *Dynamic channels* are those in which there is explicit relative motion between the transmitter and receiver. Figure 2b shows a dynamic channel with a looped path, which is one of those commonly used in our measurement platform. Dynamic UAC channels with modest TX-RX relative velocities, around or greater than 1 sea knot, easily become overspread channels in the ultrasonic band; that is, the channel coherence time is equal to or less than the effective duration of the impulse response.

Finally, Figure 3 summarizes twelve basic scenarios covered by the measurement/emulation platform for three different dimensions: scenario geometry, relative motion between communication nodes, and use or not of diversity in reception.

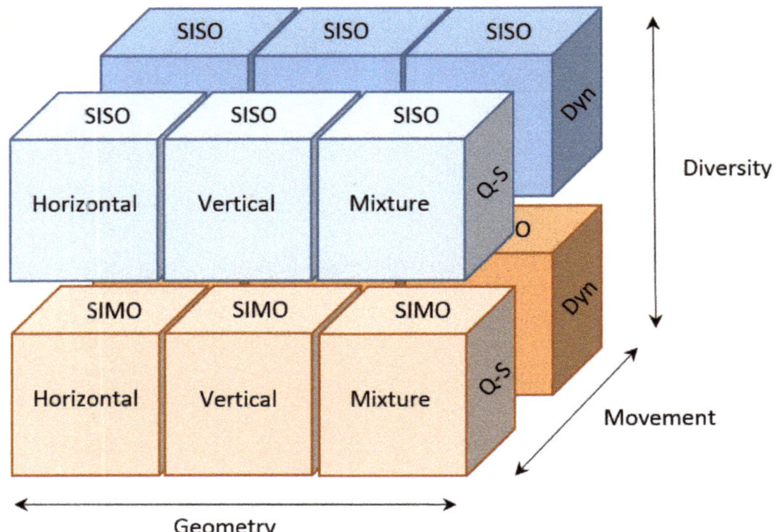

Figure 3. Summary of the UAC scenarios supported by the measurement platform according to geometry (horizontal, vertical, and mixture), movement (quasi-static and dynamic) and diversity in reception (SISO and SIMO).

3. HW of the Measurement/Emulation System

The HW of the measurement/emulation platform is shown schematically in the Figure 4. On the transmitter side, the test signal is generated by the platform SW (explained in the next section) and sent to the IOTECH Personal DAQ 3000 board that performs the digital to analog (D/A) conversion. This board is capable of generating a transmitted signal at a sampling frequency of 1 MHz and with a resolution of 16 bits. Before being injected into the projector, the signal is boosted by the classic Bruel&Kjaer 2713 power amplifier. The Bruel&Kjaer 8105 projector is quite omnidirectional and its frequency response is of the resonant type, reaching its resonance peak at around 100 kHz. Although the non-flat response of the projector must be taken into account in estimating the acoustic response of the channel. The projector response, in the band of interest, is mainly high pass and, since both the underwater channel response and the noise power spectral density are low pass [1], the overall result is that similar values of SNR are obtained throughout all frequencies. On the receiver side, the platform is capable of supporting up to a maximum of 4 RESON TC4032 hydrophones. Each pair of hydrophones use an IOTECH Personal DAQ 3000 board that has the analog to digital (A/D) conversion function. This is because the effective sampling frequency for each channel is divided by the number of channels used. Hence, in order to limit the maximum frequency of the acquired signals as little as possible, we decide to use one card for the transmitter and two for the receiver. This way the transmitter card can use a sampling frequency of up to 1 MHz and the receiver cards of up to 500 kHz per channel, which is enough for the maximum working frequency we are targeting of 128 kHz (that implies a minimum sampling frequency of 256 kHz). In contrast to the projector, the RESON TC4032 hydrophone has a fairly flat frequency response and needs a preamplifier like the RESON VP2000 before the A/D conversion. This hydrophone has a spatial response quite omnidirectional as well.

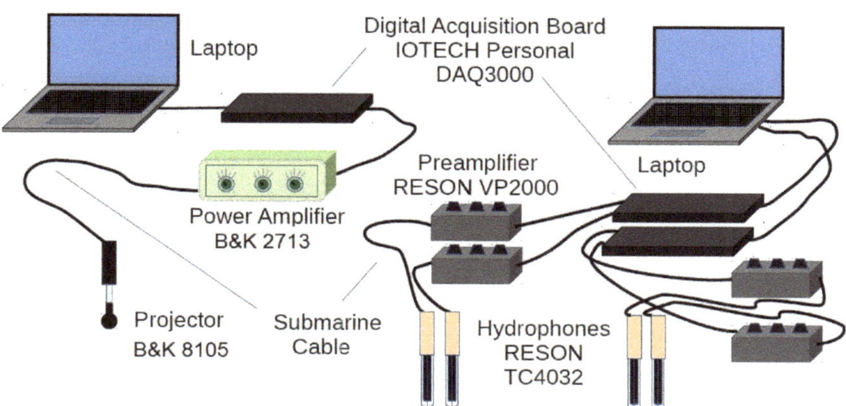

Figure 4. Schematic of the platform HW.

In Figure 5, we summarize the physical implementation of the HW platform. Figure 5a shows the transmitter unit, which is packaged in a watertight box protecting the power amplifier, and also the projector. Figure 5b presents the receiver unit for a pair of hydrophones, inside its watertight box, consisting of two preamplifiers (one for each hydrophone) and a shared acquisition board; the platform has two such units to support 4 hydrophones. Figure 5c shows the inverter power supply unit, capable of providing both 12 VDC and 220 VAC; the 220 VAC output is essential to feed the transmitter unit power amplifier. This power unit is connected to the transmitter and receiver units through watertight cables and connectors. Finally, Figure 5d presents a more detailed picture of the receiver unit, where the amplification gain knob and the cutoff frequency knobs of the preamplifier can be seen.

Figure 5. Images of the HW platform implementation: (**a**) transmitter unit, (**b**) receiver unit for 2 hydrophones, (**c**) power inverter unit and (**d**) top view of the receiver unit.

4. SW of the Measurement/Emulation System

The SW of the system has been developed to manage and control the realization of measurement campaigns. Laptop screens are not easy to see at sea outside on the deck of the ships, especially on sunny days, which makes SW operation more difficult than under normal conditions. Therefore, the interface is designed with large buttons so that the measurements can be configured with a minimum of user actions. The SW of the system consists of two applications. The first one is the transmitter app, which controls the acquisition board that generates the input signal for the amplifier connected to the projector. The second one is the receiver app, which controls the two acquisition cards in charge of acquiring the signals received by the 4 hydrophones. The configuration of a measurement requires specifying the particular conditions: distance between the transmitter and receiver boats, amplifier gains, separation between the hydrophones, swell conditions, depth, etc. The SW has been developed so that for each measurement configuration a different set of transmitted signals can be specified. To optimize the process, all these signals are transmitted sequentially, without user intervention. However, they are separated by silences, i.e., intervals in which nothing is transmitted. These silences allow the receiver app, by means of basic signal processing, to split the received signals and save each one in separate files. This greatly facilitates the measurement organization and its subsequent analysis. Despite that, everything received is recorded and stored in files, i.e., no received signal sample is discarded, because the silence periods serve later as noise measurements.

The SW of the measurement system has been developed in Microsoft Visual C++ using the Applications Program Interface (API) called DAQX. It is an API provided by the manufacturer IOTECH for its cards and valid for the Personal DAQ3000 [28]. Although the manufacturer offers other SWs to control the card, the API is the one that allows better control of the card operation and easier embedding in custom applications. As mentioned in the previous section, the cards work with sampling frequencies up to 1 MHz at the transmitter and up to 500 kHz per channel/hydrophone at the receiver.

4.1. Transmitter Application

The operation of the transmitter SW is relatively simple since it is only necessary to specify which set of signals, from the previously configured ones, are to be transmitted. The specific conditions of the measurement are not configured in the transmitter but in the receiver. By means of text files, a button is configured for each set of signals to be transmitted. The duration of each signal and the silence time among them are also configured in these files. Samples of each signal are provided by Matlab files.

Two transmission modes are considered: fixed duration and indefinite duration. In the fixed duration mode, the set of configured signals is transmitted, each with its predefined duration, ending when the last signal is sent. This is the mode used to measure quasi-static channels described in Section 2. In the indefinite duration mode, the same signal is transmitted all the time until the user cancels the transmission. The indefinite mode is interesting, for instance, to measure dynamic channels (described in Section 2) to study the effect of movement when the distance between the boat changes, and we want to analyze the effect on the same signal. To inform the user about the transmission status, the application indicates which signal is currently being transmitted from the selected set of signals or if it is in a silence period.

The most significant details in the transmitter app implementation are those related to the DAQ3000 card programming. Using the API, the card is configured in a mode that allows transmitting any signal (with the parameter *DdomDynamicWave* of the *daqDacSetOutputMode* function) and it can work either in a fixed duration mode or in an indefinite duration mode. (In the fixed duration mode, the card is configured with the *DdwmNShot* parameter of the *daqDacWaveSetMode* function, in which the number of samples to be transmitted must be specified. In the indefinite duration mode, the *DdwmInfinite* parameter is specified.) Likewise, the card internal clock is selected as the reference for the D/A conversion. The signal samples to be generated are loaded into a buffer (by using the *daqDacWaveSetBuffer* function). Once the card is configured to transmit, i.e., when the card is armed, the signal generation starts and it ends when the specified number of samples is reached in the fixed duration mode or when the card is disarmed in the indefinite duration mode if the user presses the end button. Silences between signals are simply non-transmitting states, i.e., time intervals with the card disarmed.

4.2. Receiver Application

Figure 6 presents the appearance of the receiver application interface. The buttons are fully configurable by means of text files, as well as the combination of them that defines a type of measurement. In the example shown in the figure, the first row has been configured so that determines the set of signals to be used and the second one the distance between the boats. The parameters button allows us to specify other characteristics of the measurement to be performed: power used in the transmitter, the gain of the receiver preamplifiers, depth at which the hydrophones are, swell... Another control allows us to specify the gain that the card applies to the acquired signals. During the course of measurement and with a configurable refresh rate, a graph of the received signal is displayed, either in time or in frequency using the Fast Fourier Transform (FFT). As the signals are received, the application performs a basic signal processing to estimate the received SNR, to activate a signal saturation alarm, as well as to separate the set of received signals and save each one in a different file. The saturation alarm allows us to check if the gains applied either to the transmitter amplifier, the receiver preamplifiers, or the acquisition card, are not adequate for the distance between the transmitter and the receiver. In such case, the user can readjust gains and restart the measurement.

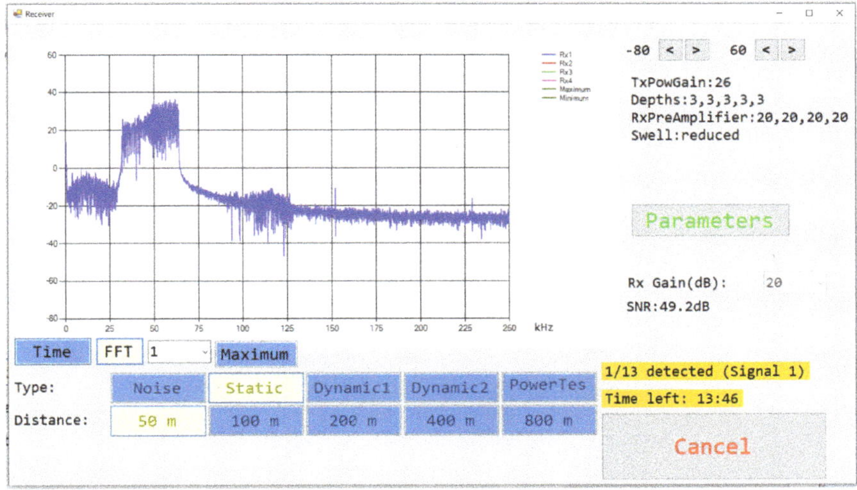

Figure 6. Appearance of Receiver application interface.

We describe the details of the receiver application implementation that are related to the DAQ3000 card programming, acquisition optimization, and the measurement-saving loop. The card is configured so that once the measurement is started it does not end until the user presses the end button which disarms the card. This way, no received signal sample is lost. The card allows defining different gains so that the analog input signal dynamic range fits the card input range, which is selectable between the larger limits, $+/-10$ V, and the smaller one, $+/-0.1$ V. The selection is done by setting an amplitude gain of 1 to 100, respectively. The SW allows the user to define which gain to apply according to the needs. The board is configured so that the reading buffer, which is a circular buffer, is managed by itself (by selecting both the *DatmDriverBuf* and *DatmCycleOn* mode). The card generates an overload event if it has to write in the buffer by overwriting samples that have not yet been read. Depending on the laptop performance, an overload event may be more or less likely. The application is designed so that, in the event of an overload, the acquisition continues and is not canceled. Nevertheless, a procedure is implemented to try to avoid overload, as described later.

The number of samples to request in each call is made configurable, although the usual value we choose for this record length is 10,000, which corresponds to a time interval of 20 ms for the sampling frequency of 500 kHz. In order to check how far we are from an overload, we calculate the waiting time of the reading samples function (samples are read from the card by the *daqAdcTransferBufData* function using the *DabtmWait* mode, whereby the function does not terminate until all requested samples are available). The percentage of this time with respect to 20ms gives a metric that allows us to anticipate possible overloads. We call *throttling* to a situation in which a high probability of overload is expected. Once the measurement is started, the application enters into a loop and performs the actions shown in the diagram of Figure 7 in each iteration. First, information on the capture status is displayed on the interface (signal being expected, saturation detection, estimated SNR, etc.). Second, the specified number of samples of the record length is obtained from the card and it is evaluated whether to get in throttling mode or not. Then, a basic signal processing is applied to the samples and, later, a short-time windowed signal is plotted on the screen, either in the time or frequency domain. Finally, the signal segment is recorded on disk.

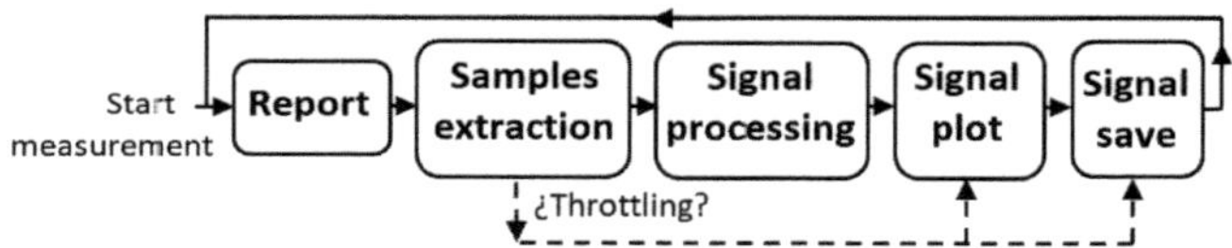

Figure 7. Measurement Loop diagram.

When the application enters in throttling mode, and to avoid overload, some actions are stopped. The frequency at which the received signal is painted on the screen is configurable, but if the application is in throttling mode, painting is stopped. On the other hand, in normal conditions the samples are recorded on disk as they are obtained, but the file is only closed when a configurable time elapses, the default value is 1 s. The file closing action is when the operating system consumes the most time. There is a trade-off between choosing a very long period of file closing that reduces the total CPU time but increases the risk of sample loss in case of a fatal error (e.g., the laptop runs out of battery). In the latter case, all samples in memory since the last time the file was closed would be lost. Into the throttling mode, although this involves more risk, the closing of files is postponed to reduce the probability of overload.

There are two kinds of tasks involved in the signal processing of the measurement loop. A first group, carried out each iteration, includes a filtering in the band of interest, from 32 to 128 kHz, and a local count of clipped samples from the maximum and minimum values registered in each segment, among others. A second group is performed after a configurable number of iterations, which is 5 by default and corresponds to an extended interval of 100 ms. They comprise the detection of the sounding signal beginning; the signal end; a saturation/clipping event. To detect the signal beginning and end, the power in the 100 ms interval is estimated. Since the set of transmitted signals are separated by silences, it is possible to detect significant power variations when changing from silence to signal or vice versa and a differential algorithm compares the power level of successive extended intervals to decide, by means of heuristic thresholds, when a signal is starting or ending. To validate a signal start detection, it is verified that the power level remains high over some time, to avoid that spurious in the received signal could lead to false start detections. The signal detection does not influence the signal recording, which is always done from the first sample, but allows to change the processing phase towards the detection of the signal end that does determine the current file closing (and automatically a new file is open to keep on saving the following samples). Regarding the detection of a saturation event, in each extended interval, the number of clipped samples is estimated by accumulating the local maxima/minima of each signal segment and if it exceeds a configurable threshold, a saturation alarm is generated. In a similar fashion to the transmitter app, the receiver app shows the evolution of the detected signal, so that the user can follow the transmission status of the selected set of sounding signals.

5. Description of the Signal Processing

In this section, we describe the technical details of the algorithms designed for channel estimation, or communication tests, from the sounding signals. This estimation procedure is an end in itself when trying to characterize the channel response from measurements and is a means when we want to demodulate the received signals to assess the performance of a given communication system.

5.1. Procedure for Channel Estimation Based on Multicarrier Signals

We denote the time-invariant impulse and frequency responses of the channel as $h(\tau)$ and $H(f)$ respectively while the sampled counterparts will be denoted as $h[n] = h(nT_s)$ for $n = 0 \ldots N-1$ and $H[k] = H(\frac{k}{NT_s})$ where T_s is the sampling frequency and N is such that NT_s is longer than the duration of $h(\tau)$. In the general case of a time-variant channel, the impulse and frequency responses will be written as $h(t, \tau)$ and $H(t, f)$ respectively and their discrete-time versions $h[m, n] = h(mNT_s, nT_s)$ and $H[m, k] = H(mNT_s, \frac{k}{NT_s})$.

One of the most widespread channel sounding systems is the correlative sounder [3,7,18] which consists in transmitting repetitively a probe signal $p[n]$, performing correlation at the receiver and storing successive channel snapshots. Common probe signals are linear frequency modulated (LFM) chirps and pseudo noise (PN) sequences due to their favorable autocorrelation properties, i.e., $\Phi[n] \equiv p[n] \circledast p[-n] \approx \delta[n]$ where $\circledast$ stands for the N-point circular convolution. The signal processing carried out by the correlative sounder can be described as follows.

Let $\tilde{p}[n]$ denote the periodic sounding signal and let N and L be the length of $p[n]$ and the number of successive transmitted copies of $p[n]$, respectively. Hence $\tilde{p}[n] = \sum_{m=0}^{L-1} p[n - mN]$. Assuming that the impulse response of the channel $h[n]$ is static (this restriction will be relaxed later), the received signal $\tilde{y}[n] = \tilde{p}[n] * h[n]$ will also be periodic. Since the convolution of a periodic signal $\tilde{p}[n]$ (with a period of N samples) with a non-periodic signal $h[n]$, with $N_h \leq N$ samples and $n = 0 \ldots N_h$, can be expressed as the concatenation of the N-point circular convolution of $p[n]$ and $h[n]$, it follows that $\tilde{y}[n] = \sum_{m=0}^{L-1} y[n - mN]$ with $y[n] = p[n] \circledast h[n]$. Next, the output of the correlation receiver can be expressed as $\tilde{s}[n] = \tilde{y}[n] * p[N - n]$, where an N-point delay has been included so that $p[N - n]$ is defined in the range $n = 0 \ldots N - 1$. Following the same reasoning as before, we can write $\tilde{s}[n] = \sum_{m=0}^{L-1} s[n - mN]$ where $s[n] = y[n] \circledast p[N - n]$ and, hence $s[n] = p[n] \circledast h[n] \circledast p[N - n] \approx \delta[n - N] \circledast h[n] = \delta[n] \circledast h[n] = h[n]$ and therefore $s[n] = \hat{h}[n]$ where $\hat{h}[n]$ is the estimated impulse response. The output of the correlator receiver can be expressed in a more convenient way as

$$\tilde{s}[n] = \sum_{m=0}^{L-1} \hat{h}[n - mN] \tag{1}$$

with

$$\hat{h}[n] = y[n] \circledast p[N - n]. \tag{2}$$

The output in (1) is composed of a concatenation of successive channel estimations $\hat{h}[n]$ each of which is computed as indicated in (2). This particular way of describing the correlator system directly leads to an alternative way of implementation using the FFT and inverse FFT (IFFT). Concretely, each period of the received signal in (2) can be obtained as

$$\hat{h}[n] = \text{IFFT}_N\{\text{FFT}_N\{y[n]\} \cdot \text{FFT}_N\{p[N-n]\}\} \tag{3}$$

The procedure is shown in the block diagram of Figure 8 where $Y_k = \text{FFT}_N\{y[n]\}$ and $P_k^* = \text{FFT}_N\{p[N-n]\}$.

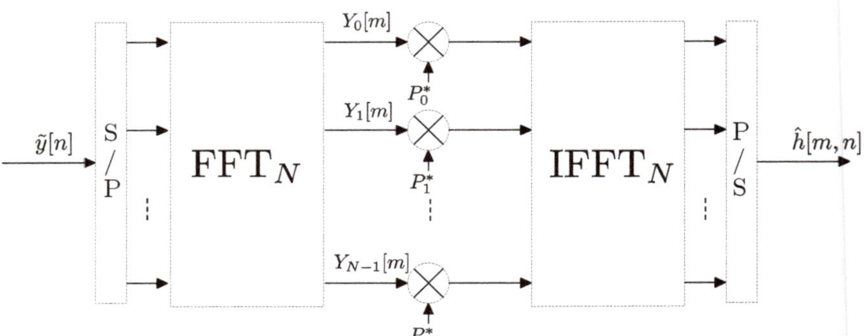

Figure 8. Block diagram of correlator receiver implemented via FFT. P/S stands for parallel to serial conversion and S/P for serial to parallel.

An alternative to the correlator receiver approach described above is the use of a multitone/multicarrier probe signal at the transmitter and a bank of filters at the receiver, which is the technique we have adopted in our measurement campaigns. In the following derivation we will show that our method is a generalization of the correlator receiver and may offer remarkable advantages. Let $\tilde{x}[n] = \sum_{k \in \mathcal{K}} X_k \cdot e^{j\frac{2\pi}{N}kn}$ be a discrete-time complex sounding signal composed of a sum of (at most N) tones or carriers equally spaced in frequency where $\mathcal{K}$ is a set of positive integers smaller than N and where X_k is the complex amplitude of the k-th carrier. As stated in [29] the peak-to-average power ratio (PAPR) of the sounding signal can be minimized if X_k is a *Zadoff-Chu* sequence, i.e., $X_k = e^{-j\phi_k}$ with $\phi_k = \pi q k^2$ where q is a constant related to the length of the sequence. Assuming a static $h[n]$, the received signal can be written as $\tilde{y}[n] = \sum_{k \in \mathcal{K}} H[k] \cdot X_k \cdot e^{j\frac{2\pi}{N}kn}$. The frequency response $H[k]$ of the channel is estimated using a bank of N filters centered at the discrete frequencies of the sounding carriers $\frac{k}{N}$, $k = 0 \ldots N-1$. The impulse response of each filter is given by $g_k[n] = g[n] \cdot e^{j\frac{2\pi}{N}kn}$, $n = 0 \ldots M-1$ where $g[n]$ is a low pass filter with cutoff frequency $\frac{1}{2N}$ and length M samples, being $M \geq N$. In ideal conditions of flat frequency response of the low pass filter $g[n]$ the output of each filter will be given by $Y_k[n] = H[k] \cdot X_k \cdot e^{j\frac{2\pi}{N}kn}$ i.e., a complex exponential with constant amplitude $H[k] \cdot X_k$. Taking a sample of $Y_k[n]$ at instant $n = N$ results in $Y_k = H[k] \cdot X_k$ and the estimation of the frequency response at index k is obtained by $\hat{H}[k] = \frac{Y_k}{X_k} = Y_k \cdot X_k^*$. Finally, an N-point IFFT is applied to obtain $\hat{h}[n]$. The procedure is shown in the block diagram of Figure 9.

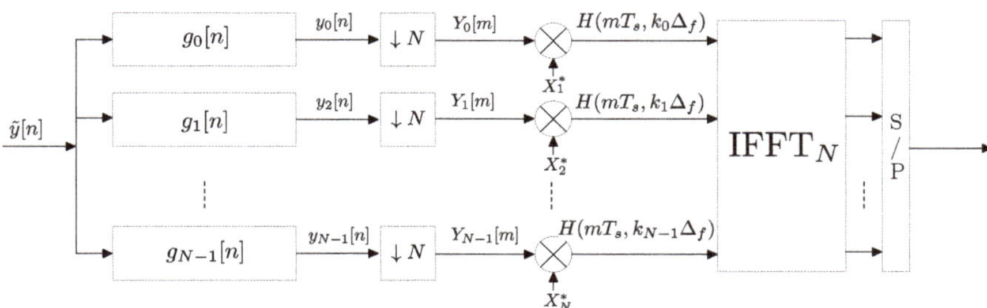

Figure 9. Block diagram of the channel estimation procedure using a bank of filters and a unit amplitude sequence at transmitter.

For the simple case of $M = N$, the bank of filters followed by decimation depicted in Figure 9 can be implemented using the FFT [30] as shown in Figure 10, where the received signal $\tilde{y}[n]$ is windowed by $g[n]$. In a more general case, if $M \geq N$, such implementation is still valid by subdividing the M-point-windowed input sequence into blocks of N points and then stacking and adding these blocks before performing the N-point FFT, which is an optimized procedure based on the linear character of the FFT (see the details in [30]).

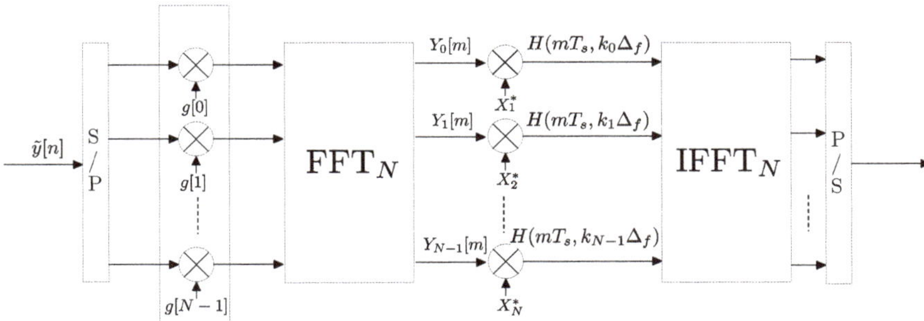

Figure 10. Block diagram of the channel estimation procedure using a multicarrier sounding signal, of unit amplitude and a bank of filters implemented via FFT.

The comparison of Figures 8 and 10 reveals that both systems are identical except for the windowing. If we consider the received signal in Figure 8 to be windowed by a rectangular window $w[n]$, $n = 0 \ldots N - 1$, we can conclude that the correlator sounder approach is a particular case of the multicarrier/bank of filters procedure. The equivalent filters of the correlator receiver have an impulse response $c_k[n] = w[n]e^{j\frac{2\pi}{N}kn}$, for $n = 0 \cdots N - 1$ and $k = 0 \cdots N - 1$ which leads to a *sinc*-shaped frequency response. See that the length of each filter is N samples.

In order to estimate the acoustic response of the channel it is necessary to compensate for the non-flat responses of both the transmitter's projector and the receiver's preamplifier. We compensate both in the frequency domain, before applying the IFFT, with the responses provided by the manufacturer or the ones obtained with a calibration procedure of the equipment setup.

One of the main advantages of the multicarrier/bank of filters method stems from the fact that the filters $g_k[n]$ can be designed with an arbitrarily close-to-ideal response since there is no restriction in their length M. The only drawbacks of using a large M are a computational burden increase and an initial delay in the decimation of the output of the filter due to the longer transient of the filters. Both are minor issues since we are dealing with an offline computation. In Figure 11, we have superimposed the frequency response of a set of consecutive filters corresponding to the correlator receiver, $c_k[n]$ (in several colors), and the bank of filters receiver $g_k[n]$ (in blue). In this latter case, the filters have been designed by Hanning-windowing the ideal impulse response and $M \gg N$.

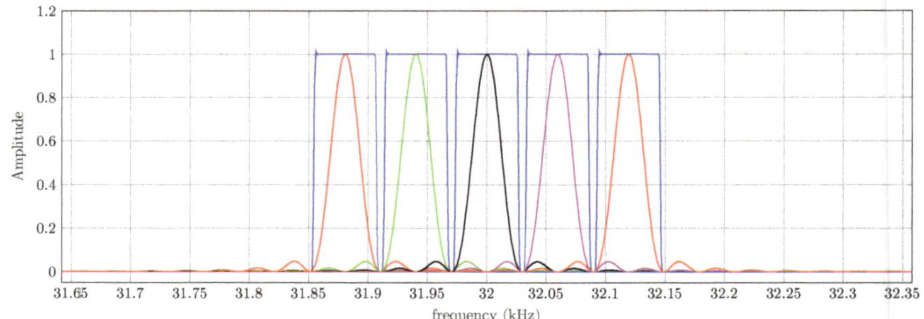

Figure 11. Frequency response of 5 consecutive filters of the bank of filters (blue) and the correlator receiver (other colors).

If the channel is time-invariant, the received carriers will remain located in their original frequencies and both systems would perform identically. However, in a real situation for UAC scenarios, the time variation of the channel response will lead to a Doppler shift and Doppler spread of the received carriers. In this case, the filters of the correlator receiver may remarkably distort each received tone and moreover, interference from adjacent bands may be significant, because neither the amplitude of the wanted signal would be maximum nor the value of the adjacent signal would be null. This fact is described in Figure 12, where a typical spectrum of received carriers is represented (the information presented in the graphics of this subsection corresponds to a measurement campaign carried out by the authors, whose details are accessible in [9]).

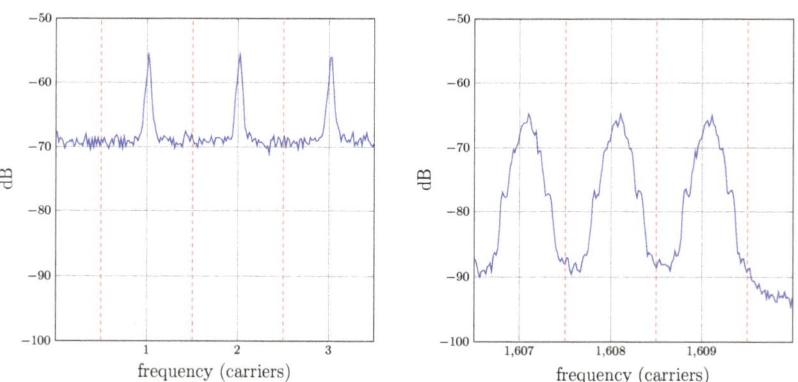

Figure 12. Received carriers spectra before resampling for low index carriers (**left**) and high index carriers (**right**). The vertical dotted lines indicate the filter band boundaries.

5.2. Doppler Effect Reduction by Resampling

The Doppler spread shown in Figure 12 can be remarkably reduced by preprocessing the received signal using resampling. This topic has been profusely mentioned in the bibliography [2–4,7,31,32] but is scarcely detailed. The procedure followed in our measurement campaigns is as follows: first, L successive estimations of the impulse response are obtained with the multicarrier/bank of filters method. The fluctuation of the initial delay on each estimation is then used to evaluate the frequency offset produced by the Doppler effect. This offset is compensated by resampling the corresponding block of the received signal. Finally, the impulse response estimation procedure is started over using the resampled received signal, which yields an improved estimation. Figure 13 shows the estimated time-variant impulse response $\hat{h}(t, \tau)$ of a channel before and after resampling. In that channel, the transmitter was 198 meters from the receiver, the hydrophone and projector were at a depth of 6 meters with respect to the surface and the seabed was at 20 m. The corresponding spectra of the received carriers before and after resampling is depicted in Figures 12 and 14, respectively. Notice that in the latter one, each spectrum is remarkably narrower and closer to the center of its band after resampling.

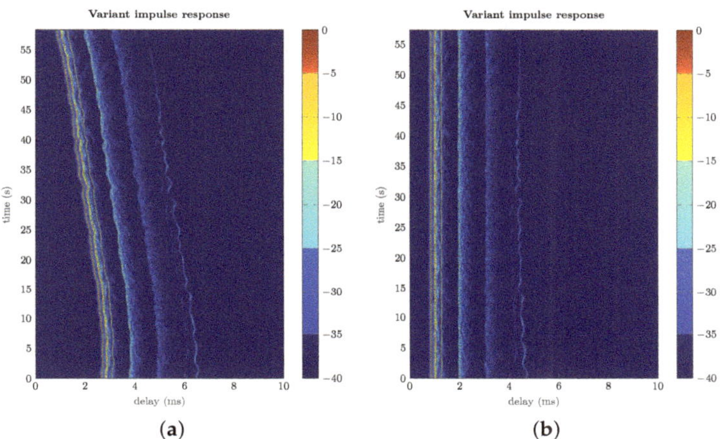

Figure 13. Time-variant impulse response estimation $\hat{h}(t, \tau)$ of a measured channel before and after resampling. (**a**) Impulse responses before resampling. (**b**) Impulse responses after resampling.

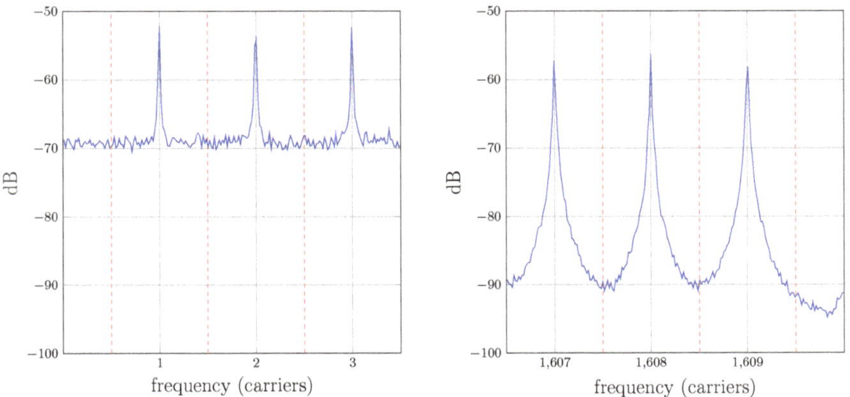

Figure 14. Received carriers spectra after resampling for low index carriers (**left**) and high index carriers (**right**). The vertical dotted lines indicate the band boundaries.

The fact that the underwater noise spectra strongly decrease with frequency [1] is also noticeable in Figure 12, where the noise floor is significantly higher for the low-frequency tones. Taking into account that these carriers are more confined in their band, we could consider an improved version of the bank of filters receiver of Figure 9 with different cutoff frequency filters, which would further reduce the noise in the estimation.

5.3. Enhancing the Measurement System Versatility

The measurement platform can work with any type of signal. For example, in the trials presented in [9], multitone signals were used for channel sounding, with the technique described above. OFDM or single-carrier signals can be employed to analyze and evaluate the performance of any underwater communications system [27]. Since measurement campaigns are generally expensive and time-consuming, it is convenient to choose the transmitted signals so that the same measurements are versatile enough to test different communication schemes. For this purpose, the following strategies have been followed. The signals are generated with non-differential constellations, for example quadriphase shift-keying (QPSK), and modulated with a pseudo-random, but deterministic and known, data sequence that is uncorrelated. If we want to study the behavior of a differential constellation, for example, differential QPSK (DQPSK), this can be achieved by interpreting the received symbols in another way (by differential decoding of the detected symbols). A similar approach can be employed to analyze the effect of using pilots in OFDM for channel estimation with a certain arrangement, more or less dense, by considering the received symbols at a set of carriers as pilots instead of data. Although we select to transmit QPSK constellations, it is also possible to test the behavior of BPSK, with this simple operation

$$\hat{x} = \text{sign}(\Re[x_R \cdot x_T^*]) \cdot x_T, \tag{4}$$

where the BPSK-equivalent demapped symbol $\hat{x}$ is estimated from an antipodal decision over the scalar product of x_R and x_T, being x_R and x_T the complex QPSK symbols at receiver and transmitter (which is indeed known), respectively.

Additionally, we adopt the strategies proposed in [33], which allows testing both any type of channel coding and interleaving schemes, and any kind of quadrature amplitude modulation (M-QAM). The first is achieved by means of a whitening process and the second by means of a so-called dithering process. In principle, the transmitted signals are generated without applying any type of channel coding. However, a post-processing of the data trials permits us to assess channel coding performance: a whitening binary sequence can be added (with modulo-2 or XOR addition) to the one obtained from the channel encoder under test, so that the resulting sequence is the pseudo-random one used in the measurements. Afterward, at the receiver, such a whitening sequence must be added to the demodulated data before applying de-interleaving and channel decoding. On the other hand, a somehow similar approach is also possible to analyze the behavior of higher order M-QAM constellations from the measurement results obtained with a simpler QPSK. For this purpose, a memoryless mapping can be used with a dithering complex sequence that is added at the transmitter after the symbol mapper and is subtracted at the receiver to the samples that feed the desired detector or M-QAM symbol demapper (this is a shortened explanation of the more detailed one that can be found in [33]).

Note that the choice of not transmitting modulated signals in measurement campaigns is also possible. Instead, one can use the same approach proposed in [34] (these authors offer a simulator and test channels available for download [35]) and make the convolution of modulated signals with the channel impulse response estimated from measurements, provided that the estimation is sufficiently accurate in time and frequency. In [27], we have presented results using this approach. However, the employ of modulated signals directly in measurement campaigns yields more realistic results. On one hand, the channel estimation is obtained using received signals that are subject to noise and residual distortion (like actual inter-carrier interference). On the other hand, the accuracy obtained in the synchronization algorithms by using periodic sounding signals and resampling, cannot be fully achieved when transmitting modulated signals.

6. Conclusions

We have described in this paper a versatile and robust measurement system to perform trials in UAC scenarios. The system comprises two subsystems prepared for on-board operation, one for the transmitter side and the other for the receiver. Both subsystems contain specific purpose HW, among which we highlight projector and hydrophones, analog front-ends, DAQ3000 acquisition boards for A/D and D/A signal conversion, and personal computers. Two SW applications have been developed for the operation, control, and management of both subsystems. Such applications carry out some on-the-fly signal processing tasks to create files where a set of sounding signals are stored after they have been transmitted through the underwater channel. Additionally, offline signal processing algorithms, which are designed to analyze the recorded data, have been explained. These set of algorithms have a two-fold purpose: some are employed to estimate the time and frequency of selective UAC channel response, and others are used to evaluate the performance of digital communication systems under realistic conditions. In particular, the need for signal resampling to compensate for the time expansion and compression it experiences through the UAC channel is addressed. Our system works in the ultrasonic frequency range and manages a wider band than the one used in other reported measurement set-ups, which supports the possibility to reach higher data rate communications. The presented work can pave the way for other research groups interested in carrying out these kinds of trials.

Author Contributions: Conceptualization, U.F.-P., J.L.-F., E.M.-N., J.F.P. and F.J.C.; Formal analysis, U.F.-P., J.L.-F., E.M.-N., J.F.P. and F.J.C.; Funding acquisition, U.F.-P. and F.J.C.; Methodology, U.F.-P., J.L.-F., E.M.-N., J.F.P. and F.J.C.; Project administration, U.F.-P. and F.J.C.; Resources, U.F.-P., E.M.-N., J.F.P. and F.J.C.; Software, U.F.-P. and E.M.-N.; Validation, U.F.-P., J.L.-F., E.M.-N., J.F.P. and F.J.C.; Writing – original draft, U.F.-P., J.L.-F., J.F.P. and F.J.C.; Writing – review and editing, U.F.-P., J.L.-F., E.M.-N., J.F.P. and F.J.C. All authors have read and agreed to the published version of the manuscript.

Funding: This work was supported in part by the Junta de Andalucía and the European Fund for Regional Development (FEDER) through project UMA18-FEDERJA-085.

Institutional Review Board Statement: Not applicable.

Informed Consent Statement: Not applicable.

Data Availability Statement: Not applicable.

Conflicts of Interest: The authors declare no conflict of interest. The funders had no role in the design of the study; in the collection, analyses, or interpretation of data; in the writing of the manuscript, or in the decision to publish the results.

References

1. Stojanovic, M.; Preisig, J. Underwater acoustic communication channels: Propagation models and statistical characterization. *IEEE Commun. Mag.* **2009**, *47*, 84–89. https://doi.org/10.1109/MCOM.2009.4752682.
2. van Walree, P.A.; Jenserud, T.; Song, H. Characterization of overspread acoustic communication channels. In Proceedings of the 10th European Conference on Underwater Acoustics, Istanbul, Turkey, 5–9 July 2010; pp. 952–958.
3. van Walree, P.A. Propagation and Scattering Effects in Underwater Acoustic Communication Channels. *IEEE J. Ocean. Eng.* **2013**, *38*, 614–631. https://doi.org/10.1109/JOE.2013.2278913.
4. Stojanovic, M. Underwater Acoustic Communications: Design Considerations on the Physical Layer. In Proceedings of the 2008 Fifth Annual Conference on Wireless on Demand Network Systems and Services, Garmisch-Partenkirchen, Germany, 23–25 January 2008; pp. 1–10. https://doi.org/10.1109/WONS.2008.4459349.
5. Singer, A.C.; Nelson, J.K.; Kozat, S.S. Signal processing for underwater acoustic communications. *IEEE Commun. Mag.* **2009**, *47*, 90–96. https://doi.org/10.1109/MCOM.2009.4752683.
6. Li, W.; Preisig, J.C. Estimation of Rapidly Time-Varying Sparse Channels. *IEEE J. Ocean. Eng.* **2007**, *32*, 927–939. https://doi.org/10.1109/JOE.2007.906409.
7. van Walree, P.A.; Otnes, R. Ultrawideband Underwater Acoustic Communication Channels. *IEEE J. Ocean. Eng.* **2013**, *38*, 678–688. https://doi.org/10.1109/JOE.2013.2253391.
8. Li, J.; Chen, F.; Liu, S.; Yu, H.; Ji, F. Estimation of Overspread Underwater Acoustic Channel Based on Low-Rank Matrix Recovery. *Sensors* **2019**, *19*. https://doi.org/10.3390/s19224976.
9. López-Fernández, J.; Fernández-Plazaola, U.; Paris, J.F.; Díez, L.; Martos-Naya, E. Wideband Ultrasonic Acoustic Underwater Channels: Measurements and Characterization. *IEEE Trans. Veh. Technol.* **2020**, *69*, 4019–4032. https://doi.org/10.1109/TVT.2020.2973495.
10. Al-Dharrab, S.; Uysal, M.; Duman, T.M. Cooperative underwater acoustic communications [Accepted From Open Call]. *IEEE Commun. Mag.* **2013**, *51*, 146–153. https://doi.org/10.1109/MCOM.2013.6553691.

11. Chitre, M.; Potter, J.; Heng, O. Underwater acoustic channel characterisation for medium-range shallow water communications. In Proceedings of the Oceans '04 MTS/IEEE Techno-Ocean '04 (IEEE Cat. No.04CH37600), Kobe, Japan, 9–12 November 2004; Volume 1, pp. 40–45. https://doi.org/10.1109/OCEANS.2004.1402892.

12. Chitre, M. A high-frequency warm shallow water acoustic communications channel model and measurements. *J. Acoust. Soc. Am.* **2007**, *122*, 2580–2586. https://doi.org/10.1121/1.2782884.

13. Sha'ameri, A.Z.; Al-Aboosi, Y.Y.; Khamis, N.H.H. Underwater Acoustic Noise Characteristics of Shallow Water in Tropical Seas. In Proceedings of the 2014 International Conference on Computer and Communication Engineering, Kuala Lumpur, Malaysia, 23–25 September 2014; pp. 80–83. https://doi.org/10.1109/ICCCE.2014.34.

14. Quazi, A.; Konrad, W. Underwater acoustic communications. *IEEE Commun. Mag.* **1982**, *20*, 24–30. https://doi.org/10.1109/MCOM.1982.1090990.

15. Aval, Y.M.; Stojanovic, M. Differentially Coherent Multichannel Detection of Acoustic OFDM Signals. *IEEE J. Ocean. Eng.* **2015**, *40*, 251–268. https://doi.org/10.1109/JOE.2014.2328411.

16. Kilfoyle, D.; Preisig, J.; Baggeroer, A. Spatial modulation experiments in the underwater acoustic channel. *IEEE J. Ocean. Eng.* **2005**, *30*, 406–415. https://doi.org/10.1109/JOE.2004.834168.

17. Li, B.; Huang, J.; Zhou, S.; Ball, K.; Stojanovic, M.; Freitag, L.; Willett, P. MIMO-OFDM for High-Rate Underwater Acoustic Communications. *IEEE J. Ocean. Eng.* **2009**, *34*, 634–644. https://doi.org/10.1109/JOE.2009.2032005.

18. Radosevic, A.; Proakis, J.; Stojanovic, M. Statistical characterization and capacity of shallow water acoustic channels. In Proceedings of the OCEANS 2009-EUROPE, Bremen, Germany, 11–14 May 2009; pp. 1–8. https://doi.org/10.1109/OCEANSE.2009.5278349.

19. Li, B.; Zhou, S.; Stojanovic, M.; Freitag, L.; Willett, P. Multicarrier Communication Over Underwater Acoustic Channels With Nonuniform Doppler Shifts. *IEEE J. Ocean. Eng.* **2008**, *33*, 198–209. https://doi.org/10.1109/JOE.2008.920471.

20. Aval, Y.M.; Wilson, S.K.; Stojanovic, M. On the Average Achievable Rate of QPSK and DQPSK OFDM Over Rapidly Fading Channels. *IEEE Access* **2018**, *6*, 23659–23667. https://doi.org/10.1109/ACCESS.2018.2828788.

21. Ribas, J.; Sura, D.; Stojanovic, M. Underwater wireless video transmission for supervisory control and inspection using acoustic OFDM. In Proceedings of the OCEANS 2010 MTS/IEEE SEATTLE, Seattle, WA, USA, 20–23 September 2010; pp. 1–9. https://doi.org/10.1109/OCEANS.2010.5663839.

22. Hegazy, R.; Kadifa, J.; Milstein, L.; Cosman, P. Subcarrier Mapping for Underwater Video Transmission Over OFDM. *IEEE J. Ocean. Eng.* **2021**, *46*, 1408–1423. https://doi.org/10.1109/JOE.2021.3076274.

23. Ochi, H.; Watanabe, Y.; Shimura, T.; Hattori, T. Experimental Results of Short Range Wideband Acoustic Communication Using QPSK and 8PSK. In Proceedings of the OCEANS 2008—MTS/IEEE Kobe Techno-Ocean, Kobe, Japan, 8–11 April 2008; pp. 1–5. https://doi.org/10.1109/OCEANSKOBE.2008.4531085.

24. Diamant, R.; Tan, H.P.; Lampe, L. LOS and NLOS Classification for Underwater Acoustic Localization. *IEEE Trans. Mob. Comput.* **2014**, *13*, 311–323. https://doi.org/10.1109/TMC.2012.249.

25. Headrick, R.; Freitag, L. Growth of underwater communication technology in the U.S. Navy. *IEEE Commun. Mag.* **2009**, *47*, 80–82. https://doi.org/10.1109/MCOM.2009.4752681.

26. Cañete, F.J.; López-Fernández, J.; García-Corrales, C.; Sánchez, A.; Robles, E.; Rodrigo, F.J.; Paris, J.F. Measurement and Modeling of Narrowband Channels for Ultrasonic Underwater Communications. *Sensors* **2016**, *16*, 256. https://doi.org/10.3390/s16020256.

27. Cobacho-Ruiz, P.; Cañete, F.J.; Martos-Naya, E.; Fernández-Plazaola, U. OFDM System Design for Measured Ultrasonic Underwater Channels. *Sensors* **2022**, *22*, 5703. https://doi.org/10.3390/s22155703.

28. IOtech. DaqX Software. Available online: http://support.elmark.com.pl/iotech/desktop/DaqLab_Users_Manual.pdf (accessed on 23 April 2022).

29. Hans-Jurgen Zepernick, A.F. *Pseudo Random Signal Processing: Theory and Application*; Wiley-Blackwell: Chichester, UK, 2013.

30. Crochiere, R.; Rabiner, L. *Multirate Digital Signal Processing*; Prentice-Hall Signal Processing Series: Advanced monographs; Prentice-Hall: Hoboken, NJ, USA, 1983.

31. Daoud, S.; Ghrayeb, A. Using Resampling to Combat Doppler Scaling in UWA Channels With Single-Carrier Modulation and Frequency-Domain Equalization. *IEEE Trans. Veh. Technol.* **2016**, *65*, 1261–1270. https://doi.org/10.1109/TVT.2015.2409560.

32. Liu, X.; Qiao, G.; Ma, L.; Zheng, N.; Zhao, Y. Non-Uniform Doppler Compensation Method for Staggered Multitone Filter Bank Multicarrier in the Underwater Acoustic Channel. In Proceedings of the 2021 OES China Ocean Acoustics (COA), Harbin, China, 14–17 July 2021; pp. 586–590. https://doi.org/10.1109/COA50123.2021.9520061.

33. Yang, S.; Deane, G.B.; Preisig, J.C.; Sevüktekin, N.C.; Choi, J.W.; Singer, A.C. On the Reusability of Postexperimental Field Data for Underwater Acoustic Communications R&D. *IEEE J. Ocean. Eng.* **2019**, *44*, 912–931. https://doi.org/10.1109/JOE.2019.2925921.

34. van Walree, P.A.; Socheleau, F.X.; Otnes, R.; Jenserud, T. The Watermark Benchmark for Underwater Acoustic Modulation Schemes. *IEEE J. Ocean. Eng.* **2017**, *PP*, 1–12. https://doi.org/10.1109/JOE.2017.2699078.

35. van Walree, P.A.; Otnes, R.; Jenserud, T. The Watermark Acoustic Modem Benchmark. 2016. Available online: http://www.ffi.no/watermark (accessed on 10 August 2022).

Article

Autoencoder-Based Signal Modulation and Demodulation Methods for Sonobuoy Signal Transmission and Reception

Jinuk Park [1], Jongwon Seok [2,*] and Jungpyo Hong [2,*]

1   School of Electrical Engineering, Korea Advanced Institute of Science and Technology, Daejeon 34141, Korea
2   Department of Information and Communication Engineering, Changwon National University, Changwon 51140, Korea
*   Correspondence: jwseok@changwon.ac.kr (J.S.); hansin@changwon.ac.kr (J.H.)

Abstract: Sonobuoy is a disposable device that collects underwater acoustic information and is designed to transmit signals collected in a particular area to nearby aircraft or ships and sink to the seabed upon completion of its mission. In a conventional sonobuoy signal transmission and reception system, collected signals are modulated and transmitted using techniques such as frequency division modulation or Gaussian frequency shift keying. They are received and demodulated by an aircraft or a ship. However, this method has the disadvantage of a large amount of information being transmitted and low security due to relatively simple modulation and demodulation methods. Therefore, in this paper, we propose a method that uses an autoencoder to encode a transmission signal into a low-dimensional latent vector to transmit the latent vector to an aircraft or vessel. The method also uses an autoencoder to decode the received latent vector to improve signal security and to reduce the amount of transmission information by approximately a factor of a hundred compared to the conventional method. In addition, a denoising autoencoder, which reduces ambient noises in the reconstructed outputs while maintaining the merit of the proposed autoencoder, is also proposed. To evaluate the performance of the proposed autoencoders, we simulated a bistatic active and a passive sonobuoy environments. As a result of analyzing the sample spectrograms of the reconstructed outputs and mean square errors between original and reconstructed signals, we confirmed that the original signal could be restored from a low-dimensional latent vector by using the proposed autoencoder within approximately 4% errors. Furthermore, we verified that the proposed denoising autoencoder reduces ambient noise successfully by comparing spectrograms and by measuring the overall signal-to-noise ratio and the log-spectral distance of noisy input and reconstructed output signals.

Keywords: sonobuoy; autoencoder; denoising; signal transmission and reception; modulation and demodulation

Citation: Park, J.; Seok, J.; Hong, J. Autoencoder-Based Signal Modulation and Demodulation Methods for Sonobuoy Signal Transmission and Reception. *Sensors* **2022**, 22, 6510. https://doi.org/10.3390/s22176510

Academic Editors: Sylvain Girard, Haixin Sun and Xuebo Zhang

Received: 29 July 2022
Accepted: 26 August 2022
Published: 29 August 2022

1. Introduction

Sonobuoy is a combination of sonar and buoy, and refers to a device that collects underwater information through sound waves. Sonobuoy is a disposable device that is dropped from the maritime patrol to the area of interest. It is designed to transmit the collected underwater signal to the maritime patrol via wireless communication and sink to the sea when the mission is completed. Sonobuoy is divided into passive and active types according to the detection method and the detection range, operating time, and service life vary widely for each product model [1]. Among them, the transmission bit rate of signals is used from hundreds of kbps (kilobits per second) to tens of Mbps (Megabits per second) [2].

The sonobuoy may be operated in a monostatic sonobuoy when the active sonobuoy (CASS) or DICASS (Directional CASS) is used alone. It may be operated in a bistatic and multi-static sonobuoy with an explosive sound source or a combination of active and passive sonobuoys [3]. In general, since a bistatic target detection has different positions of a transmitter and a receiver, the detection area is wider and confidentiality is guaranteed compared to monostatic target detection [4]. Figure 1 presents a research conceptual diagram schematically illustrating a bistatic target detection environment using active and passive sonobuoys in anti-submarine warfare (ASW). Since the system of the electronic unit of the sonobuoy cannot perform complex signal processing, sonobuoy inevitably transmits the collected underwater signal to the maritime patrol plane or the ship through wireless communication. The existing signal modulation and demodulation methods used in wireless communication include frequency division multiplexing [5] and frequency shift keying (FSK) [6]. Such a signal modulation and demodulation method have a disadvantage in that a high bit rate is required due to the large amount of information to be transmitted because it is the entire acoustic signal. In addition, since the frequency band of the modulated signal is easily analyzed, the modulation scheme is relatively easy to predict and is highly likely to be demodulated, resulting in low security.

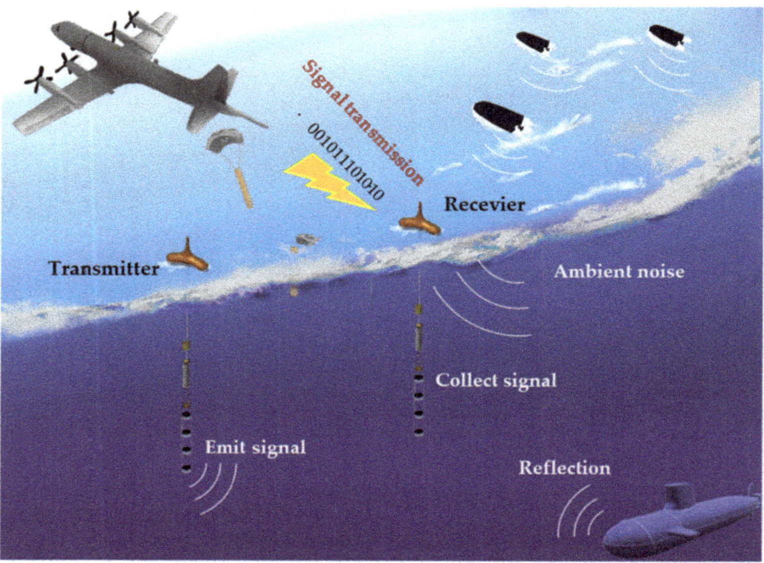

Figure 1. A research conceptual diagram of the bistatic sonobuoy signal transmission and reception.

Conversely, using deep neural nets recently, there have been great achievements in various fields, i.e., speech recognition, visual object recognition, object detection, and natural language processing [7]. In particular, autoencoder is an unsupervised learning-based feature extraction technique that can obtain high-level features of input signals by learning and using unlabeled data. It is more practical because it can be applied to a wider range of data than supervised learning, which is expensive to obtain labeled data. The autoencoder mainly consists of encoder and decoder parts, where the encoder yields high-level features, usually called codes or latent variables. These represent input signals compressively well, and the decoder is trained to restore the codes as close as possible to the original inputs. Autoencoder is similar to principal component analysis (PCA) in that it compresses inputs into latent variables by reducing the dimension of data in the encoding process. However, autoencoder is a nonlinear generalization of PCA and presents superior performance to PCA in general [8]. Normally, the structure of the autoencoder is stacked by multiple layers to extract high-level features. Denoising, sparse, and variational autoencoders are developed for further performance improvements in the feature representation [8–18]. According to [10,11], denoising autoencoders learn more key high-dimensional features in the training process by performing denoising tasks on noisy input signals and are known to be superior to the traditional autoencoder [10,11]. Usually, traditional and denoising autoencoders are implemented as under-complete models in which the input dimension is larger than the hidden layer, and as the autoencoder is stacked, the input of the layer is compressed and the number of activations (outputs of each layer) is reduced. Conversely, sparse autoencoders are implemented as over-complete models, which have a larger dimension of hidden layer than the dimension of input, unlike traditional and denoising autoencoders [12,13]. Sparse autoencoders are other methods for extracting interesting structures of data by imposing "sparsity", which means most nodes are inactive and active nodes exist very rarely, on nodes of layers. However, the disadvantage of sparse autoencoders is the computational complexity of which the activation value must be calculated in advance in order to add sparsity to the cost function [12]. An efficient algorithm to iteratively solve this problem has been proposed [13]. In addition, variational autoencoders are the ones that involve the notion of probability [14,15]. Variational autoencoders are stochastic generative models that model the probability distribution of parameters, whereas the other autoencoders are deterministic discriminative models that model the value itself of parameters [15]. The common and ultimate goals of the above-mentioned methods are to extract high-dimensional features or representations of the input and to improve the performance of tasks (mainly classification) with the features. The results of the autoencoders are used independently or in combination with other methods, i.e., support vector machine (SVM), convolutional neural network (CNN), and Gaussian mixture model (GMM), mainly as a front-end for parameter initialization of supervised learning [16–18].

In this paper, a novel approach to apply the under-complete structure of the autoencoder to sonobuoy signal modulation and demodulation for signal transmission and reception in order to decrease the amount of information to be transmitted and increase security. Our contributions are two-fold. First, we propose a method that modulates the transmission signal to a low-dimensional latent vector using an autoencoder to transmit the latent vector to an aircraft or vessel and demodulates the received latent vector to reduce the amount of transmission information and improve the security of the signal. Second, a denoising autoencoder that reduces ambient noises in the reconstructed outputs while maintaining the merit of the proposed autoencoder is also proposed.

2. Conventional Sonobuoy Signal Modulation and Demodulation Methods

2.1. Frequency Division Multiplexing

Frequency division multiplexing is a method used to transmit multi-channel signals to a single channel, and multi-channel signals are transferred to different frequency bands within a multiplexer. The modulated signals are combined into a single-channel signal through simple addition and then transmitted. Directional frequency analysis and recording (DIFAR) uses frequency division multiplexing and the overall structure is in Figure 2. As shown in Figure 2, DIFAR requires a high bit rate because it transmits the entire signal combined into a single channel, and it is less secure because the transmitted signal is easily distinguishable in the frequency domain.

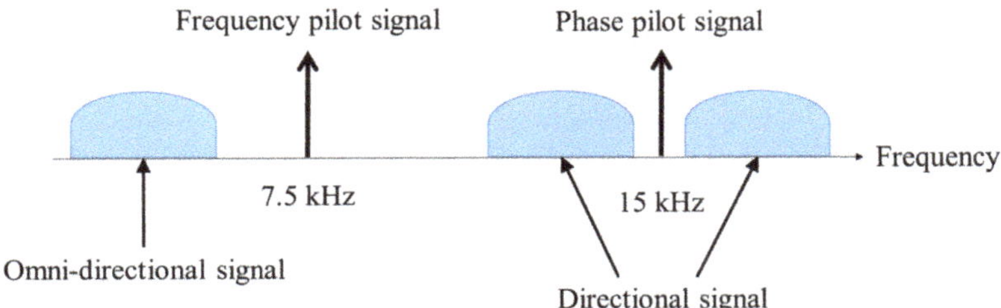

Figure 2. Frequency division multiplexing of DIFAR for signal modulation and demodulation.

2.2. Frequency Shift Keying

Frequency shift keying transmits information as a frequency change of a carrier signal such as a sine wave. For the simplest binary frequency variation modulation (binary FSK), two signals of different frequencies are used to transmit binary information of 0 and 1. In addition, a Gaussian filter is used for frequency conversion, called Gaussian frequency shift keying (Gaussian FSK), which is a signal transmission method used in various sonobuoy models.

3. Proposed Autoencoder-Based Signal Modulation and Demodulation Methods

3.1. Autoencoder-Based Signal Modulation and Demodulation Method (General Form)

Autoencoder is a widely used structure in the field of deep learning. It trains the output value of the model to be the same as the input value and forms a symmetrical structure. The structure is largely divided into an encoder, a decoder, and a bottleneck section between the encoder and the decoder. The dimension of the latent vector in the bottleneck section extracted through the encoding is generally much lower than the dimension of the input, and the latent vector reflects the compressed implications of the data. Therefore, the autoencoder mainly serves as a feature extractor using the characteristics of latent vectors to obtain initial weights of other models [16–18].

In this paper, using the autoencoder structure, latent vectors are extracted by the encoder installed within the sonobuoy and transmitted. The signal processor in the marine patrol reconstructs the original signal from the received latent vectors. The schematic description and the entire training process of the proposed autoencoder-based method are shown in Figure 3 and Algorithm 1, respectively.

Algorithm 1: Pseudocode for autoencoder training algorithm

1: AE training $(e, b, l, E, \Delta, \theta)$
2: $\mathbf{x} = [x_1, x_2, \ldots, x_n] \in R^{N*1}$ is the input signal, in which $x_i \in [-1, \; 1]$ $(1 \leq i \leq N)$ is a single acoustic signal sample
3: e is the number of epochs
4: l is the learning rate
5: b is the batch size
6: E is encoder network
7: Δ is decoder network
8: θ is the network parameters
9:    **for** 0 to e **do**
10:      $\hat{\mathbf{x}} = \Delta(E(\mathbf{x}))$
11:      $L = \sum\limits_{i=1}^{b} (\mathbf{x} - \hat{\mathbf{x}})^2$
12:      $L = mean(L)$
13:      g = gradient of θ
14:      **for** θ in θ **do**
15:        $\theta = \theta - l * g$
16:      **end**
17: **end**

The purpose of this study is to secure the security of the signal using fewer bits than the conventional method when transmitting the sonobuoy signal. The autoencoder judged it to be a suitable model that satisfies both of these purposes. Since the autoencoder trains the same input and output values, it is consistent with the concept of demodulating the transmission signal again. Furthermore, since the latent vectors in the bottleneck section have information that can be restored to the original signal, they can be demodulated to the original signal if only the latent vector and trained demodulator are present, even if the entire signal is not transmitted. By using this, it is possible to transmit signals even in adverse communication environments with a very low bit rate compared to the conventional sonobuoy signal transmission technique. In addition, since all of these transmission and reception processes can only be demodulated by having a basically trained autoencoder model, a third party cannot demodulate into an original signal even if it acquires a transmitting latent vector.

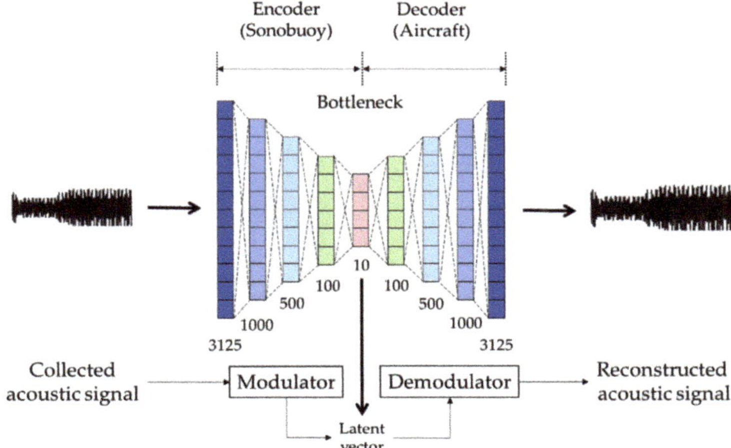

Figure 3. The schematic description of the proposed autoencoder-based signal modulation and demodulation. The trained encoder and decoder are installed on sonobuoy and aircraft, respectively. The main goal of the proposed method is to reconstruct the input signal as close as possible with the latent vector.

3.2. Denoising Autoencoder-Based Signal Modulation and Demodulation Method (Optional Form)

As can be seen in Figure 1, ambient noise exists in the underwater environment [19]. Ambient noise acts as one of the major causes of performance degradation in underwater target detection and identification using signals acquired by sonar and sonobuoy [20]. Therefore, noise reduction algorithms are usually applied as preprocessing to prevent unnecessary performance degradation [20]. In this section, we propose a denoising autoencoder method that can perform the additional ambient noise reduction while maintaining the advantages of the general autoencoder-based method proposed in the above section.

The overall structure of the proposed denoising autoencoder is shown in Figure 4. The denoising autoencoder, like the autoencoder in Figure 3, consists of an encoder, a bottleneck section, and a decoder. However, the output is different in that it yields a noise-removed signal, and the training method for this is different. The entire training process is described in detail in Algorithm 2. The denoising autoencoder is trained with noise-corrupted input data at various signal-to-noise ratios (SNRs) in order to restore the signal of interest even if the input data is distorted or contains a noise. Therefore, using the denoising autoencoder, ambient noise reduction is possible in the transmission and reception process without a separate noise reduction method.

As can be seen in Figure 4, since the network structure is in the under-complete form, the size of the layer is decreasing. As such, it can be transmitted using only a few bits compared to the conventional signal transmission and reception technique; the security of the signal can be guaranteed. In addition, it is possible to obtain an ambient noise-reduced output.

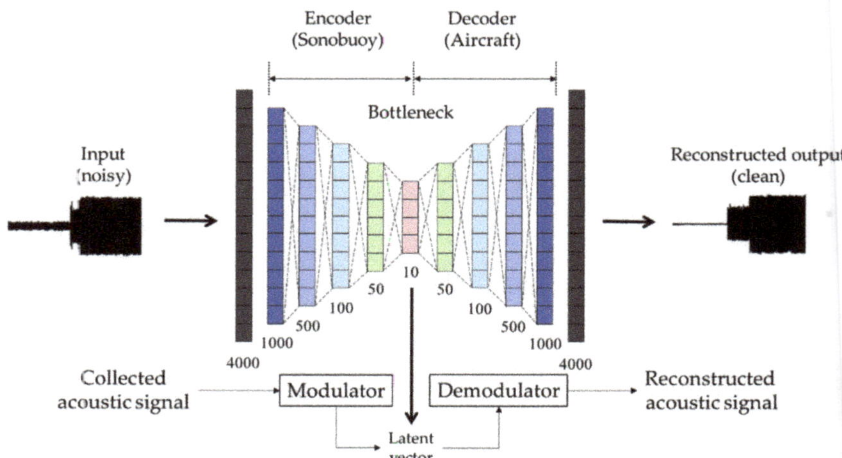

Figure 4. This is the schematic description of the proposed denoising autoencoder-based signal modulation and demodulation. This is an optional form of the general form in Figure 3 for ambient noise reduction. The trained encoder and decoder are installed on sonobuoy and on aircraft, respectively. The main goals of the proposed method are to reconstruct the input signal as close as possible with the latent vector and to reduce ambient noise.

Algorithm 2: Pseudocode for denoising autoencoder training algorithm

---

1: DAE training $(e, b, l, E, \Delta, \theta)$
2: $\mathbf{x} = [x_1, x_2, \ldots, x_n] \in R^{N*1}$ is the clean input and $\mathbf{z} = [z_1, z_2, \ldots, z_n] \in R^{N*1}$ is the noise input in which $x_i \in [-1, \ 1]$ and $z_i \in [-1, \ 1]$ $(1 \leq i \leq N)$ are a single acoustic signal sample
3: e is the number of epochs
4: l is the learning rate
5: b is the number of batches
6: E is encoder network
7: Δ is decoder network
8: θ is the network parameters
9:    **for** 0 to e **do**
10:      **for** $j = 1$ to b **do**
11:      r is SNR between $\mathbf{x}_j$ and $\mathbf{z}_j$ in dB scale $\in [0, \ 5, \ 10, \ 15]$
12:      $\mathbf{y}_j = \mathbf{x}_j + r\mathbf{z}_j$
13:      $\hat{\mathbf{y}}_j = \Delta(E(\mathbf{y}_j))$
14:      $L = \sum_{i=1}^{b} (\mathbf{x}_j - \hat{\mathbf{y}}_j)^2$
15:      $L = mean(L)$
16:      $\mathbf{g}$ = gradients of θ
17:       **for** θ, g in $\theta, \mathbf{g}$ **do**
18:        $\theta = \theta - l * g$
19:      **end**
20:    **end**
21: **end**

---

4. Experiments with Simulated Data

In this paper, the performance of the autoencoder was verified in a bistatic active sonobuoy environment and the performance of the denoising autoencoder was verified in a passive sonobuoy environment. However, both autoencoder and denoising autoencoder are basically applicable to both active and passive sonobuoy environments.

4.1. Experiments for Evaluation of Autoencoder in a Bistatic Active Sonobuoy Environment

4.1.1. Experimental Setup

In this paper, we generate bistatic simulation data in an underwater environment to verify the proposed method. The transmitting signals are generated in two forms: continuous wave (CW) and linear frequency modulation (LFM). The positions of the transmitter and receiver were fixed, and the maximum distance between a target and sonobuoys was limited to 9 km. The target maneuver range was set between 50 m and 150 m. Other detailed conditions for simulation data generation are summarized in Table 1.

Table 1. Simulation environmental setup.

	CW	LFM
Center frequency	3500 Hz, 3600 Hz, 3700 Hz, 3800 Hz	
Bandwidth	-	400 Hz
Pulse duration	0.1 s, 0.5 s, 1 s	
Sampling frequency	31,250 Hz	

Data were generated in a scenario of receiving a pulse signal reflected from a target. Scenarios in which the target location is randomly set are stored as files that are about 10 s long. The total training data were about 50 h, and the evaluation set for evaluating whether the training was converged was about 3 h of data separately from the training data. All data were generated by applying ray tracing [21] as in Figure 5, and the sound velocity profile used for the ray tracing is shown in Figure 6.

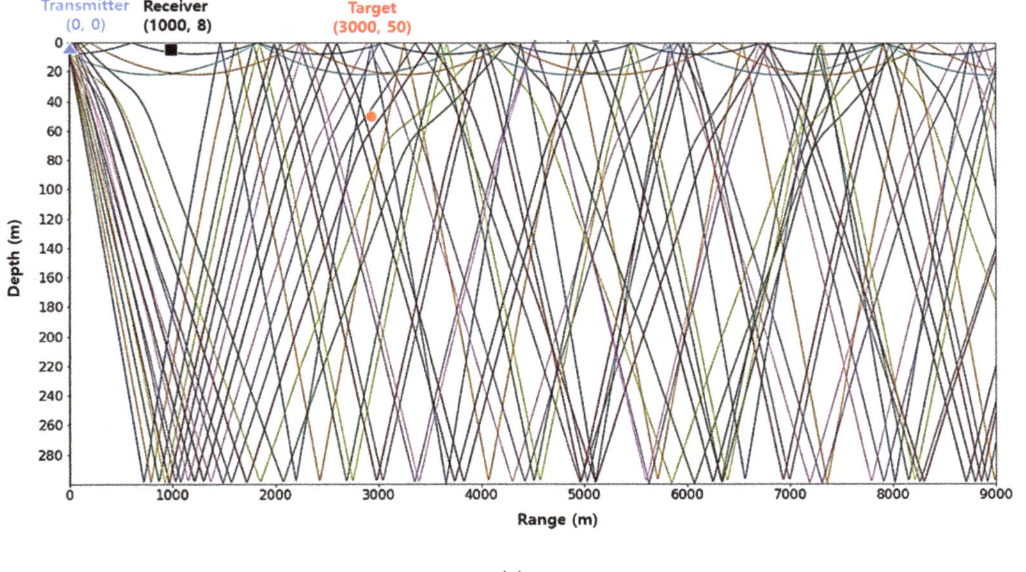

(a)

Figure 5. *Cont.*

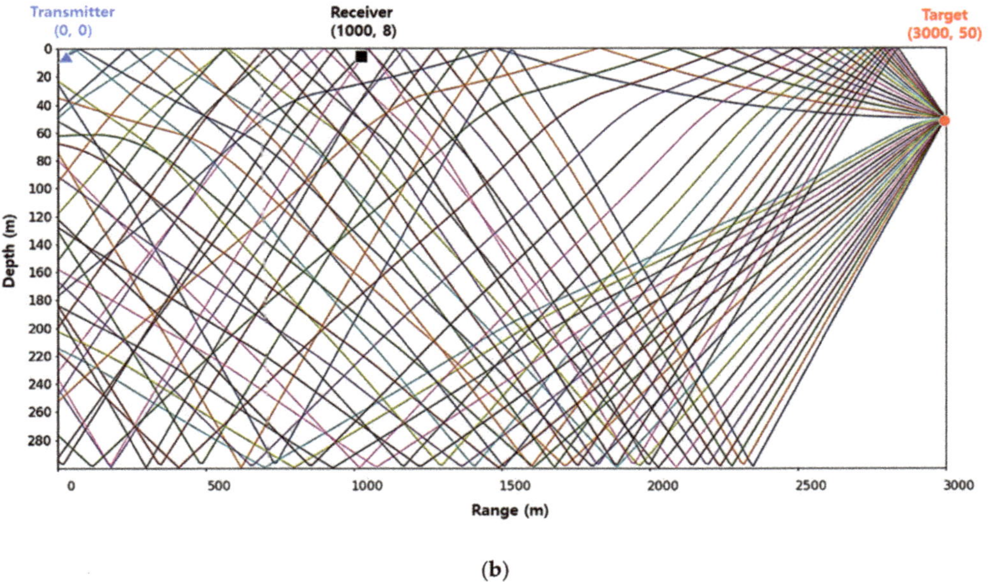

(b)

Figure 5. Sound propagation paths samples [range (m), depth (m)] (triangle: transmitter [0,8], square: receiver [1000,8], circle: target [3000,50]): (**a**) Sound propagation paths sample (from the transmitter to the target); (**b**) Sound propagation paths samples (from the target to the receiver).

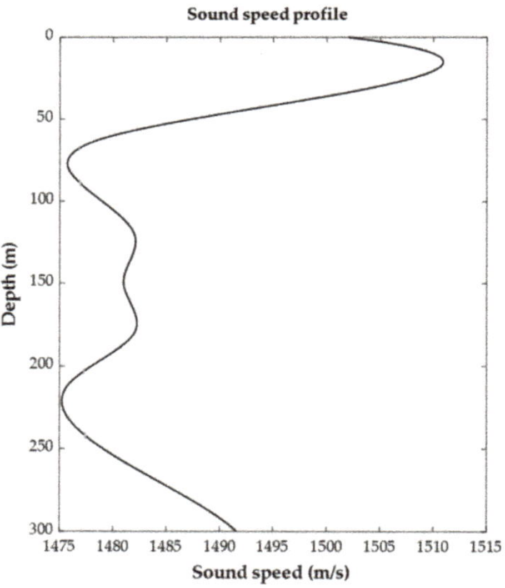

Figure 6. Sound speed profile of the simulated bistatic sonobuoy environment.

Since the sonobuoy signal cannot be modeled with a simple single-layer autoencoder structure, this experiment used a stacked autoencoder to train. Table 2 shows the parameter setting of the model used in the experiment. The parameters were empirically set, and due to the nature of the research data, the result cannot be confirmed only by the loss value, so it should be determined through the restored sample. Although the width and depth of the training layer are not optimal variables, we have failed to model the distribution of complicated input signals when the depth of the layer is too low.

Table 2. Model structure of the baseline autoencoder model.

Encoder (Dim)	Decoder (Dim)
Noisy input (3125)	Latent vector (10)
Linear (3125–1000)	Linear (10–100)
ReLU	ReLU
Linear (1000–500)	Linear (100–500)
ReLU	ReLU
Linear (500–100)	Linear (500–1000)
ReLU	ReLU
Linear (100–10)	Linear (1000–3125)
ReLU	Tanh
Latent vector (10)	Output (3125)

Unlike general audio signals, underwater acoustic signals are very sparse in the frequency domain, thus, we used the time domain acoustic signal as an input for training, which means end-to-end training. Input signals were normalized in the range of −1 to 1 per file, and if the input size of the model was too small, it could not reflect the pulse of one cycle properly, resulting in discontinuity, so it was put into the model in 0.1 s (3125 samples). Between linear layers, ReLU was used as an activation function to reflect the nonlinearity of the input signal, and in the last layer of the decoder, that was used as an activation function to restore the signal to the range from –1 to 1. As a loss function, a mean square error (MSE) between the input and the output signals was calculated for each sample, and the adam optimizer [22] with a learning rate of 0.001 was used.

4.1.2. Experimental Results

To evaluate the original signal restoration performance of the autoencoder, we measured the MSE of the original and the restored spectrograms. The signal used for the evaluation generated 60 s of data not used for model training, and the average energy in the frequency domain was 0.0074. The performances of three autoencoder models (autoencoder I, autoencoder II, and autoencoder III) were measured and summarized in Table 3. The experimental results presented that MSE, which represents a difference from the original signal, had 4.08%, 3.88%, and 3.22%, respectively, compared to the energy average of the original signal. The performance of the model consisting of eight linear layers was the best.

Table 3. Comparison of autoencoder models in MSE. Autoencoder III refers to model in Table 2. M denotes 10^6.

Model	MSE	The Number of Parameters
Autoencoder I	0.00302	6.27 M
Autoencoder II	0.00287	6.66 M
Autoencoder III	0.00238	7.36 M

Comparing the spectrograms depicted in Figure 7, it can be confirmed that the original signal can be restored by a low-dimensional latent vector with a small artifact. In addition to the frequency band where the echo signal exists, it can be seen that signals such as harmonics are seen or signals such as noise are also present in the signal-absent interval. In the case of the signal-absent interval, it is considered that the magnitude of the data in the interval of the original signal is too small. This results in a noise-like signal in the band in which the echo signal exists due to the bias value of the autoencoder model. In addition, since the proposed method trains in the form of end-to-end without a separate feature extraction process, small value differences in the time domain may appear as noise in the high-frequency band even if the value of the loss function decreases. Nevertheless, both types of signals have been restored very similarly in the frequency bands where echo signals exist. Artifacts generated in the signal absent intervals and bands are negligible compared to the energy of the target signal of interest.

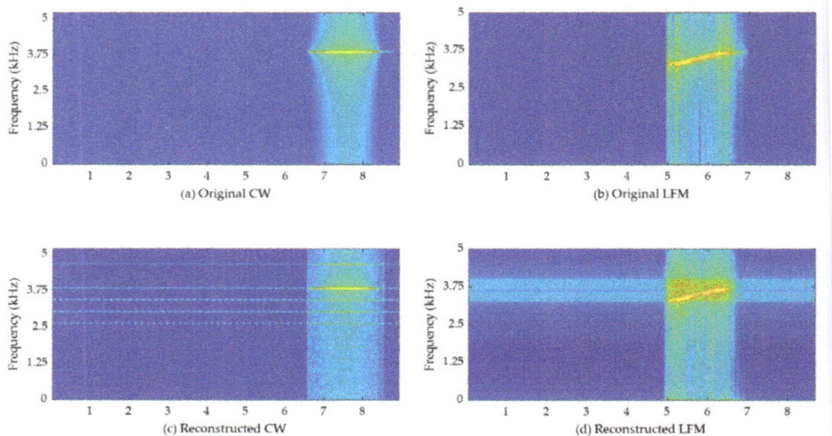

Figure 7. Spectrogram comparison between original and reconstructed signals: (**a**) Original CW; (**b**) Original LFM; (**c**) CW reconstructed by the proposed autoencoder; (**d**) LFM reconstructed by the proposed autoencoder.

In order to demodulate the original signal, in addition to the latent vector, two values used for normalization must be transmitted. Therefore, if the 10-dimensional latent vector and the size value used for normalization are quantized to 16 bits, the amount of information in the proposed method is about 1.92 kbps. To compare the reconstruction performance, we measured MSE according to the number of quantized bits per sample used in the conventional method [1]. Figure 8 shows the reconstruction error of the conventional and the proposed methods depending on the amount of information required for encoding and decoding. Here, Autoencoder III in Table 3 is used for comparison.

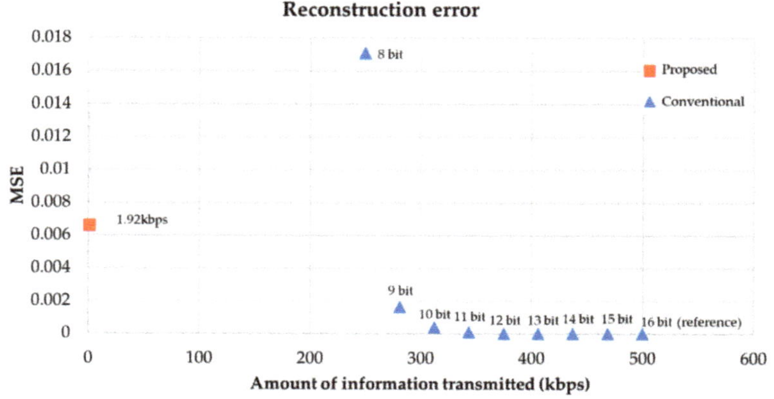

Figure 8. Reconstruction error (MSE) according to the amount of information transmitted. A signal quantized to 16 bits was regarded as a reference signal without errors.

Figure 8 presents that the MSE value of the proposed method (red square) is located between the MSEs of the conventional method (blue triangle) using 8 bits and 9 bits per sample. Considering that the conventional method quantizes to 14 or 16 bits [1], the proposed method relatively presents a large MSE. Therefore, further study to reduce the MSE of the proposed method is necessary. Nevertheless, the reason why we insist that the proposed method is superior to the conventional method is the noticeable reduction in the amount of information transmitted. Consider the cases of quantizing a sampling frequency of 31,250 Hz to 16 bits and 8 bits per sample. These cases require 500 kbps (31,250 samples × 16 bits) and 250 kbps (31,250 samples × 8 bits), respectively. In this paper, we generated a 10-dimensional latent vector every 0.1 s, and store the two-dimensional information that we used for normalization in the training process. Therefore, the proposed method only requires transmitting vectors of 120 dimensions for transmitting a signal of 1 s, not the total number of samples (e.g., 31,250 samples for 31,250 Hz). Compared to the conventional method using 16-bit quantization, the amount of transmission information of the proposed method is 260 times smaller than that of the conventional method. Furthermore, assuming that the amount of transformation information of the conventional method presenting similar MSE performance in Figure 8 is approximately 250 kbps, the amount of transmission information of the proposed method is 130 times smaller than that of the conventional method. This means that the proposed method can encode and decode sonobuoy signals with 130 times less information, which is due to the nature of the latent vector, generated from autoencoder, being represented as a compressed, very high-dimensional feature vector of the input signal. Additionally, owing to the inherent characteristics of the autoencoder, the latent vector can obtain high security that cannot be decrypted without the decoder of the corresponding autoencoder.

4.2. Experiments for Evaluation of Denoising Autoencoder in a Passive Sonobuoy Environment

4.2.1. Experimental Setup

In order to verify the performance of the denoising autoencoder, a DIFAR passive sonobuoy detection environment was simulated using the MATLAB Phased-Array System Toolbox [23,24], and detailed experimental conditions are shown in Table 4.

Table 4. Simulation setup for evaluating the denoising autoencoder.

	Tonal Signal
Depth	30, 60, 120, 300 (m)
Detection range	0~15 km
Center frequency	1000~2400 Hz
Sampling frequency	40,000 Hz

It is assumed that a tonal signal is continuously generated at a target located randomly in the detection range. The signal generated at the target reaches the receiver in consideration of the Doppler effect, the reflection loss of sound waves, and the spreading loss. One file was about 60 s long, a total of 50 h of data were used for training, and 3 h of data were used for evaluation and test.

Like the autoencoder in Table 3, the denoising autoencoder model consists of an encoder and a decoder with multiple layers, and the detailed model structure is shown in Table 5. The input signal was normalized in the range from −1 to 1, chopped by 0.1 s, and inserted into the model for network training. The difference from Table 3 is slight in the number of input samples caused by the different sampling rate, and there is one more layer.

Table 5. Model structure of the baseline denoising autoencoder model.

Encoder (Dim)	Decoder (Dim)
Noisy input (4000)	Latent vector (10)
Linear (4000–1000)	Linear (10–50)
ReLU	ReLU
Linear (1000–500)	Linear (50–100)
ReLU	ReLU
Linear (500–100)	Linear (100–500)
ReLU	ReLU
Linear (100–50)	Linear (500–1000)
ReLU	ReLU
Linear (50–10)	Linear (1000–4000)
ReLU	Tanh
Latent vector (10)	Output (4000)

As described in Algorithm 2, the input of the encoder synthesized the clean signal and the white noise with an SNR from 0 to 15 dB. The output of the decoder was trained to reduce the MSE with the clean signal. Through this training method, the noisy signal may be restored as a clean signal for various SNR environments.

4.2.2. Experimental Results

To evaluate the performance of the denoising autoencoder, the spectrogram was subjectively analyzed, and the overall SNR and log spectral distance (LSD) of inputs and outputs were objectively measured [25]. We summarized the noise reduction performance of the proposed denoising autoencoder and the Wiener filter-based sonobuoy noise reduction method [26] in Table 6.

Table 6. Overall SNR and LSD results at various SNRs. All measures are in dB.

	Input SNR	0 dB	5 dB	10 dB	15 dB
Overall SNR	Noisy input	2.24	4.01	6.56	10.35
	Conventional method [26]	4.35	7.13	10.73	15.14
	Reconstructed output	*10.91*	*12.96*	*12.80*	*15.90*
LSD	Noisy input	7.54	7.16	6.78	5.96
	Conventional method [26]	7.05	6.56	6.04	5.52
	Reconstructed output	*3.64*	*3.61*	*3.68*	*3.22*

In the spectrogram of Figure 9, most of the noises present in the noisy input have been removed from the restored signals. The noise around the signal band remains; however, the level of the remaining noise is negligible.

In addition, an overlayed frequency analysis of time segments of clean, noisy, and reconstructed signals at the 2.4–2.6 s interval in Figure 9 is inserted in Figure 10. Normalization was performed using each maximum value of clean, noisy, and reconstructed signals for power spectrum comparison. In Figure 10, it can be seen that the spectra of the clean signal and reconstructed output are very close by removing the noises distributed in the entire band.

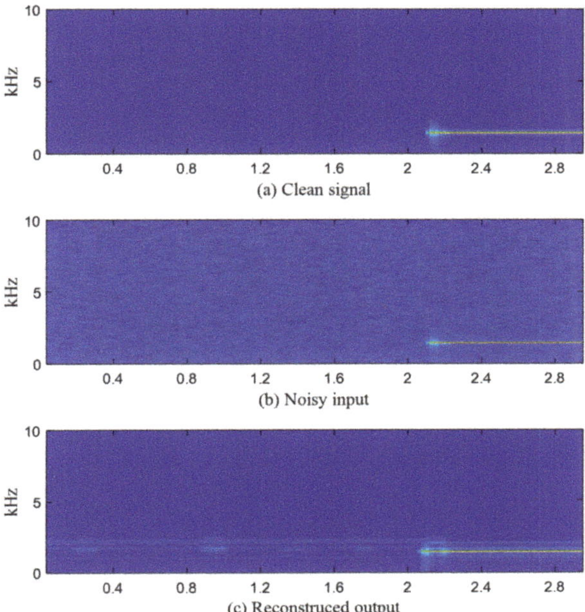

Figure 9. Spectrogram comparison: (**a**) Clean signal; (**b**) Noisy autoencoder input at 5 dB SNR; (**c**) Reconstructed autoencoder output.

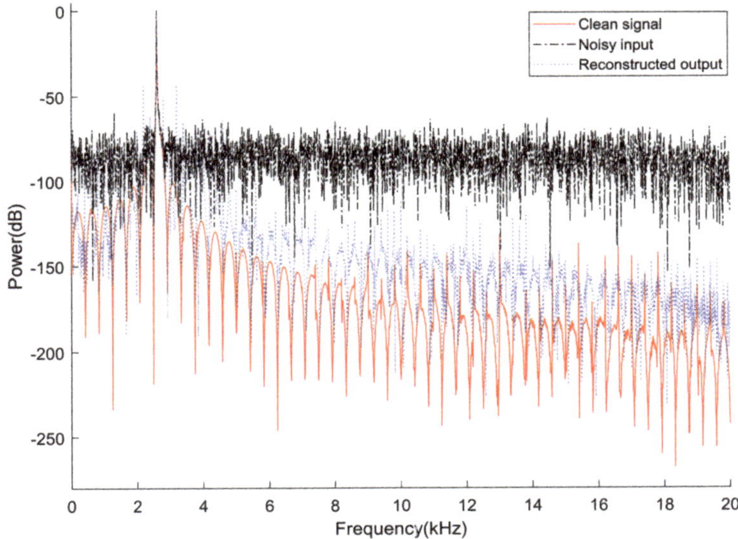

Figure 10. An overlayed frequency analysis of time segments of clean, noisy, and reconstructed signals from 2.4 s to 2.6 s in Figure 9. The "power (dB)" means the square of magnitude of each spectrum represented in decibels. In addition, normalization was performed using each maximum value of clean, noisy, and reconstructed signals for power spectrum comparison. (Red solid: clean signal, black dash-dot: noisy input, and blue dot: reconstructed output).

Furthermore, in order to objectively evaluate the noise reduction performance of the denoising autoencoder, overall SNR and LSD for 50 min of test data were measured and summarized in Table 6. Table 6 presented that the proposed method is superior to the conventional method satisfying higher overall SNR and lower LSD simultaneously for all SNR conditions. Through this, it was confirmed that the transmission/reception technique using the denoising autoencoder successfully performs ambient noise removal in the transmission/reception stage without using a separate noise reduction method.

5. Conclusions

In this paper, novel sonobuoy signal transmission and reception methods using autoencoders are proposed. Through evaluation, we confirmed that the original signal could be restored from a low-dimensional latent vector by using the proposed autoencoder with approximately 4% errors. We also proposed that the autoencoder shows similar reconstruction performance only using 130 times less information than the conventional method. Furthermore, we verified that the proposed denoising autoencoder successfully reduces ambient noise by comparing spectrograms and by measuring the overall SNR and the LSD of noisy input and reconstructed output signals. The proposed method demonstrates superior denoising performance satisfying higher overall SNR and lower LSD simultaneously for all SNR conditions than the conventional denoising method. However, studies to improve the reconstruction performance by reducing the MSE and to verify the proposed method with real sonobuoy data are necessary and these remain as future works.

Author Contributions: Conceptualization, J.S. and J.H.; Data curation, J.P.; Funding acquisition, J.H.; Investigation, J.P.; Methodology, J.H.; Project administration, J.S. and J.H.; Software, J.P.; Writing—original draft, J.P.; Writing—review & editing, J.P. and J.H. All authors have read and agreed to the published version of the manuscript.

Funding: This work was supported by Advanced materials and components laboratory project for defense industry, DCL2020L, funded by Korea Research Institute for Defense Technology Planning and Advancement (KRIT).

Institutional Review Board Statement: Not applicable.

Informed Consent Statement: Not applicable.

Data Availability Statement: Not applicable.

Conflicts of Interest: The authors declare no conflict of interest.

References

1. Lee, J.; Han, S.; Kwon, B. Development of communication device for sound signal receiving and controlling of sonobuoy. *J. KIMS Technol.* **2021**, *24*, 317–327. [CrossRef]
2. ROASW System: Remote Operation ASW. Available online: https://www.naval-technology.com/products/roasw-system-remote-operation-asw/ (accessed on 25 August 2022).
3. Holler, R.A. The evolution of the sonobuoy from World War II to the Cold War. *U.S. Navy J. Underw. Acoust.* **2014**, *62*, 322–346.
4. Cox, H. Fundamentals of bistatic active sonar. In *Underwater Acoustic Data Processing*; Chan, Y.T., Ed.; Springer: Dordrecht, The Netherlands, 1989; pp. 3–24.
5. Cimini, L. Analysis and simulation of a digital mobile channel using orthogonal frequency division multiplexing. *IEEE Trans. Commun.* **1985**, *33*, 665–675.
6. Watson, B. *FSK: Signals and Demodulation*; Watkins-Johnson Company: Palo Alto, CA, USA, 1980; Volume 7, pp. 1–13.
7. LeCun, Y.; Bengio, Y.; Hinton, G.E. Deep learning. *Nature* **2015**, *521*, 436–444. [CrossRef] [PubMed]
8. Hinton, G.E.; Salakhutdinov, R.R. Reducing the dimensionality of data with neural networks. *Science* **2006**, *313*, 504–507. [CrossRef] [PubMed]
9. Bengio, Y. Learning Deep Architectures for AI. *Found. Trends Mach. Learn.* **2009**, *2*, 1–127. [CrossRef]
10. Vincent, P.; Larochelle, H.; Lajoie, I.; Bengio, Y.; Manzagol, P.A. Stacked denoising autoencoders: Learning useful representations in a deep network with a local denoising criterion. *J. Mach. Learn. Res.* **2010**, *11*, 3371–3408.
11. Vincent, P.; Larochelle, H.; Bengio, Y.; Manzagol, P.A. Extracting and Composing Robust Features with Denoising Autoencoders. In Proceedings of the 25th International Conference on Machine Learning (ICML), Helsinki, Finland, 5–9 July 2008; pp. 1096–1103. [CrossRef]

12. Ng, A.Y. Sparse autoencoder. *Lect. Notes* **2011**, *72*, 1–19.
13. Lee, H.; Battle, A.; Raina, R.; Ng, A.Y. Efficient Sparse Coding Algorithms. In Proceedings of the 19th Neural Information Processing Systems (NIPS), Vancouver, BC, Canada, 4–7 December 2006; pp. 801–808.
14. Kingma, D.P.; Welling, M. Auto-Encoding Variational Bayes. In Proceedings of the 2nd International Conference on Learning Representations (ICLR), Banff, AB, Canada, 14–16 April 2014.
15. An, J.; Cho, S. Variational Autoencoder based Anomaly Detection using Reconstruction Probability. *Spec. Lect. IE* **2015**, *2*, 1–18.
16. Masci, J.; Meier, U.; Ciresan, D.; Schmidhuber, J. Stacked Convolutional Auto-Encoders for Hierarchical Feature Extraction. In Proceedings of the 21th International Conference on Artificial Neural Networks (ICANN), Espoo, Finland, 14–17 June 2011; pp. 52–59.
17. Du, B.; Xiong, W.; Wu, J.; Zhang, L.; Zhang, L.; Tao, D. Stacked convolutional denoising auto-encoders for feature representation. *IEEE Trans. Cybern.* **2017**, *47*, 1017–1027. [CrossRef] [PubMed]
18. Zong, B.; Min, M.R.; Cheng, W.; Lumezanu, C.; Cho, D.; Chen, H. Deep Autoencoding Gaussian Mixture Model for Unsupervised Anomaly Detection. In Proceedings of the 6th International Conference on Learning Representations (ICLR), Vancouver, BC, Canada, 30 April–3 May 2018.
19. Dahl, P.H.; Miller, J.H.; Cato, D.H.; Andrew, R.K. Underwater ambient noise. *Acoust. Today* **2007**, *3*, 23. [CrossRef]
20. Ashraf, H.; Jeong, Y.; Lee, C.H. Underwater ambient-noise removing GAN-based on magnitude and phase spectra. *IEEE Access* **2021**, *9*, 24513–24530. [CrossRef]
21. Urick, R.J. *Principles of Underwater Sound*, 3rd ed.; Peninsula Publishing: Los Altos, CA, USA, 1983; pp. 147–197.
22. Kingma, D.P.; Ba, L.J. Adam: A Method for Stochastic Optimization. In Proceedings of the 3rd International Conference on Learning Representations (ICLR), San Diego, CA, USA, 7–9 May 2015.
23. Fernandes, J.d.C.V.; de Moura Junior, N.N.; de Seixas, J.M. Deep learning models for passive sonar signal classification of military data. *Remote Sens.* **2022**, *14*, 2648. [CrossRef]
24. Chen, H.; Gentile, R. Phased array system simulation. In Proceedings of the 2016 IEEE International Symposium on Phased Array Systems and Technology (PAST), Waltham, MA, USA, 18–20 October 2016.
25. Loizou, P.C. *Speech Enhancement*; CRC Press: Boca Raton, FL, USA, 2007.
26. Hong, J.; Bae, I.; Seok, J. Wiener filtering-based ambient noise reduction technique for improved acoustic target detection of directional frequency analysis and recording sonobuoy. *J. Acoust. Soc. Korea* **2022**, *41*, 192–198.

Article

Influence of Temporal and Spatial Fluctuations of the Shallow Sea Acoustic Field on Underwater Acoustic Communication

Zhichao Lv [1,*], Libin Du [1], Huming Li [1], Lei Wang [1], Jixing Qin [2], Min Yang [3,*] and Chao Ren [4]

[1] College of Ocean Science and Engineering, Shandong University of Science and Technology, Qingdao 266590, China; dulibin@sdust.edu.cn (L.D.); lihuming78@163.com (H.L.); wangleiimg@foxmail.com (L.W.)

[2] State Key Laboratory of Acoustics, Institute of Acoustics, Chinese Academy of Sciences, Beijing 100190, China; qjx@mail.ioa.ac.cn

[3] North China Sea Marine Technical Support Center, State Oceanic Administration, Qingdao 266061, China

[4] Acoustic Science and Technology Laboratory, Harbin Engineering University, Harbin 150001, China; qingtianxiayu@hrbeu.edu.cn

* Correspondence: lvzhichao@hrbeu.edu.cn (Z.L.); yangmin@ncs.mnr.gov.cn (M.Y.)

Abstract: In underwater acoustic communication (UAC) systems, the channel characteristics are mainly affected by spatiotemporal changes, which are specifically manifested by two factors: the effects of refraction and scattering caused by seawater layered media on the sound field and the random fluctuations from the sea floor and surface. Due to the time-varying and space-varying characteristics of a channel, the communication signals have significant variations in time and space. Furthermore, the signal shows frequency-selective fading in the frequency domain and signal waveform distortion in the time domain, which seriously affect the performance of a UAC system. Techniques such as error correction coding or space diversity are usually adopted by UAC systems to neutralize or eliminate the effects of deep fading and signal distortion, which results in a significant waste of limited communication resources. From the perspective of the sound field, this study used experimental data to analyze the spatiotemporal fluctuation characteristics of the signal and noise fields and then summarized the temporal and spatial variation rules. The influence of the system then guided the parameter configuration and network protocol optimization of the underwater acoustic communication system by reasonably selecting the communication signal parameters, such as frequency, bandwidth, equipment deployment depth, and horizontal distance.

Keywords: underwater acoustic communication; channel characteristics; spatiotemporal fluctuation

Citation: Lv, Z.; Du, L.; Li, H.; Wang, L.; Qin, J.; Yang, M.; Ren, C. Influence of Temporal and Spatial Fluctuations of the Shallow Sea Acoustic Field on Underwater Acoustic Communication. *Sensors* **2022**, *22*, 5795. https://doi.org/10.3390/s22155795

Academic Editor: Sylvain Girard

Received: 21 May 2022
Accepted: 28 July 2022
Published: 3 August 2022

Publisher's Note: MDPI stays neutral with regard to jurisdictional claims in published maps and institutional affiliations.

1. Introduction

At present, using sound waves is the only method for transmitting data over long distances in seawater. Underwater acoustic communication has become an indispensable part of data transmission technology for exploring, developing, and protecting the ocean. The complexity of UAC systems is mainly manifested in the time-varying and space-varying channels. The signal-to-noise ratio (SNR) as a measurement parameter of system performance design is also an important indicator for evaluating the quality of underwater acoustic communication. The spatiotemporal variation range of the SNR can be used to describe the spatiotemporal fluctuation characteristics of underwater acoustic communication signals.

Due to the complexity and time variation of the marine environment, the SNR of communication signals varies widely in time and space. Therefore, the degeneration in the UAC system performance would be caused by two reasons: the difficulties in optimizing the synchronization signal detection threshold and determining the location of the UAC system equipment. However, from the perspective of the sound field, it is possible to describe the spatiotemporal variation from the signal field and noise field, and then analyze the fluctuation characteristics of the underwater acoustic communication channel. Environmental factors, such as ocean currents, tides, and internal waves, with large spatial-temporal fluctuations have not been considered.

Since the 1950s, researchers have gradually paid attention to signal interference in shallow-water sound fields, and they mainly analyzed the spatial-temporal characteristics of sound fields based on ray acoustic theory and normal wave theory [1–3]. Ray acoustics researchers focused on the Loe mirror effect in optics, which assumes that the interfering sound rays are approximately parallel, and they derived a partially analytical solution and discovered vertical distribution characteristics of the sound field [4]. In the 1980s, in view of the normal wave theory and far-field assumption, scholars analyzed the interference sound field and put forward the conception of waveguide invariants. However, previous theoretical analyses of the interference phenomenon, which are based on restricted scenarios and simplified assumptions, were detrimental to the establishment of a universal analysis model. This study mainly focused on the influence of short-range shallow sea interference on signal fluctuations. Compared with the normal wave method, the ray acoustic method is superior given its clear concept and simple calculation; therefore, the ray acoustic method was adopted to analyze the spatial distribution of the sound field.

From the perspective of the sound field, the influencing factors of the spatial variation range of the SNR are not only the spatial-temporal distribution characteristics of the signal field caused by the signal interference and interface fluctuations but also the spatial-temporal distribution characteristics of the system noise field. The operating performance of a UAC system is significantly affected by the noise of the system and the marine environment. The marine environmental noise was first measured in two studies [5,6]. After analyzing a large amount of measured data, it was found that marine environmental noise is mainly composed of wind-induced noise, ship noise, and biological noise; reference [7] gives a marine environmental noise spectrum and notes that the low-frequency noise components mainly come from the machinery of ships, while high-frequency noise is mainly wind-induced wave noise. Due to the continuous development of signal acquisition technology and the increase in marine research investment, research on marine environmental noise has become a popular topic of discussion [8–15]. On the basis of different sound field propagation theories, researchers proposed various marine environmental noise models. The classic models include the C/S model, K/I fast-field model, and P/K model [16–18]. Using these models, scholars have conducted in-depth studies on calculation accuracy, calculation speed, orientation, and boundary conditions, and concluded a series of results, which promoted the development and application of exploration of the noise field [19–22]. Zhou Jianbo et al. considered wind-induced waves as the noise source and used the transmission theory method instead of the traditional Monte Carlo method to construct a noise field model. They analyzed the spatial noise distribution and concluded that high-frequency noise had fluctuations in intensity at the offshore surface [23]. Avrashi, G. et al. considered the problem of carrier frequency offset estimation in OFDM underwater acoustic communication and analyzed the causes of changing environmental impacts [24]. Z.L. et al. analyzed wave fluctuation on underwater acoustic communication using measured data collected with USV [25]. X.Z. et al. used quantile–quantile (Q-Q) plots to analyze real marine environmental data, interpreting the impulsive property of ocean ambient noise in shallow waters [26]. X.Z. et al. applied Loffeld's bistatic formula to SAS image processing, which provided a more accurate approximation of the spectrum compared to that based on phase center approximation [27]. An, J. et al. propose underwater acoustic (UWA) communications using a generalized sinusoidal frequency modulation (GSFM) waveform, which makes full use of the time and frequency variation laws of the marine environment in experimental data [28]. Zhang, Y. et al. proposed a deep-learning-based orthogonal frequency division multiplexing receiver for underwater acoustic communications to process marine environmental data through neural networks [29].

The stratum structure of the ocean space determines the multi-channel coherent structure characteristics of the ocean sound field. The motion of the transmitter end, the interface, and the receiving sensor affect the spatiotemporal fluctuation characteristics of the sound field, which shows that the channel response function is time-varying and space-varying. The feature of time-varying and space-varying channels is the key point to manage to achieve effective and stable UAC systems. In this study, the signal field and noise field were evaluated by establishing a model and obtaining data through experiments, and the spatial-temporal distribution of signals was summarized to provide theoretical support for the design of UAC systems.

2. Spatial Distribution Characteristics of the Signal and Noise Fields

2.1. Investigation of the Spatial Distribution of the Signal Field

In a uniformly shallow sea, the characteristics of the medium do not change significantly with the depth; therefore, the sound field can be studied using ray theory. The accuracy of ray theory for calculating the sound field is correlated with the number of sound rays examined in the study. The larger the number of sound rays considered, the less the sound rays reflected from the bottom of the water contribute to the sound field. Thus, the direct sound and the first-order reflection rays of the sea surface are mainly considered. The sound pressure field is

$$p(r, z, t) = \frac{1}{R_{00}} exp[i(kR_{00} - \omega t)] - G_{10} \frac{1}{R_{10}} exp[i(kR_{10} - \omega t)] \tag{1}$$

The direct sound path $R_{00} = \sqrt{r^2 + (z - z_s)^2}$ and the first surface reflection sound path $R_{10} = \sqrt{r^2 + (z + z_s)^2}$, where r is the horizontal distance between the sound source and the receiving hydrophone and z_s is the depth of sound source placement; G_{10} is the absolute value of the surface reflection coefficient. Separating out the time variable gives

$$p(r, z) = \frac{1}{R_{00}R_{10}} exp(ikR)\{R_{10}exp[ik(R_{00} - R)] - G_{10}R_{00}exp[ik(R_{10} - R)]\} \tag{2}$$

where

$$R_{00} - R = \sqrt{r^2 + z^2}\left(\sqrt{1 + \frac{z_s^2 - 2z_s z}{r^2 + z^2}} - 1\right) \tag{3}$$

When the value of r is greater than 3 times the depth of the sea, i.e., $(z_s^2 - 2z_s z)/(r^2 + z^2) \ll 1$, Formula (3) can be approximated:

$$R_{00} - R \approx \sqrt{r^2 + z^2}\left(\sqrt{1 + 2 \cdot \frac{z_s^2 - 2z_s z}{2(r^2 + z^2)} + \left[\frac{z_s^2 - 2z_s z}{2(r^2 + z^2)}\right]^2} - 1\right) = \frac{z_s^2 - 2z_s z}{2(r^2 + z^2)} \tag{4}$$

Similarly:

$$R_{00} - R \approx \frac{z_s^2 + 2z_s z}{2(r^2 + z^2)} \tag{5}$$

Substituting (4) and (5) into Equation (2), we obtain

$$p = \frac{1}{R_{00}R_{10}} exp\left[ik\left(R + \frac{z_s^2}{2\sqrt{r^2+z^2}}\right)\right] \\ \times \left\{(R_{10} - G_{10}R_{00})cos\left[k\left(\frac{z_s z}{\sqrt{r^2+z^2}}\right)\right] \\ - i(R_{10} - G_{10}R_{00})sin\left[k\left(\frac{z_s z}{\sqrt{r^2+z^2}}\right)\right]\right\} \tag{6}$$

Then, the mean square sound pressure in the sound field is

$$\overline{p^2} = \frac{1}{2}\frac{1}{(r^2+z^2+z_s^2)^2 - 4z_s^2 z^2}\left[(1 + G_{10}^2)(r^2 + z^2 + z_s^2) + 2(1 - G_{10}^2)z_s z \\ - 2G_{10}\sqrt{(r^2 + z^2 + z_s^2)^2 - 4z_s^2 z^2}cos(\frac{2kz_s z}{\sqrt{r^2+z^2}})\right] \tag{7}$$

When focusing on the sound field distribution in the central area of the water body, there is an approximate value of $z \approx z_s/2$; after substitution into Equation (7), this gives

$$\overline{p^2} = \frac{1}{2}\frac{1}{\left(r^2+\frac{5}{4}z_s^2\right)^2-z_s^4}\left[\left(1+G_{10}^2\right)\left(r^2+\frac{5}{4}z_s^2\right)+\left(1-G_{10}^2\right)z_s^2 \right. \\ \left. -2G_{10}\sqrt{\left(r^2+\frac{5}{4}z_s^2\right)^2-z_s^4}\cos\left(\frac{2kz_sz}{\sqrt{r^2+z^2}}\right)\right] \tag{8}$$

If the signal is a broadband signal, the following formula can be obtained:

$$\overline{p^2}_{\Delta f} = \frac{1}{\Delta f}\int_{f_0-0.5\Delta f}^{f_0+0.5\Delta f}\overline{p(f)^2}df \tag{9}$$

Further derivation can be written as follows:

$$\overline{p^2} = \frac{1}{2}\frac{1}{\left(r^2+\frac{5}{4}z_s^2\right)^2-z_s^4}\left[\left(1+G_{10}^2\right)\left(r^2+\frac{5}{4}z_s^2\right)+\left(1-G_{10}^2\right)z_s^2 \right. \\ \left. -2G_{10}\sqrt{\left(r^2+\frac{5}{4}z_s^2\right)^2-z_s^4}\,\frac{\sin\theta}{\theta}\cos\left(\frac{2kz_sz}{\sqrt{r^2+z^2}}\right)\right] \tag{10}$$

where $\theta = 2\pi z_s z\cdot\Delta f/\left(c\sqrt{r^2+z^2}\right)$ and $k_0 = 2\pi f_0/c$. It is worth noticing that when the signal has a single frequency, i.e., Δf approximates 0, $\sin\theta/\theta$ approaches 1; furthermore, when the signal bandwidth increases, θ increases and $\sin\theta/\theta$ approaches 0. Given this, when the signal has sufficient bandwidth, the fluctuation of the sound field of the signal can be effectively smoothed. When $z \approx z_s/2$, the approximation can be given as

$$\theta \approx \frac{\pi z_s^2\cdot\Delta f}{c\sqrt{r^2+z^2}} \approx \frac{\pi}{c}\frac{z_s^2}{r}\cdot\Delta f \tag{11}$$

According to Formula (11), it can be obtained that the fluctuation range of the signal is proportional to the distance r at the transmitting and receiving ends and inversely proportional to the center frequency f, the bandwidth Δf, and the modem depth z_s.

2.1.1. Simulation Testing

In order to verify the theoretical distribution of the signal field, the signal sound intensity fluctuations of different center frequencies, bandwidths, and horizontal distances of the transceiver were simulated and examined.

1. Simulation of the signal interference at different center frequencies

Simulation parameters: the seabed was absolutely hard and flat; water depth: 100 m; sound source deployment depth: 95 m (near the seabed); horizontal distance of the transmitting and receiving end: 200 m; signal bandwidth: 10 Hz; frequencies were 500 Hz, 2 kHz, and 5 kHz.

Figure 1 shows the signal interference diagrams at different frequencies. The frequencies from (a) to (c) were 500 Hz, 2 kHz, and 5 kHz. The abscissa is the signal sound pressure level and the ordinate is the depth. Comparing the figures, it can be seen that as the center frequency of the signal doubled, the vertical fluctuation range of the sound intensity decreased by a factor of half; the higher the mark frequency, the smaller the spatial fluctuation. In underwater acoustic communication, the center frequency of the signal should be appropriately increased within the tolerance range of the high-frequency absorption and attenuation of the signal.

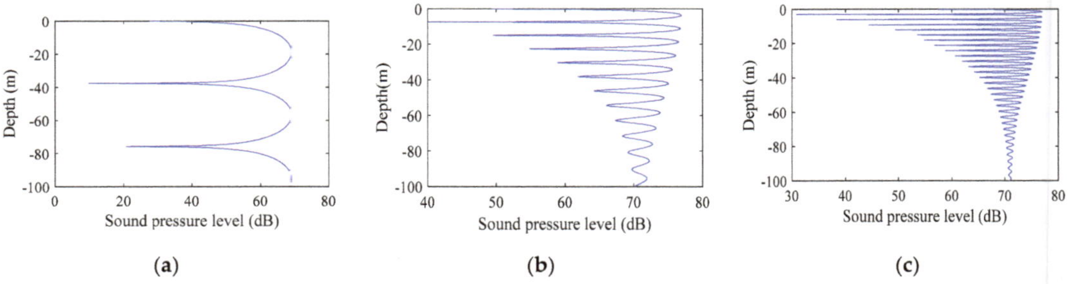

Figure 1. Signal interferograms at different frequencies: (**a**) 500 Hz, (**b**) 2000 Hz, and (**c**) 5000 Hz.

2.   Simulation of signal interference at different horizontal distances

Simulation parameters: the seabed was absolutely hard and flat; water depth: 100 m; sound source deployment depth: 95 m (near the seabed); frequency: 5 kHz; mark bandwidth: 10 Hz; the horizontal distances of the transceiver end were 100 m, 300 m, and 1000 m.

Figure 2 shows the signal interference patterns at different horizontal distances. The distances from (a) to (c) were 100 m, 300 m, and 1000 m, where the abscissa is the mark sound pressure level and the ordinate is the depth. Comparing the figures, the range of vertical fluctuations in sound intensity increased exponentially as the horizontal distance increased, which was consistent with the inference obtained using Formula (11). In the case of the same bandwidth, as the signal frequency and the horizontal distance increased, the sound intensity fluctuation was inversely proportional to the signal frequency and proportional to the horizontal distance; as the depth increased, the sound intensity fluctuation decreased.

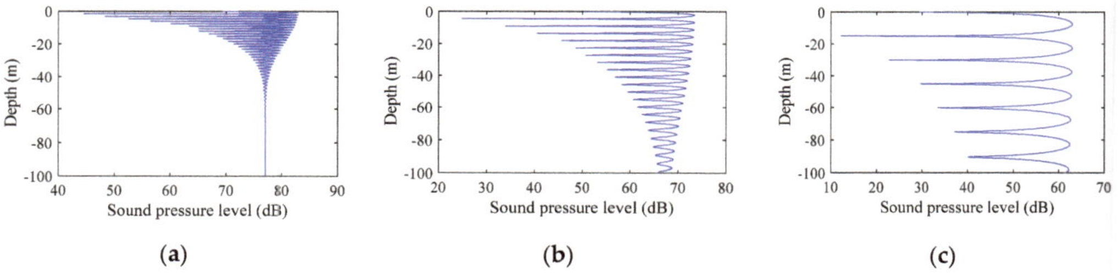

Figure 2. Signal interferograms at different horizontal distances: (**a**) 100 m, (**b**) 300 m, and (**c**) 1000 m.

3.   Interference simulation of different bandwidth signaling

Simulation parameters: the seabed was absolutely hard and flat; water depth: 100 m; sound source deployment depth: 95 m (near the seabed); frequency: 1 kHz; horizontal distance of the transmitting and receiving end 200 m; mark bandwidths were 1 Hz, 10 Hz, and 100 Hz.

Figure 3 shows the signal interference diagrams at different bandwidths. The bandwidths from (a) to (c) were 1 Hz, 10 Hz, and 100 Hz, where the abscissa is the signal sound pressure level and the ordinate is the depth. It can be seen from the comparison of the figures that as the bandwidth increased, the vertical fluctuation range of the sound intensity decreased exponentially until it tended to be stable in the end.

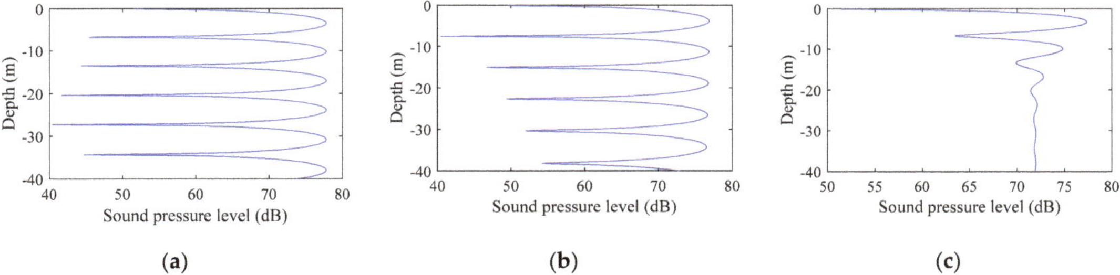

Figure 3. Signal interferograms at different bandwidths: (**a**) 1 Hz, (**b**) 10 Hz, and (**c**) 100 Hz.

In summary, from the simulation results shown in Figures 1–3, it can be seen that the signal fluctuation range was proportional to the distance between the sending and receiving ends and was inversely proportional to the signal center frequency, bandwidth, and modem lowering depth, which were related to Formulas (3)–(11) and were consistent with the inferences given.

2.1.2. Analysis of Experimental Data

The bottom of the Yellow Sea is relatively flat and the sea conditions are relatively stable in the autumn, which is suitable for sound field analysis. With a view to verify the spatial distribution of the sound field obtained from the simulation experiments, a sound field analysis experiment ExQD_1701 was performed in the Yellow Sea in the autumn of 2017.

The depth of this experimental sea area was precisely 40 m. A signal-launching ship, which used a UW350 type transmitting transducer with a working frequency range of 20 Hz–20 kHz, was used for the transmission. The transmitting transducer was cylindrical with a diameter of 0.2 m and a length of exactly 1 m. The net weight was exactly 100 kg with its own hoisting device. The schematic diagram of the experiment is shown in Figure 4. The Xiangyang Hong 81 experimental ship was utilized to be the signal-receiving ship with five sub-arrays of the same specification for signal reception with a pitch of 1 m. When the signal-transmitting ship reached the preset position, it could transmit signals with different frequencies. The real-time positions of the signal-transmitting ship and the signal-receiving ship were recorded using GPS, which was used to calculate the relative distance between the transmitting and receiving ends.

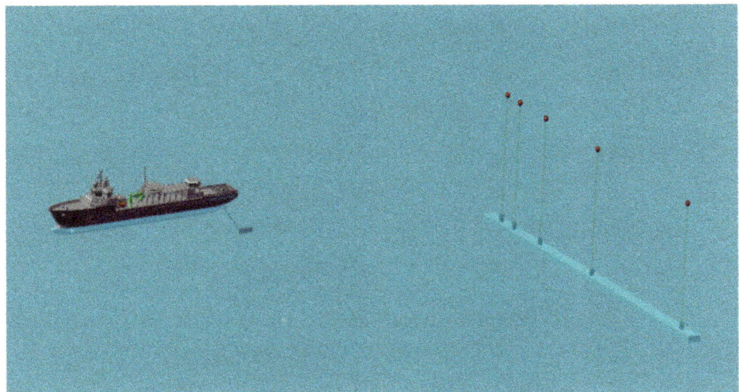

Figure 4. ExQD_1701 test.

1.   Spatial fluctuation of the low-frequency signal field

The former simulation experiments demonstrated that when the signal frequency was low, the fluctuation was large. First, the experiment processed and analyzed the low-frequency signal data below 1 kHz. During the experiment, single-frequency signals of 95 Hz and 400 Hz were transmitted, and the sound source level of the transmitting transducer was stable. In order to summarize the spatial distribution of signals, Figures 5 and 6 display the vertical distribution of sound pressure levels for single-frequency signals of 95 Hz and 400 Hz when the relative distances between the transmitting and receiving ends were different. The abscissa is the received sound pressure level and the ordinate is the water depth. By comparing different distances, frequencies, and depths, the spatial fluctuation characteristics of the signal could be summarized. When the signal frequency was low, the sea trial results fit well with the simulation results. As for the normal wave, the fluctuation law could be described as: the modes of the normal wave excited at different frequencies were different. The higher the frequency was, the greater the number of modes, and the more complicated the signal fluctuation law. This rule can be used to guide the equipment placement of low-frequency remote UAC systems.

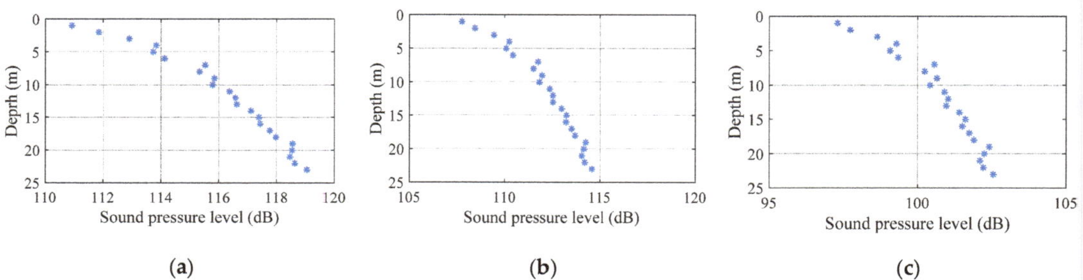

(a) **(b)** **(c)**

Figure 5. Spatial distribution at different distances (95 Hz): (**a**) vertical distribution at 1 km, (**b**) vertical distribution at 4 km, and (**c**) vertical distribution at 8 km.

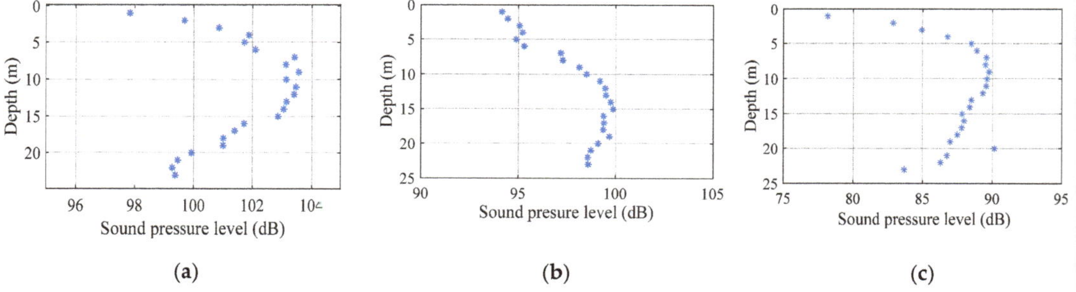

(a) **(b)** **(c)**

Figure 6. Spatial distribution at different distances (400 Hz): (**a**) vertical distribution at 1 km, (**b**) vertical distribution at 4 km, and (**c**) vertical distribution at 8 km.

2. Spatial fluctuation of the high-frequency signal field

The spatial fluctuations of high-frequency signals were also analyzed. The communication frequency band was selected from 5 kHz to 20 kHz with a steady energy level used in underwater acoustic communication experiments. The transmitting signals were single-frequency signals with frequencies of 12 kHz and 20 kHz. The sound source level of the transmitting transducer was stable. In order to discover the law of spatial fluctuations, Figures 7 and 8 demonstrate the vertical distribution of the sound pressure level when the relative distances between the sending and receiving ends were different for single-frequency signals with frequencies of 12 kHz and 20 kHz, respectively. The abscissa in the figure is the obtained sound pressure level and the ordinate is the water depth. Through the comparison of different distances, frequencies, and depths, it was discovered that when the signal frequency was above 1 kHz, the signal wavelength was short; meanwhile, the environment was greatly affected during the propagation, which was difficult to study qualitatively. The farther the horizontal distance was, the larger the vertical fluctuation range, while the deeper the equipment deployment depth, the larger the vertical fluctuation range and the sound intensity of the near-sea surface signal was slightly lower than the signal strength in water. With the increase in frequency, the signal fluctuation is reduced; however, when the signal frequency was too high, i.e., the signal wavelength was short, which was greatly affected by the scattering and reflection of surface fluctuations, and the signal absorption loss was greater than when the frequency of the signal was low. Therefore, a single-frequency signal appeared to have a violent spatial distribution. When the frequency was as high as 20 kHz, there was a 25 dB intensity difference in the vertical distribution at a horizontal distance of 8 km.

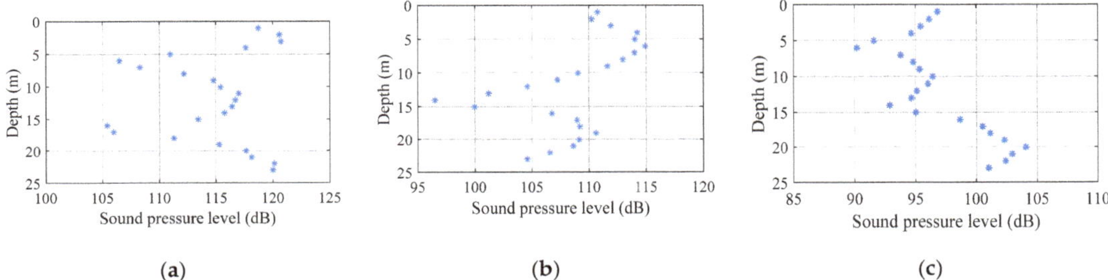

Figure 7. Spatial distribution at different distances (12 kHz): (**a**) vertical distribution at 1 km, (**b**) vertical distribution at 4 km, and (**c**) vertical distribution at 8 km.

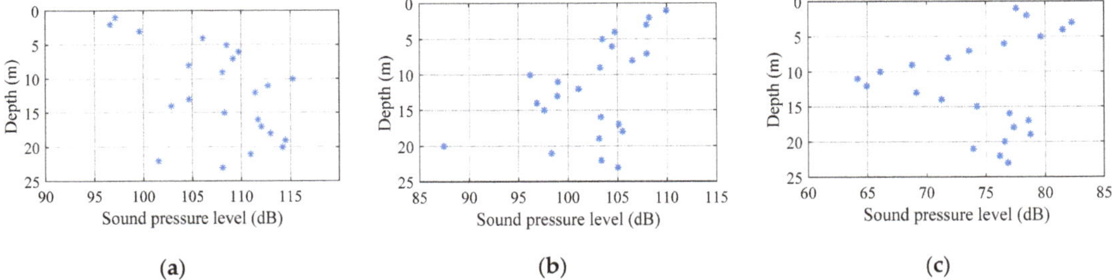

Figure 8. Spatial distribution at different distances (20 kHz): (**a**) vertical distribution at 1 km, (**b**) vertical distribution at 4 km, and (**c**) vertical distribution at 8 km.

2.2. Analysis of the Spatial Distribution of the Noise Field

The time distribution characteristics of the noise field are mainly targeted at the commonly used high-speed underwater acoustic communication frequency bands of 5 kHz–20 kHz. According to the sound field analysis, in addition to the spatial distribution of the signal field due to the interference and the interface fluctuation, the influencing factors of the spatial variation range also have the spatial distribution of the system noise field. This study investigated the spatial distribution characteristics of the noise field in the 5 kHz–20 kHz frequency band with a general volume noise model [22]. It was assumed that all noise sources were uniformly distributed on an infinite plane; then, the spatial correlation coefficient of the marine environmental noise was simulated. In the ExQD_1701 experiment, the curve of the spatial correlation coefficient with respect to depth for a 10 kHz signal for 10 h is shown in Figure 9. The black dotted line is the theoretical value given by the general model of volume noise, and the spatial correlation of noise was small at high frequencies. Due to the correlation, calculations were performed between 30 array elements at different vertical depths and the no. 1 surface array element; consequently, the spatial correlation coefficient curve was clearly revealed. It was found that the curve matched the theoretical value given by the volume noise model. In subsequent high-frequency noise experimental data processing, the vertical correlation between array elements could be ignored.

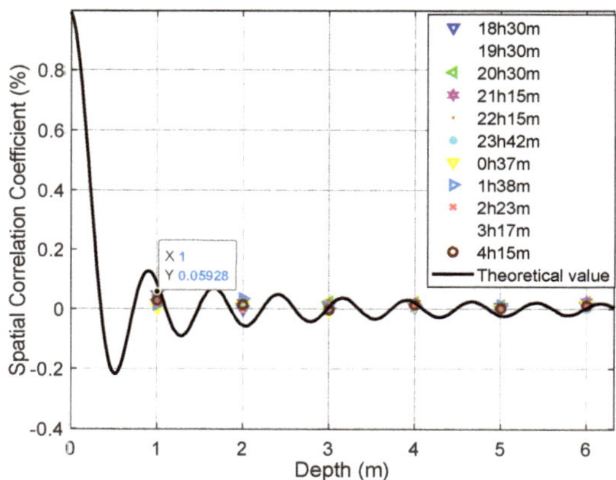

Figure 9. Correlation coefficient with depth (10 kHz, 10 h).

In order to discuss the spatial distribution of wind and wave noise, the environmental noise was collected by using an array in ExQD_1701. The seabed in the experimental sea area was approximately the same level. The water depth was 40 m and the wind speed during the experiment was approximately 3 m/s. There were no other vessel activities within 5 km. Figure 10a indicates the noise field distribution of the marine environment at different depths, and Figure 10b shows the noise field distribution with the ship's self-noise. The no. 1 array element was an offshore array element, and the no. 30 array element was a near-seabed array element. The ship's self-noise had a greater impact on the frequency band below 5 kHz, which gradually decreased with the increase in depth and slightly fluctuated at high frequencies. Underwater acoustic communications often utilize high-frequency bands, with an associated 5 dB of noise fluctuations. The spatial distribution of environmental noise was not obvious and it was mainly because the surface noise source was a surface source composed of multiple noise sources. During the propagation process, the noise signals overlapped and neutralized each other with relatively small spatial fluctuation.

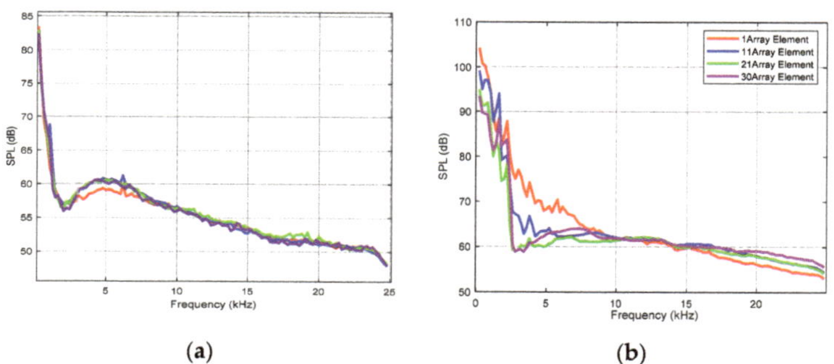

Figure 10. Environmental noise and vertical distribution of ship self-noise: (**a**) environmental noise and (**b**) ship noise.

Figure 11 demonstrates the vertical distribution of noise at different frequencies measured using an array suspended from the side of the ship when the auxiliary ship was still working. Compared with the noise in Figure 10b, it can be considered that the noise was below 4 kHz. The noise in the frequency band was mainly the self-noise of the receiving ship. Due to this frequency band, the noise had a more obvious vertical distribution at each frequency point. As the depth increased, the noise power spectral density gradually increased, and the difference in noise spectral level could accumulate to approximately 20 dB. When the frequency was higher than 4 kHz, the noise mainly came from the surface waves, and the surface fluctuations contributed more significantly to the high-frequency noise field strength of 1 kHz–10 kHz. In the experiment, the array element closest to the surface was placed about 2 m underwater. When the water depth reached 5 m, the noise variation curve had no obvious fluctuations in the depth range covered by the hydrophone array. Thus, after the device was placed at a certain depth, the contribution of high-frequency noise to the fluctuation of the communication signal could be ignored.

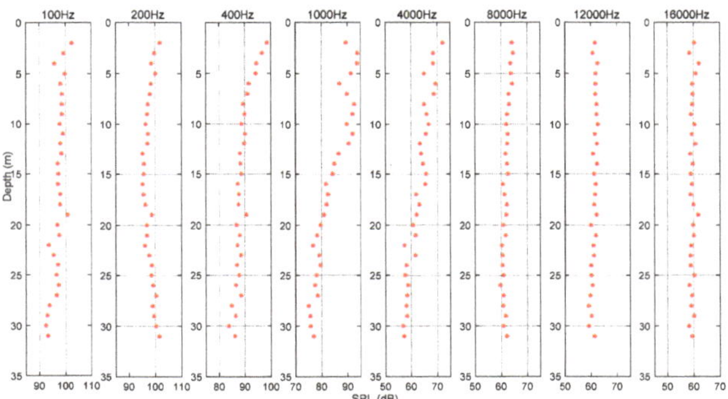

Figure 11. Different frequency noise field intensities with depth (40 m).

According to the analysis of the experimental results, the spatial distribution of the signal field and the noise field was basically consistent with the simulation results, and its vertical distribution showed that the intensity of the near-sea surface signal was lower than that of other depth signals.

3. Analysis of the Time Fluctuation of Underwater Acoustic Communication Signals

The time window of a UAC system is smaller than other systems. Therefore, this study mainly analyzed the impact of small-scale spatial-temporal fluctuations caused by environmental parameters, such as wind and waves, on underwater acoustic communication. Moreover, environmental factors, such as ocean currents, tidal waves, and internal waves, with large spatial-temporal fluctuations were not considered.

3.1. Statistics of the Time Fluctuation of Low-Frequency Signal Fields

When an acoustic signal propagates in a shallow sea channel, it also has an undulating effect that changes over time, which corresponds to a time-varying channel in underwater acoustic communication. In this section, based on the spatial fluctuations of the signal field, the temporal fluctuations of the signal field were studied. The experimental ExQD_1701 data was statistically analyzed. In order to fully consider the selective frequency fading of the signal and ignore the effect of bandwidth on the time fluctuation of the signal field, the analysis used single-frequency signals. The time fluctuations in the low-frequency signal field and the high-frequency signal field were examined.

1.   Spatial fluctuation of the high-frequency signal field

First, we analyzed the time fluctuation of the low-frequency signal. The frequencies of the transmitted signal were 95 Hz and 400 Hz, and the sound source level of the transmitted transducer was stable. Figures 12 and 13 show the spurious color maps of the spatiotemporal distribution of the single-frequency signals (95 Hz and 400 Hz) at different distances. The vertical axis represents the water depth; the horizontal axis is the time when the signal was collected. With the increase in the horizontal distance between the transmitting and receiving ends, the signal strength increased significantly with time, which verified the conclusion of the theoretical calculations. The variation law of the signal intensity in the vertical direction was also consistent with the previous experimental results. When the frequency remained unchanged, the time fluctuation of the near-sea surface signal was larger than the sea floor fluctuation with the increase of the horizontal distance. With the increase in frequency, the number of normal wave modes of the signal field increased and the time fluctuation became stable.

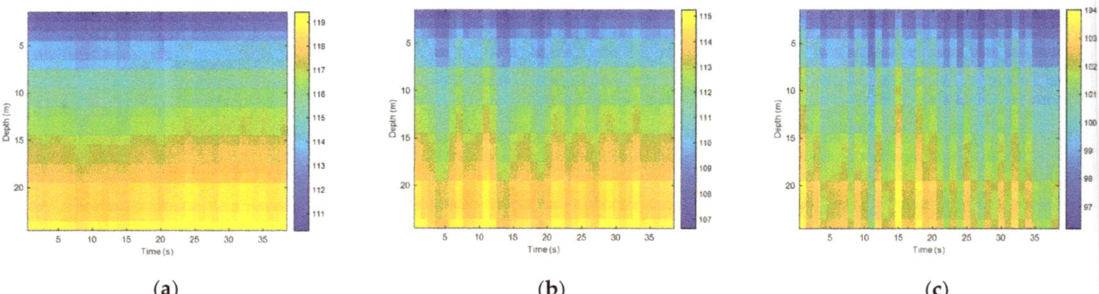

(a)　　　　　　　　　　(b)　　　　　　　　　　(c)

Figure 12. Temporal and spatial distribution at different horizontal distances (95 Hz): (**a**) spatiotemporal distribution map at 1 km, (**b**) spatiotemporal distribution map at 4 km, and (**c**) spatiotemporal distribution map at 8 km.

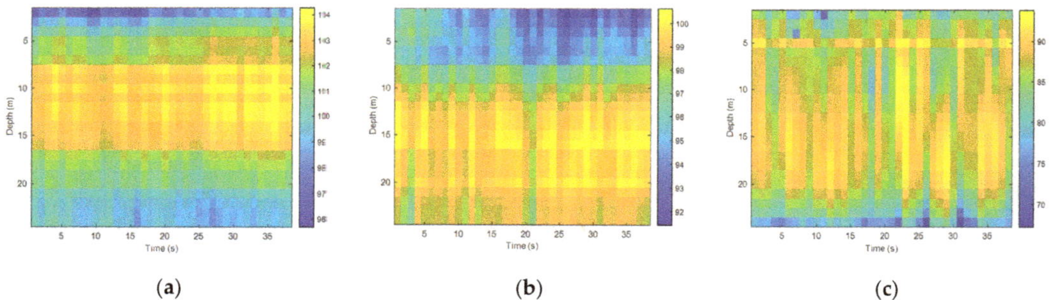

(a)　　　　　　　　　　(b)　　　　　　　　　　(c)

Figure 13. Temporal and spatial distribution at different horizontal distances (400 Hz): (**a**) spatiotemporal distribution map at 1 km, (**b**) spatiotemporal distribution map at 4 km, and (**c**) spatiotemporal distribution map at 8 km.

2.　Statistics of the time fluctuation of high-frequency signal fields

Analyzing the time fluctuations of high-frequency signals and selecting the 5 kHz–20 kHz communication frequency band were commonly used in underwater acoustic communication experiments, with the transmitted single frequency signals at 12 kHz and 20 kHz being used. The sound source level of the transmitting transducer was stable and it was the same as the processing flow of the time fluctuation of the low-frequency signal field. Figures 14 and 15 show the spurious color maps of the spatiotemporal distribution of the single-frequency signals (12 kHz and 20 kHz) at different distances. The abscissa represents the time when the signal was collected, and the ordinate is the water depth, which indicates the distribution of the hydrophones from the surface to the sea floor. With the increase in distance, the time distribution of high-frequency signals became more pronounced, which showed that the channel structure stabilization time became shorter in underwater acoustic communication. As the frequency increased, the wavelength of the acoustic wave became shorter, the signal field became more complex under the influence of scattering and interference caused by interface fluctuations, and the fluctuation law of single-frequency signals appeared to be less significant.

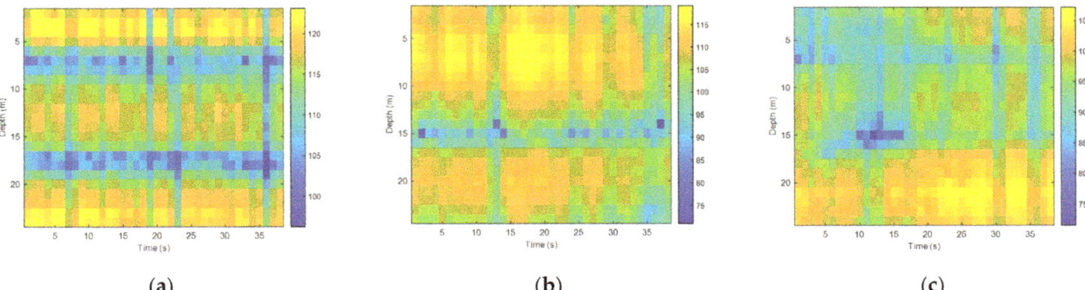

(a) **(b)** **(c)**

Figure 14. Temporal and spatial distribution at different horizontal distances (12 kHz): (**a**) spatiotemporal distribution map at 1 km, (**b**) spatiotemporal distribution map at 4 km, and (**c**) spatiotemporal distribution map at 8 km.

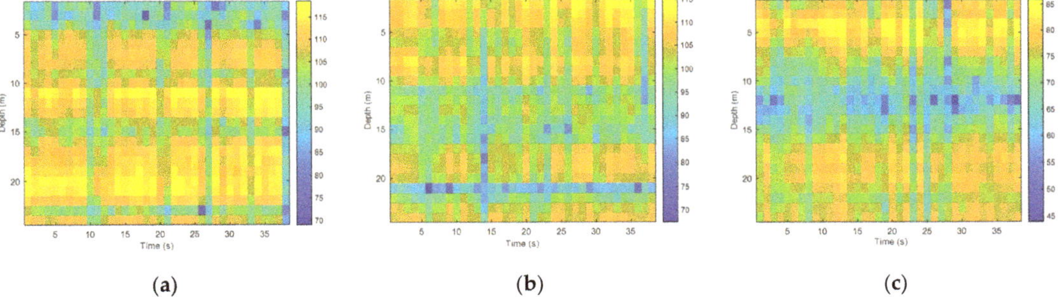

(a) **(b)** **(c)**

Figure 15. Temporal and spatial distribution at different horizontal distances (20 kHz): (**a**) spatiotemporal distribution map at 1 km, (**b**) spatiotemporal distribution map at 4 km, and (**c**) spatiotemporal distribution map at 8 km.

3.2. Analysis of the Time Distribution Characteristics of Noise Fields

The time distribution characteristics of the noise fields were mainly targeted at the commonly used high-speed underwater acoustic communication frequency bands of 5 kHz–20 kHz. According to the foregoing, the vertical elements of the noise distribution were weakly correlated to each other. Therefore, the array elements at 5 m, 15 m, and 25 m were selected as research objects to study the noise time distribution in the surface, seabed, and water column situations. In the ExQD_1701 experiment, the environmental noise data in the experimental stage was randomly taken for 600 s, and a 1 s Hanning window was applied to intercept the data at an overlap rate of 0.66. Therefore, a total of 1762 sample points were obtained. Statistics of narrow-band noise distribution at different frequencies are shown in Figure 16. In the high-frequency range, the environmental high-frequency noise was mainly distributed in the 60 dB to 70 dB range. As the frequency increased, the noise intensity gradually decreased; the noise intensity also gradually decreased as the depth increased. This was found to be in accordance with the theoretical model.

In order to study the time distribution characteristics of noise further, the probability distribution of the single-frequency noise intensity at a depth of 15 m (Figure 16) was selected to explore the time distribution characteristics of noise.

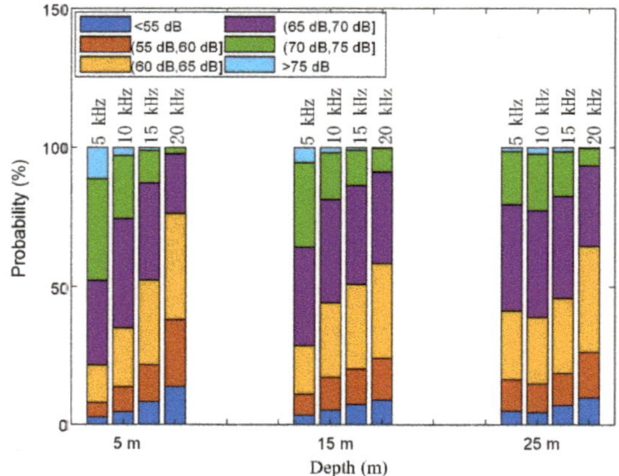

Figure 16. Noise distribution at different frequencies and different depths.

In Figure 17, subfigure (a) shows the time probability distribution of the 5 kHz narrowband noise with a mean of 68.59 dB and a standard deviation of 5.08 dB, subfigure (b) shows the time probability distribution of the 10 kHz narrowband noise with a mean of 66.16 dB and a standard deviation of 5.01 dB, subfigure (c) shows the time probability distribution of the 15 kHz narrowband noise with a mean of 65.30 dB and a standard deviation of 5.10 dB, and subfigure (d) shows the 20 kHz narrowband noise time probability distribution with a mean of 64.39 dB and a standard deviation of 4.97 dB. A comparison of these figures demonstrated that the probability of noise intensity basically followed the Gaussian distribution with a relativity small fluctuation on the time scale. As shown in Figure 17, the mean decreased with increasing frequency while the standard deviation remained stable.

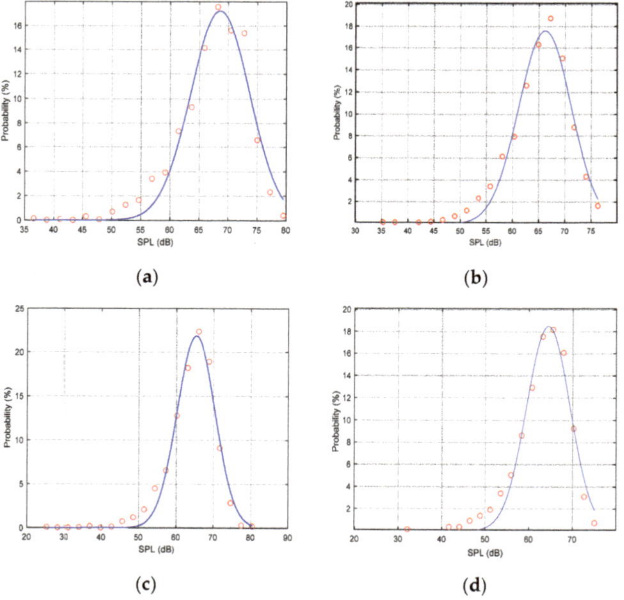

Figure 17. Statistical distribution of the noise over time: (**a**) 5 kHz, (**b**) 10 kHz, (**c**) 15 kHz, and (**d**) 20 kHz.

According to the analyses of the above experimental results, it can be concluded that the time distribution of the high-frequency noise field was relatively stable. The analysis results can provide a theoretical guide and data support for the underwater acoustic communication quality evaluation model.

4. Analysis of the Spatiotemporal Variation Range of the Signal-to-Noise Ratio in Underwater Acoustic Communication

Based on the spatial-temporal distribution characteristics of the sound field obtained in the previous section, the spatial-temporal variation range of the SNR was analyzed from the experimental data. The data was selected from the Yellow Sea Acoustic Communication Experiment ExDQ_1702. The external field experimental parameters were: water with a depth of 15 m, communication signal frequency band of 8 kHz–16 kHz, transducer placement with a depth of 5 m, and five arrays for each receiving array. There were 32 hydrophones in each array and the distance between the hydrophones was exactly 0.5 m, the horizontal communication distance was 3.5 km, the experimental sea state was approximately two levels, and the vertical distribution of the entire bandwidth SNR was counted.

Figure 18 indicates the vertical fluctuation of the SNR of different arrays. Because the receiving array was limited by the size of the receiving ship, the relative horizontal distance between the arrays was not large and the gap between the SNR was small. During the underwater acoustic communication experiment, the auxiliary ship of the receiving ship was continuously working and displayed certain random fluctuations on the surface. As the receiving ship fluctuated up and down on the sea, the SNR of the surface array element was significantly lower than that of the underwater array element. With the increase in water depth, the variation range of the SNR gradually decreased and basically remained stable after 5 m underwater.

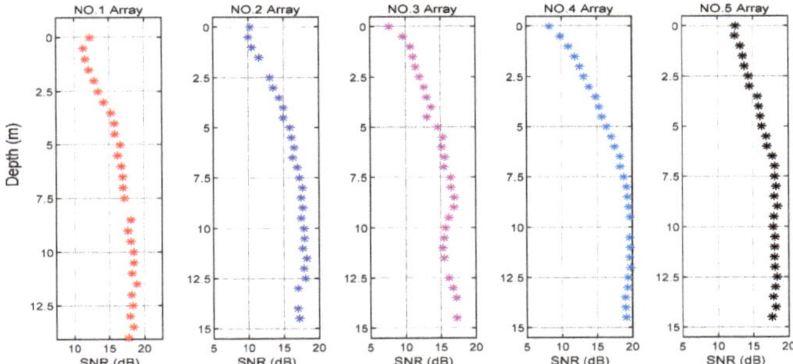

Figure 18. Signal-to-noise ratio of different arrays as a function of depth.

Figure 19 demonstrates the statistical range of the space-time variation of the SNR of array 5 over 31.5 s. The black hexagon indicates the mean of the SNR within the changing range, which is in line with Figure 18. From the figure, it was relatively small and fluctuated dramatically, which matched the trend of the spatial and temporal distributions of the signal field and noise field of the single-frequency signal. However, the experimental signal was further processed to cope with the slight inconsistency of the fluctuation range. Signals with center frequencies of 9 kHz and 15 kHz and a length of 31.5 s were selected. The sampling frequency of the system was 50 kHz and the number of sample points was about 1,574,520. The processing results are shown in Figures 20 and 21. In these figures, the abscissa is the SNR and the ordinate is the depth, and the plots show the variation ranges of the SNR within the signal time of 31.5 s at different depths, with the blue point representing the mean value. With the increase in the processing bandwidth, the spatial-temporal variation range of the SNR was significantly reduced, and the SNR of the surface array elements in the vertical direction was significantly lower than that in the water array elements, which verified the laws obtained using the theory and simulation experiments.

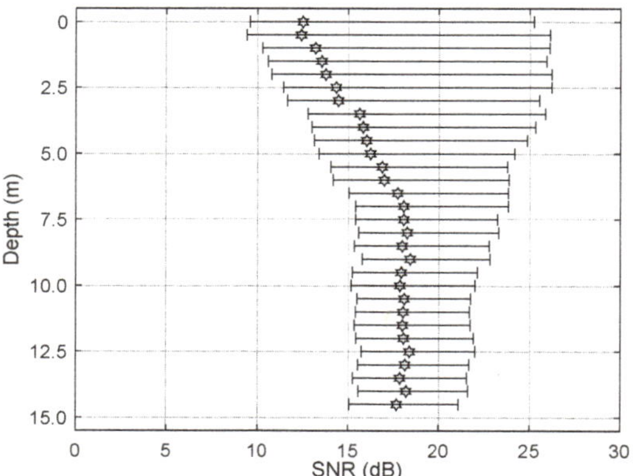

Figure 19. Signal-to-noise ratio fluctuation chart (fifth array).

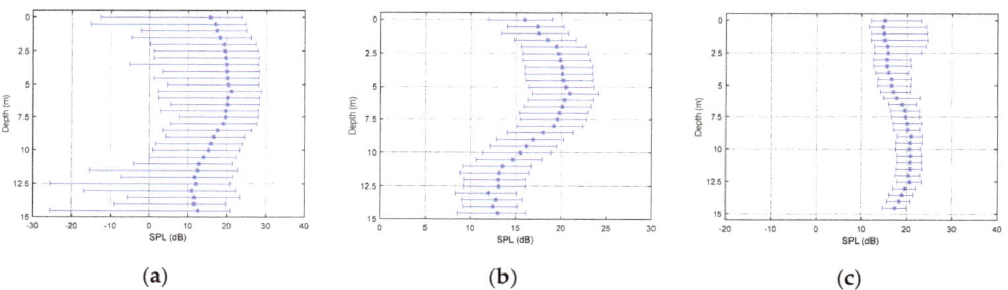

(a) **(b)** **(c)**

Figure 20. Signal-to-noise ratio vertical distribution (9 kHz): (**a**) bandwidth at 1 Hz, (**b**) bandwidth at 10 Hz, and (**c**) bandwidth at 500 Hz.

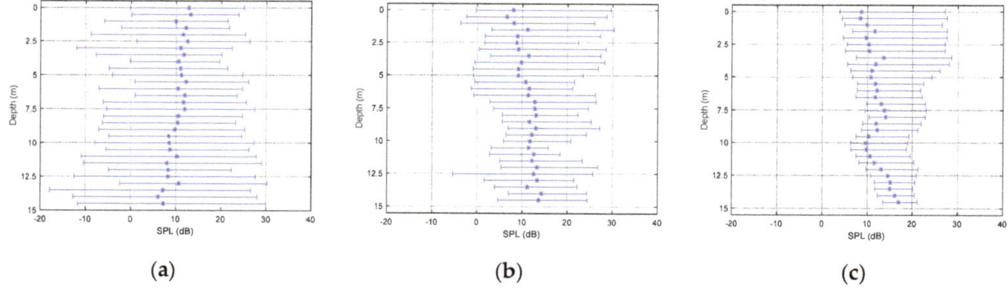

(a) **(b)** **(c)**

Figure 21. Signal-to-noise ratio vertical distribution (15 kHz): (**a**) bandwidth at 1 Hz, (**b**) bandwidth at 10 Hz, and (**c**) bandwidth at 500 Hz.

Figure 22 shows the simulation experiments performed under the fixed coding method using the channel parameters in the ExDQ_1702 experiment and the bit error rate corresponding to different signal-to-noise ratios (SNRs) and mapping methods. It can be seen that as the SNR increased, the bit error rate decreased. When the SNR in the experiment is less than 10 dB, the system should avoid using the 16QAM mapping method with a bit error rate higher than 0.01 and select other mapping methods according to the actual effective rate requirements.

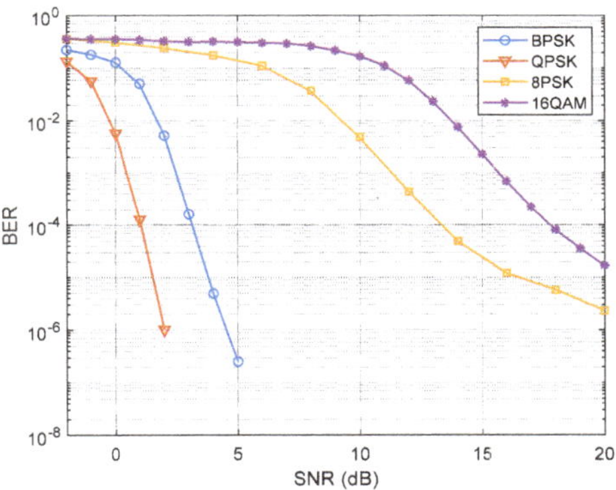

Figure 22. Bit error rate variation simulation results.

5. Conclusions

Aiming at solving problems of the serious spatiotemporal fluctuations of the signal caused by the time-varying channel structure of the shallow sea, this study took identifying the spatiotemporal variation range of the SNR of underwater acoustic communication as the research goal, which was explored in the sound field from two aspects: the signal field and the noise field. The investigation of the temporal and spatial distribution of the signal field considered signal interference effects caused by surface reflection and scattering and theoretically deduced the variation of signal intensity fluctuations with horizontal distance, signal frequency, bandwidth, and deployment depth, which was further verified through both simulations and the Yellow Sea trial. The investigation of the noise field mainly considered the spatiotemporal distribution of high-frequency wind-induced noise and ship noise. Through processing and analysis of the experimental data, it was found that the time fluctuation of noise basically conformed to the Gaussian distribution, the spatial distribution consistency was high, and the near-sea surface noise was slightly higher than the bottom noise. Combining the analysis of the spatiotemporal distribution characteristics of the signal field and the noise field, it was found that when the signal frequency was high, the bandwidth was large; furthermore, when the horizontal distance between the transceivers was small and the depth of the receiver was deep, the spatial-temporal variation range of the SNR of the UAC system was relatively small. These characteristics were verified using sea trial data, and the derived law will be used to guide the parameter configuration and network protocol optimization of the UAC systems.

The correlation radius of the acoustic signal decreased with increasing frequency. The vertical correlation radius was smaller than the horizontal correlation radius. Therefore, in shallow sea acoustic communication, vertical arrays should be used for receiving to enhance the SNR and improve system reliability.

Author Contributions: Z.L. conceived of the study, designed the study, and wrote the manuscript. M.Y. and L.D. provided guidance on the ideas and mathematical treatment. H.L. and C.R. compiled the experimental data. J.Q. and L.W. assisted in the editing of manuscripts. All authors have read and agreed to the published version of the manuscript.

Funding: This research was funded by the National Natural Science Foundation of China (Grant No. 11874061), the Youth Innovation Promotion Association CAS (No. 2021023), and Shandong Province "Double-Hundred Talent Plan (WST2020002)".

Institutional Review Board Statement: Not applicable.

Informed Consent Statement: Not applicable.

Data Availability Statement: Not applicable.

Conflicts of Interest: The authors declare no conflict of interest.

References

1. Hui, J.; Sheng, X. *Underwater Acoustic Channel*; National Defense Industry Press: Arlington, VA, USA, 2007.
2. Li, Q.; Wang, L.; Wei, C.; Li, Y.; Ma, X.; Yu, H. Theoretical analysis and experimental results of interference striation pattern of underwater target radiated noise in shallow water waveguide. *Chin. Ournal Acoust.* **2011**, *21*, 73–80.
3. Wang, H.; Zhao, Y.; Wu, B.; Chen, Y.; Chen, H. Estimation of source parameters based on underwater acoustic interference pattern in shallow water. In Proceedings of the IEEE International Conference on Signal Processing, Communications and Computing, Xi'an, China, 14–16 September 2011; pp. 1–4.
4. Chen, S.; Zhao, L.; Cao, J. Analytical study on acoustic interference pattern in shallow water. *Acta Acust.* **2017**, *37*, 3–16.
5. Dahl, P.H.; Miller, J.H.; Cato, D.H.; Andrew, R.K. Underwater Ambient Noise. *Acoust. Today* **2007**, *3*, 23–33. [CrossRef]
6. Urick, R.J. Ambient Noise in the Sea. *J. Acoust. Soc. Am.* **1984**, *86*, 194.
7. Wenz, G.M. Acoustic Ambient Noise in the Ocean: Spectra and Sources. *J. Acoust. Soc. Am.* **1962**, *34*, 1936–1956. [CrossRef]
8. Ross, N.C.; Andrea, P. Low frequency deep ocean ambient noise trend in the Northeast Pacific Ocean. *J. Acoust. Soc. Am.* **2011**, *129*, 161–165.
9. Chapman, R.; Price, A. Trends in low-frequency deep ocean ambient noise levels: New results from old data. *J. Acoust. Soc. Am.* **2010**, *127*, 1782. [CrossRef]
10. Mcdonald, M.A.; Hildebrand, J.A.; Wiggins, S.M. Increases in deep ocean ambient noise in the Northeast Pacific west of San Nicolas Island, California. *J. Acoust. Soc. Am.* **2006**, *120*, 711–718. [CrossRef] [PubMed]
11. Kibblewhite, C.A.; Shooter, J.A.; Watkins, S.L. Examination of attenuation at very low frequencies using the deep-water ambient noise field. *J. Acoust. Soc. Am.* **1998**, *60*, 1040–1047. [CrossRef]
12. Shooter, A.J.; Demary, T.E.; Wittenborn, A.F. Depth Dependence of Noise Resulting from Ship Traffic and Wind. *IEEE J. Ocean Eng.* **1990**, *15*, 292–298. [CrossRef]
13. Wille, P.C.; Geyer, D. Measurements on the origin of the wind-dependent ambient noise variability in shallow water. *J. Acoust. Soc. Am.* **1984**, *75*, 173–185. [CrossRef]
14. Ross, D. Mechanics of underwater noise. *J. Acoust. Soc. Am.* **1989**, *86*, 1626. [CrossRef]
15. Piggott, C.L. Ambient Sea Noise at Low Frequencies in Shallow Water of the Scotian Shelf. *J. Acoust. Soc. Am.* **1964**, *36*, 2152. [CrossRef]
16. Kuperman, W.A.; Ingenito, F. Spatial correlation of surface generated noise in a stratified ocean. *J. Acoust. Soc. Am.* **1998**, *67*, 1988–1996. [CrossRef]
17. Perkins, J.S. Modeling ambient noise in three-dimensional ocean environments. *J. Acoust. Soc. Am.* **1993**, *93*, 739–752. [CrossRef]
18. Cron, B.F.; Sherman, C.H. Spatial-Correlation Functions for Various Noise Models. *Acoust. Soc. Am. J.* **1962**, *34*, 1. [CrossRef]
19. Li, T. Research on the Spatial Correlation of the Acoustic Vector Field for Surface-Generated Noise in Shallow Water. Master's Thesis, Harbin Engineering University, Harbin, China, 2010.
20. Huang, Y.; Li, T.; Guo, J. Spatial Correlation of Surface Noise Received by Acoustic Vector Sensors in a Horizontally Stratified Medium. *J. Harbin Eng. Univ.* **2010**, *31*, 975–981.
21. Shang, H. Research on the Correlation of Ambient Noise Based on Acoustic Vector Hydrophone. Master's Thesis, Harbin Engineering University, Harbin, China, 2008.
22. Huang, Y.; Yang, S. Spatial Correlation Coefficients of Sound Pressure for General Ambient Noise Models. *Tech. Acoust.* **2007**, *26*, 1004–1005.
23. Zhou, J.; Piao, S.; Liu, Y.; Zhu, H. Ocean Surface Wave Effect on the Spatial Characteristics of Ambient noise. *Acta Phys. Sin.* **2017**, *66*, 121–131.
24. Avrashi, G.; Amar, A.; Cohen, I. Time-varying carrier frequency offset estimation in OFDM underwater acoustic communication. *Signal Process.* **2022**, *190*, 108299. [CrossRef]
25. Lv, Z.; Bai, Y.; Jin, J.; Wang, H.; Ren, C. Analysis of wave fluctuation on underwater acoustic communication based USV. *Appl. Acoust.* **2020**, *175*, 107820. [CrossRef]
26. Zhang, X.; Ying, W.; Yang, B. Parameter estimation for class a modeled ocean ambient noise. *J. Eng. Technol. Sci.* **2018**, *50*, 330–345. [CrossRef]
27. Zhang, X.; Wu, H.; Sun, H.; Ying, W. Multireceiver SAS imagery based on monostatic conversion. *IEEE J. Sel. Top. Appl. Earth Obs. Remote Sens.* **2021**, *14*, 10835–10853. [CrossRef]
28. An, J.; Ra, H.; Youn, C.; Kim, K. Experimental Results of Underwater Acoustic Communication with Nonlinear Frequency Modulation Waveform. *Sensors* **2021**, *21*, 7194. [CrossRef] [PubMed]
29. Zhang, Y.; Li, C.; Wang, H.; Wang, J.; Yang, F.; Meriaudeauc, F. Deep learning aided OFDM receiver for underwater acoustic communications. *Appl. Acoust.* **2022**, *187*, 108515. [CrossRef]

Article

Underwater Acoustic Signal Detection Using Calibrated Hidden Markov Model with Multiple Measurements

Heewon You [1], Sung-Hoon Byun [2,3] and Youngmin Choo [4,*]

1   Department of Ocean Systems Engineering, Sejong University, Seoul 05006, Korea; youhw0905@sju.ac.kr
2   Korea Research Institute of Ships and Ocean Engineering (KRISO), Daejeon 34103, Korea; byunsh@kriso.re.kr
3   KRISO Campus, Korea University of Science and Technology, Daejeon 34103, Korea
4   Department of Defense Systems Engineering, Sejong University, Seoul 05006, Korea
*   Correspondence: ychoo@sejong.ac.kr

Abstract: It is important to find signals of interest (SOIs) when operating sonar systems. A threshold-based method is generally used for SOI detection. However, it induces a high false alarm rate at a low signal-to-noise ratio. On the other side, machine-learning-based detection is performed to obtain more reliable detection results using abundant training data, costing intensive time and labor. We propose a method with favorable detection performance by using a hidden Markov model (HMM) for sequential acoustic data, which requires no separate training data. Since the detection results from HMM are significantly affected by the random initial parameters of HMM, the genetic algorithm (GA) is adopted to reduce the sensitivity of the initial parameters. The tuned initial parameters from GA are used as a start point for the subsequent Baum–Welch algorithm updating the HMM parameters. Furthermore, multiple measurements from arrays are exploited both in determining the proper initial parameters with GA and updating the parameters with the Baum–Welch algorithm. In contrast to the standard random selection of the initial point with single measurement, a stable initial point setting by the GA ensures improved SOI detections with the Baum–Welch algorithm using the multiple measurements, which are demonstrated in passive and active acoustic data. Particularly, the proposed method shows the most confidential detection in finding weak elastic surface waves from target, compared to existing methods such as conventional HMM.

Keywords: sonar signal detection; hidden Markov model; genetic algorithm

Citation: You, H.; Byun, S.-H.; Choo, Y. Underwater Acoustic Signal Detection Using Calibrated Hidden Markov Model with Multiple Measurements. *Sensors* **2022**, *22*, 5088. https://doi.org/10.3390/s22145088

Academic Editors: Haixin Sun and Xuebo Zhang

Received: 17 June 2022
Accepted: 3 July 2022
Published: 6 July 2022

Publisher's Note: MDPI stays neutral with regard to jurisdictional claims in published maps and institutional affiliations.

1. Introduction

Sonar systems with arrays comprising multiple sensors have been used to detect signals of interest (SOIs). However, various noises from vessels, submarines, and fish schools, etc., exist in the ocean and are measured by the sensors along with the SOIs. Therefore, detection methods are required to discriminate the desired signals from the noises.

Threshold-based detection schemes such as energy detection [1] or constant false alarm rate (CFAR) detection [2] are generally used in finding the SOIs. In energy detection, the energy of measured data is compared with a predefined threshold value. CFAR detection is a scheme that uses an adaptive threshold based on the relationship between a specified cell (sample under test) and adjacent auxiliary data. The threshold-based detections do not require prior information regarding the marine environment and exhibit low computational complexity. However, the detection performance is inferior in low signal-to-noise ratio (SNR) owing to the simple decision rules for SOIs. Hence, sophisticated detection methods using algorithms from machine learning (ML) have been proposed [3–6].

Owing to technological developments, various ML schemes have been applied in detecting the SOIs passively or actively [3–7], which treat the SOI detection as classification problems. To distinguish a target from a clutter in active sonar systems, a perceptual-based signal features from the human auditory system are exploited [3]. To suppress interference from background noise in recognizing underwater sound signals, a denoising autoencoder are used with random forest [4]. Furthermore, signal detection methods using various convolution neural networks have been used actively [5–7]. Although ML-based detections or classifications have remarkably enhanced performance, they require abundant training data for a given task, which have time and labor costs.

A hidden Markov model (HMM), a ML algorithm, has been applied widely to speech and text recognition with sequence data [8–11]; it estimates hidden states (or hidden information) of samples in the sequence data by using probabilities (HMM parameters) explaining the hidden states, which are extracted from the given data themselves. HMM has been applied to sequential data measured by radar and sonar systems to detect the corresponding target signals (i.e., SOIs) [12–14]. For a track-before-detection strategy in noisy environments, HMM was used for radar target detection to avoid the usage of threshold-based detection [12]. In bioacoustics, the vocalizations of Bryde's whales were automatically identified by using HMM, which enables the SOI detection even when ship noise interfering with the whale sounds is present [13]. In [14], to enhance detection performance in passive sonar data, a pregrouping of acoustic signals (i.e., samples are preliminarily clustered into "signal" and "noise") was incorporated with HMM.

In the current study, we attempt to detect SOIs in sonar data such as scattered signals from targets with low false alarms using limited measurements. Thus, we propose an HMM-based detection requiring no separate training data. Since the detection results from HMM are significantly affected by random initial parameters of HMM, a genetic algorithm (GA) is utilized to reduce the sensitivity of the initial parameters to the detection [15]. Furthermore, multiple measurements from array are exploited in the HMM-based detection to enhance accuracy and stability in finding the SOIs within sonar data. Section 2 presents problems in underwater signal detection and describes the parameters in the HMM. In Section 3, the proposed scheme using the HMM is explained comprehensively. The detection performance of the proposed scheme is investigated using synthetic passive and real active sonar data (Sections 4 and 5). Finally, the conclusions are provided in Section 6.

2. Problem Description

Here, an HMM-based detection scheme is proposed to detect SOIs with less false alarms and without separate training data by exploiting sequential acoustic data with pre-established probability models in the HMM. The conventional HMM is modified to accommodate multiple measurements from an array, which enhance the detection accuracy and robustness by correlated SOIs over sensors in the array.

The HMM has been widely adopted in speech and text recognition involving sequential data [8–11]. In the HMM framework, samples in the sequential data have hidden states, which are estimated from observed signals. Sonar data can be sequenced based on the regularity of the SOIs, and the HMM can be applied to underwater acoustic signal detection using time-domain sonar data after quantization is performed. Two states exist in sonar data, i.e., signal and noise states.

Figure 1 shows the structure of the HMM. The HMM (or probability models in the HMM) can be expressed as $\theta = [\pi, \mathbf{A}, \mathbf{B}]$, where π, $\mathbf{A}$, and $\mathbf{B}$ are the initial state distribution, transition matrix, and emission matrix, respectively. The initial state distribution, π, indicates the probability distribution over the states at the initial time. The states, as time progresses, are connected by the first-order Markov chain [16], and their transitions are represented by the transition matrix $\mathbf{A}$ (relevant to dotted line), whose element (i, j) represents the probability of the state changing from the ith state at the present time to the jth state at the next time. The probability of a specific observation at a certain state (emission probability) is represented by emission matrix $\mathbf{B}$ (relevant to dashed line). The sizes of the transition and emission matrix are $M \times M$ and $M \times N$, respectively, when the number of states is M and the values in sonar data are quantized with N (in the current study, $M = 2$ and $N = 150$).

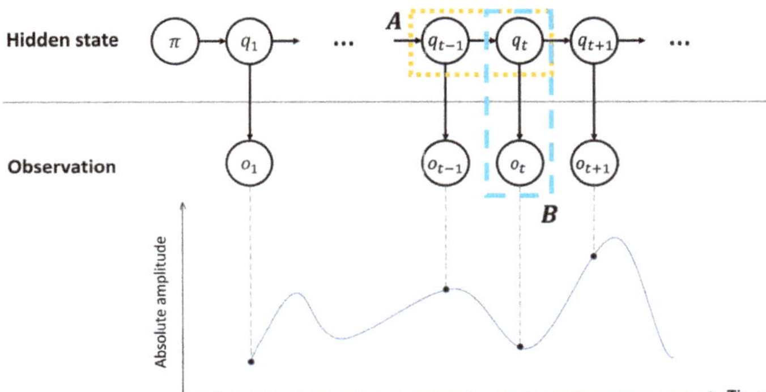

Figure 1. The structure of the HMM. The HMM explains observations o_t by using three probability models. An initial state distribution π is the probability distribution over the states at the initial time. The state change (marked with dotted line) is accounted for by the first-order Markov chain, and the previous state affects the present state. An observation at a specific time appears probabilistically, which depends on the state (marked with dashed line). HMM finds the optimal probability models for observations (Baum–Welch algorithm), which are subsequently used to identify the hidden states with the observations (Viterbi algorithm).

In the HMM framework, optimal probability models (also, referred to as optimal parameters) are obtained with best explaining the observations by the Baum–Welch algorithm in HMM [17] ($\theta^* = \underset{\theta}{\operatorname{argmax}} P(\mathbf{o}|\theta)$). Then, the hidden states are revealed by the Viterbi algorithm, which uses the estimated parameters with observations as follows [18,19]: $\underset{q}{\operatorname{argmax}} P(\mathbf{q}|\mathbf{o}, \theta^*)$. From the perspective of detection, the SOIs are identified by samples possessing the "signal" state.

During parameter estimation using the HMM, the Baum–Welch algorithm was used to determine θ^* from randomly selecting the initial values. However, the estimates strongly depend on the initial values owing to local optimal points; HMM-based detection results for the same data can be different owing to the random initial values differing along with the applications of HMM. Hence, several studies have been conducted to determine the appropriate initial values for the parameters [14,15,20,21]. In relevant studies pertaining to the detection of underwater SOIs [14], k-means and pre-grouping were used to derive the proper initial values instead of the random ones. This detection scheme is referred to as expectation-maximization (EM)-Viterbi Algorithm (VA) as in [14]. Although the EM-VA provides an accurate detection of SOIs in environments containing transient noise, its performance degrades when ambient noise is present, as discussed in Section 4. Furthermore, when multiple SOIs exist, SOIs having magnitudes similar to or less than the magnitude of noise are overlooked by the EM-VA.

In the EM-VA [14], a single measurement (acoustic data at a single sensor) was used as observation. In the sonar system, multiple measurements can be acquired by sensor arrays. Unlike previous studies using HMM [14,22,23], here, multiple measurements were exploited not only to determine the reliable initial values using the genetic algorithm (GA) but also to update parameters using the Baum–Welch algorithm; these are described comprehensively in the following section.

3. HMM Calibration and Parameter Adjustment Using Multiple Measurements

The Baum–Welch algorithm is sensitive to the initial values of the parameters, and its estimations strongly depend on the initial values. To determine the appropriate initial values, multiple measurements from arrays in the sonar system are used to ensure accurate and robust SOI detection. The detection process, which involves various algorithms, is illustrated in Figure 2. First, the initial values for the HMM, which exhibit the most effectively the observations of multiple measurements in terms of probability, are determined using the GA (Section 3.1). Next, the parameters in the calibrated HMM are adjusted by the Baum–Welch algorithm using multiple measurements (Section 3.2). Finally, the hidden states are derived for the multiple measurements using the Viterbi algorithm, and they indicate the SOIs.

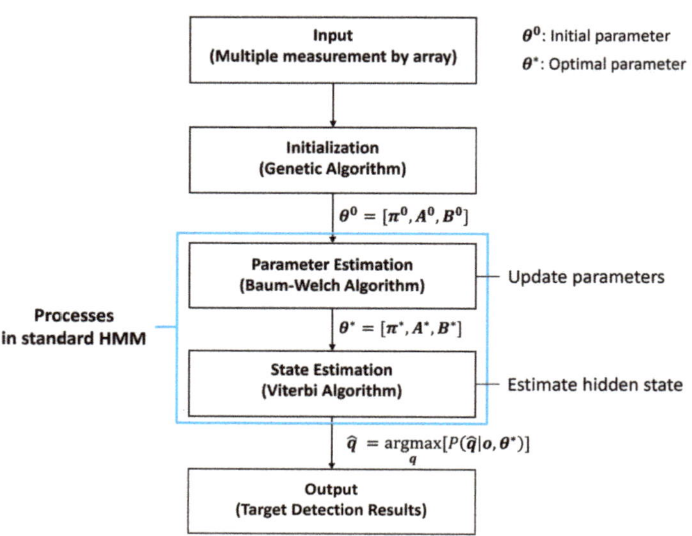

Figure 2. Block diagram of detection process in hidden Markov model (HMM). Multiple measurements were exploited to determine the best initial values for the HMM parameters using the genetic algorithm. They were updated with the Baum–Welch algorithm using the multiple measurements. Then, hidden states of the multiple measurements were revealed via the Viterbi algorithm using the HMM parameters.

3.1. Initialization: Calibrating HMM

We used a GA, which is inspired by the natural selection process, to estimate the initial values of the parameters. The GA is a representative scheme for solving optimization problems, where genes (candidates for the solution) in a population at the current generation produce genes in a population at the next generation using crossover, mutation, and selection (evolution) to approach the solution [24]. In particular, mutation in the GA prevents a solution from being a local optimum.

In the current study, genes in the GA are parameters with distinct values (θ_g^p, where the subscript g and superscript p indicate the gene and generation numbers, respectively) and the appropriate initial values for the measurements are determined by evolving θ_g^p in the GA. The criterion for the appropriateness is calculated using the fitness function $P_g^p = lnP(\mathbf{o}|\theta_g^p)$, where $o = [o_1, ..., o_T]$ is observed in the time domain, o_t is a quantized acoustic signal v_n, and T is the total number of observations with a signal length. $P(\mathbf{o}|\theta_g^p)$ is the likelihood function, which stochastically explains the observation based on the specified parameters and is calculated using the forward or backward algorithm in the Baum–Welch algorithm. In this study, the forward algorithm was adopted for the calculation with probability $P(o_1, ..., o_t, q_t = s_m|\theta)$, which was obtained from $\alpha_t(m') = b_{m'}(o_t = v_n) \sum_{m=1}^{M} \alpha_{t-1}(m) a_{m,m'}$. q_t is the observed state at time t and is one of the state types s_m (i.e. "signal" or "noise"; hence, $M = 2$); $a_{m,m'}$ ((m, m') element of the transition matrix) and $b_{m'}(o_t = v_n)$ ((m', n) element of the emission matrix) are the probabilities of state transition from state m to state m' and the observation of $o_t = v_n$ at state m', respectively.

A new gene was created for the next population by selectively using θ_g^p among the current population; θ_g^p with a high probability was used preferentially. This process was repeated until a terminal condition was satisfied.

The signals received at the sensor array contain the SOIs. Hence, the parameters for the measurements are expected to be the same. The shared parameters were estimated to increase the detection accuracy and robustness, and the fitness function for the single measurement was modified to accommodate the multiple measurements with the assumption of their stochastic independence. A product of likelihood functions for the measurements was used for the fitness function [15], as follows:

$$P_g^p = \sum_{k=1}^{K} lnP(\mathbf{o}^{(k)}|\theta_g^p) \tag{1}$$

The superscript k in parenthesis represents the measurement number; K represents the total number of the measurements, which is the same as the number of sensors in the arrays. The optimal values from the GA using (1) are the initial values for the parameter in the HMM, which is referred to as the calibrated HMM herein.

3.2. Parameter Adjustment Using Baum–Welch Algorithm with Multiple Measurements

Optimal parameters (θ^*) were derived using the Baum–Welch algorithm [17], which uses the parameters from GA (θ^0) as a starting point. The Baum–Welch algorithm has an iterative loop composed of an expectation (E) step and a maximization (M) step. Hidden variables from the E-step are used to update old parameters in the M-step, and they are denoted as follows [17]:

$$\gamma_t = P(q_t = s_m|\mathbf{o}, \theta^q), \tag{2}$$

$$\xi_t(m, m') = P(q_t = s_m, q_{t+1} = s_{m'}|\mathbf{o}, \theta^q), \tag{3}$$

where θ^q is the parameter after q iterations; the superscript indicates the iteration number. γ_t and ξ_t are the probabilities of state s_m at time t, and the joint states of s_m at time t and $s_{m'}$ at time $t+1$ for $\mathbf{o}$ and θ^q, respectively.

When applying the Baum–Welch algorithm with a single measurement, the parameters are updated using the hidden variables as follows [17]:

$$\pi_m^{q+1} = \gamma_1(m), \qquad 1 \leq m \leq M \tag{4}$$

$$a_{m,m'}^{q+1} = \frac{\sum_{t=1}^{T-1} \xi_t(m, m')}{\sum_{t=1}^{T-1} \gamma_t(m)}, \qquad 1 \leq m \leq M, 1 \leq m' \leq M, \tag{5}$$

$$b_m^{q+1}(v_n) = \frac{\sum_{t=1}^{T} \mathbb{I}_{o_t=v_n} \gamma_t(m)}{\sum_{t=1}^{T} \gamma_t(m)}, \qquad 1 \leq m \leq M, 1 \leq n \leq N, \tag{6}$$

where $\pi_m^{q+1}, a_{m,m'}^{q+1}$, and $b_m^{q+1}(v_n)$ are elements of the initial state distribution, transition matrix, and emission matrix at $q+1$ iterations, respectively. $\mathbb{I}_{o_t=v_n}$ is an indicator function, which equals one when $o_t = v_n$. Otherwise, it is zero. T is the signal length of observation (or measurement) $\mathbf{o}$. The initial state distribution of (4) is obtained from γ_1. The transition probability of (5) is a conditional probability that accounts for the state changing from the mth state at the present time to the m'th state at the next time. It is the ratio of the sum of joint probabilities of s_m at time t and $s_{m'}$ at time $t+1$ to the sum of probabilities of s_m at time t; the sums are conducted in the time domain. The emission probability of (6) conforms to its definition (i.e., the probability of a specific observation quantity v_n at state s_m) by counting $\gamma_t(m)$ with the observation o_t matching v_n among all $\gamma_t(m)$. Equations (2)–(6) are used repeatedly until the parameters converge or the iteration reaches a predefined number.

To exploit the commonality (i.e., the shared parameters) of the multiple measurements from the array, the parameters are updated during the iterations of the Baum–Welch algorithm as follows [17]:

$$\pi_m^{q+1} = \frac{1}{K} \sum_{k=1}^{K} \gamma_1^{(k)}(m), \qquad 1 \le m \le M \tag{7}$$

$$a_{m,m'}^{q+1} = \frac{\sum_{k=1}^{K} \sum_{t=1}^{T^{(k)}-1} \xi_t^{(k)}(m,m')}{\sum_{k=1}^{K} \sum_{t=1}^{T^{(k)}-1} \gamma_t^{(k)}(m)}, \qquad 1 \le m \le M, 1 \le m' \le M, \tag{8}$$

$$b_m^{q+1}(v_n) = \frac{\sum_{k=1}^{K} \sum_{t=1}^{T^{(k)}} \mathbb{I}_{o_t^{(k)}=v_n} \gamma_t^{(k)}(m)}{\sum_{k=1}^{K} \sum_{t=1}^{T^{(k)}} \gamma_t^{(k)}(m)}, \qquad 1 \le m \le M, 1 \le n \le N, \tag{9}$$

Hidden variables $\gamma_t^{(k)}$ and $\xi_t^{(k)}$ are calculated using the kth measurement $\mathbf{o}^{(k)}$ in the E-step; $T^{(k)}$ is the signal length of $\mathbf{o}^{(k)}$ and is set as a constant of T in the current study. Equations (7)–(9) are obtained by modifying (4)–(6) with an additional summation over the spatial domain based on multiple measurements by the array. The initial state distribution of (7) is the average of $\gamma_1^{(k)}$ over the spatial domain. Similar to (5) and (6), the transition probability of (8) is the ratio of the sum of the joint probabilities $\xi_t^{(k)}(m,m')$ and the sum of the corresponding marginal probabilities $\gamma_t^{(k)}(m)$; the sums are conducted in the space and time domains. The emission probability of (9) is calculated by counting $\gamma_t^{(k)}(m)$, with the observation o_t matching v_n among $\gamma_t^{(k)}(m)$.

Multiple measurements from the array are beneficial to the HMM because they provide additional samples that are in proportion to the number of sensors for estimating the conditional probabilities, as shown in (7)–(9). The probabilities from rich data are more reliable and result in stable and accurate signal detections.

Next, the hidden states for each measurement are revealed using the Viterbi algorithm [18,19], based on observations as well as shared parameters θ^* from the Baum–Welch algorithm. $\hat{\mathbf{s}}^{(k)}$, which comprises hidden states as time progresses at the kth measurement, is derived using the Viterbi algorithm by maximizing $P(\mathbf{q}^{(k)} = \mathbf{s}^{(k)}|\mathbf{o}^{(k)}, \theta^*)$; it indicates the SOIs in the measurement. The suggested process is abbreviated as the GA-HMM.

Although the parameters can be determined using the GA or Baum–Welch algorithm separately, the two optimization schemes are used sequentially in SOI detection for a superior estimation of parameters; the GA derives an unbiased initial point for the Baum–Welch algorithm (Figure 3a), and a desired optimal point is subsequently determined from the initial point (Figure 3b). The detection performance afforded by the Baum–Welch algorithm alone is sensitive to the random initial values of the parameter (or random initial point), which are updated consecutively using the Hill-Climbing [25,26] and hence can fall into a local optimum point next the neighboring random initial point. Using only the GA incurs a high computational cost for parameter convergence. Furthermore, noise hinders the GA from converging near global optimal points. The parameters cannot converge even after sufficient generations; hence, the detection performance based on the GA deteriorates.

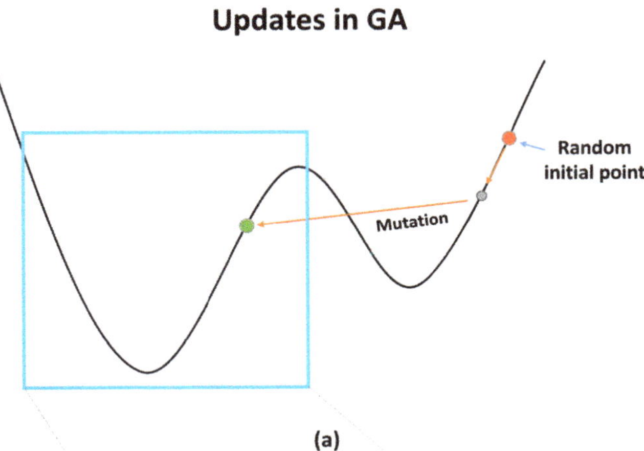

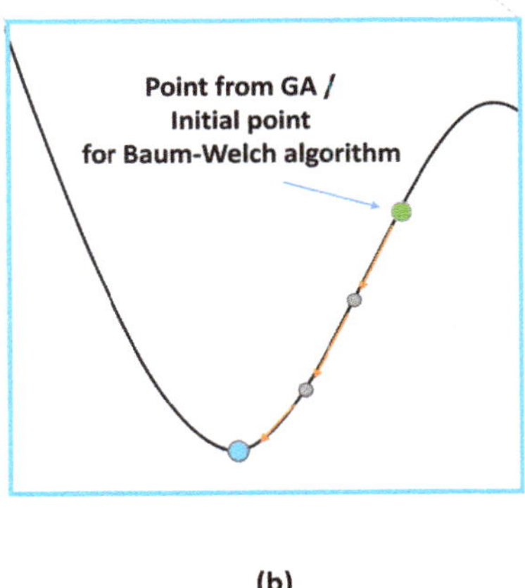

Figure 3. Parameters in HMM evaluated via sequential usage of GA and Baum–Welch algorithm: (**a**) GA yielded unbiased initial point for Baum–Welch algorithm in subsequent stage; (**b**) Baum–Welch algorithm yielded global optimal point from unbiased initial point.

4. Analysis of GA-HMM Using Synthetic Data

The detection performance of the GA-HMM was analyzed by comparing its detection results with those of other schemes. The effects of the fine initial point from the GA were demonstrated with synthetic data.

4.1. Numerical Environment

To analyze the GA-HMM, synthetic data were generated while considering the acoustic signals measured using the sonar systems. Each synthetic datum with a signal length of 0.3 s was discretized with a sampling frequency of 500 Hz and contained 150 samples ($T = 150$). Here, the SOI in the synthetic data was a 50 Hz three-cycle sine wave comprising 30 samples, and it was contaminated by additive white Gaussian noise (Figure 4a). Although the starting point of the SOI did not affect the detection performance, the 50th sample of the synthetic data was used as the starting point to ease the visual inspection of the detection results.

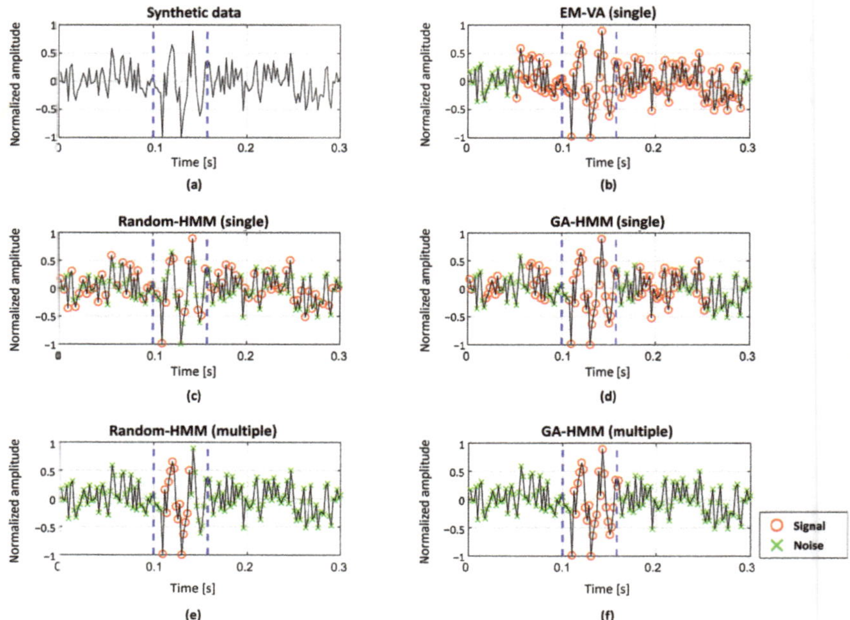

Figure 4. (**a**) Representative example of synthetic data with SNR of 8 dB. Detection results from (**b**) EM-VA, (**c**) single-measurement Random-HMM, (**d**) single-measurement GA-HMM, (**e**) multiple measurement Random-HMM, and (**f**) multiple-measurement GA-HMM. The Interval for SOI is indicated by vertical dashed lines. Symbols "o" and "x" represent "signal" and "noise" states, respectively.

Noise with various magnitudes were added to the clean synthetic to investigate the detection performance according to SNRs. Additionally, different numbers of the synthetic data were used for the detection to demonstrate properties of multiple measurements in finding the SOI.

In the current study, an observation value of HMM is an absolute value of the acoustic signal quantized with uniform intervals of 150 ($N = 150$) after normalization with its absolute maximum. To obtain the fine initial point using the GA, 200 randomly generated parameters were used as genes in the first-generation population. A score for the appropriateness was calculated for the genes using the fitness function presented in (1), and the genes with high scores had a high probability of being selected for generating the next genes with crossover. The probability of mutation was set to 0.01. The most feasible parameter after 10 generations ($p = 10$) was used for the fine initial point. In the GA, the transition probability between the same (or different) states had a lower (or upper) bound of 0.5, owing to rare transitions between different states, which occurred at the 50th (from noise to signal) and 80th samples (from signal to noise) among 150 samples in the simulation.

Subsequently, the Baum–Welch algorithm commenced from the fine initial points and terminated when the parameters converged or the iteration reached a predefined number (in this study, $Q = 500$).

The Viterbi algorithm, which is applied to the sequential samples in the observations with estimated parameters, implies the states at the samples with a value of 0 or 1 ($M = 2$). The variance of samples with the same state was calculated. The state with a higher (or lower) variance was assigned to the "signal" state (or "noise" state) as in [14]; here, samples with values of 0 and 1 correspond to the "noise" and "signal" states, respectively. Quantities for the hyperparameters in the GA-HMM, including the quantization number and upper and lower bounds, were determined empirically.

4.2. Detection Performance Analysis of GA-HMM

Figure 4 shows representative examples of detection results obtained by the GA-HMM, EM-VA, and Random-HMM; the HMM using a random initial point for the Baum–Welch algorithm is referred to as the Random-HMM here in. The SOI is indicated by vertical dashed lines. The SOI was detected under a harsh condition without using a matched filter enhancing the SNR (passive sonar signal detection). Noise comparable to the SOI (SNR = 8 dB) restricts the use of threshold-based detection schemes. Therefore, sophisticated schemes were used. In this study, the GA-HMM and Random-HMM were applied to perform detections using single or multiple measurements based on (1) and (7)-(9). Meanwhile, the EM-VA used a single measurement to identify the SOI because it cannot accommodate multiple measurements [14].

When using the EM-VA, most of the samples in the synthetic data were identified as the SOI by noise, and false alarms occurred, as shown in Figure 4b. Figure 4c,d show the detection results obtained from the Random-HMM and GA-HMM based on a single measurement. Many noise samples were misclassified as SOIs, increasing false alarm rates (FAR) owing to some inappropriate initial values in Random-HMM. This problem was mitigated using the GA-HMM, which determined the parameters using the fine initial point. However, considerable false alarms remained. Therefore, multiple measurements comprising 30 synthetic data were used, as shown in Figure 4e,f, to reduce false alarms. As a result, the Random-HMM using multiple measurements achieved significantly reduced the FAR of the SOI sample. On the other hand, the GA-HMM using multiple measurements exhibited the highest recall with less false alarms, thereby demonstrating its superior detection performance compared with the considered schemes.

Table 1 summarizes the recall, FAR, and computation time of the schemes based on the average detection results for 100 trials at a fixed SNR of 8dB. In this study, recall is defined based on the ratio of the number of correctly identified SOI samples to the total number of SOI samples, and the FAR is defined based on the ratio of the number of misidentified noise samples to the total number of noise samples. Although the EM-VA exhibited a high recall, it incorrectly identified noise persistently. In particular, the noise near the SOI tended to be identified as a "signal", and it resulted in the highest FAR as shown in Figure 4b. The Random-HMM using single measurement overlooks the SOI and misclassified noise because of unstable detection from the random initial point; thus, it resulted in an inferior recall and FAR. While these problems were alleviated using the single-measurement GA-HMM, it still exhibited a considerable FAR. Therefore, all single-measurement schemes exhibited unsatisfactory detection performance owing to excessive false alarms. Noise could not be distinguished from the SOI, thereby resulting in high FARs in the scarce measurement. Therefore, multiple measurements were used to mitigate these problems.

Table 1. Recall, false alarm rates, and computation time of the investigated schemes.

Scheme	Recall	False Alarm Rate	Computation Time
EM-VA (single)	0.86	0.63	0.83 s
Random-HMM (single)	0.60	0.48	0.08 s
GA-HMM (single)	0.65	0.43	6.76 s
Random-HMM (multiple)	0.69	0.27	11.02 s
GA-HMM (multiple)	0.88	0.06	12.10 s

The considered schemes were implemented at a computer with an intel(R) Core (TM) i9-9900K CPU, and the corresponding computational times were measured (Table 1). Although the proposed scheme showed the hugest computational burden, it can be applied to acoustic measurements during experiments and detect SOIs in semi-real-time, owing to its computational time in the order of 10 s.

The multiple-measurement Random-HMM exhibited improved performance in terms of both recall and FAR (moderate recall with significantly reduced FAR) because it exploited the consistency of the SOI in the multiple measurements when updating the parameters and was less affected by erratic noise. Detection performance was significantly improved by using multiple measurements to determine better initial points with GA and update parameters with the Baum–Welch algorithm. As a result, the multiple-measurement GA-HMM exhibited the highest recall and lowest FAR, indicating that most of the samples were identified correctly.

To investigate the detection performance of the schemes for various noise magnitudes, synthetic data with various SNRs were generated. The recalls and FARs from schemes were displayed according to the SNRs, as in Figure 5, where they were averaged over 100 trials at each SNR. At low SNR, EM-VA had high recall and FAR, and most of samples were identified as "signal". With the increment in the SNR, the false alarm significantly reduced with lower recall by overlooking SOI samples more frequently. The other schemes using single measurement resulted in increased recalls and decreased FARs as the SNR increased. In particular, the initial point obtained using the GA improved the single-measurement performance, which also improved as the SNR increased. The schemes based on single measurement could not provide reliable detections (even at high SNRs) because of their insufficient recalls (EM-VA and Random-HMM) or high FARs (GA-HMM and Random-HMM). As shown previously in Figure 4, the multiple measurements improved the detection performance by evaluating the conditional probabilities in (7)–(9) more confidently. The recall (or FAR) of the multiple-measurement Random-HMM improved gradually as the SNR increased and reached 0.75 (or approximately 0.2) at a high SNR of 13. The detection performance improved considerably by the multiple-measurement GA-HMM, whose classification accuracy was accelerated by the increase in the SNR and became almost perfect at the appropriate SNR. Additional methods for obtaining fine initial points such as the GA are important in HMM-based detection because the initial points significantly affect the parameter estimation in the Baum–Welch algorithm of HMM.

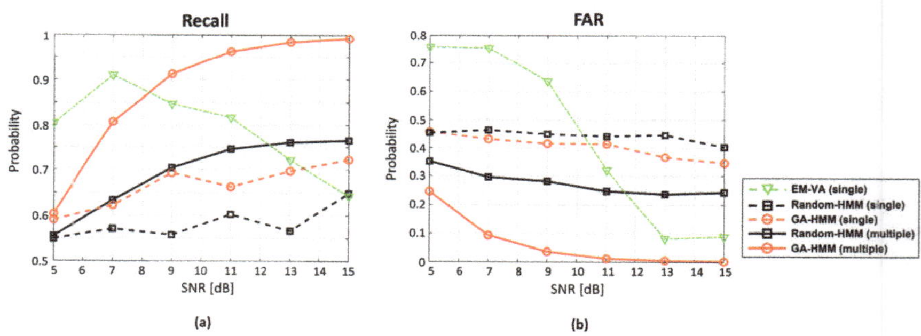

Figure 5. Detection performance of scheme according to SNRs: (**a**) Recall; (**b**) FAR. Multiple measurements consist of 30 synthetic data.

Figure 6 shows the detection results from the Random-HMM and GA-HMM analyzed based on the measurement number (sensor number in array) at a fixed SNR of 8 dB. Although their performances improved in proportion to the measurement number, the GA-HMM with a fine initial point exhibited superior accuracy in terms of detection regardless of the measurement number. The GA-HMM exploited multiple measurements more effectively than the Random-HMM because it used them in the Baum–Welch-algorithm-based update as well as the GA-based initial point identification. The performance differences increased until the measurement number reached 30. Despite the slow performance enhancement after 30 measurements, the multiple measurements afforded accurate and robust SOI detection.

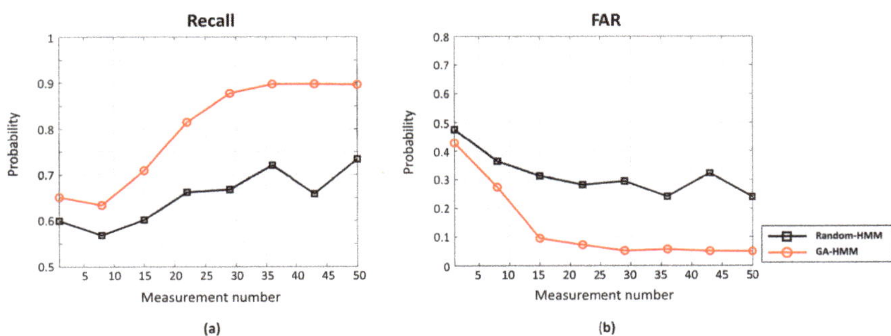

Figure 6. Detection Performance of scheme according to measurement number at fixed SNR of 8 dB: (**a**) Recall; (**b**) FAR.

5. Application of GA-HMM to Measured Acoustic Data

The feasibility of the multiple-measurement GA-HMM was investigated by analyzing acoustic data from a water tank experiment, which included intense specular echoes and weak elastic waves from shell targets (SOIs). The detection results for the real data were compared with those obtained using the EM-VA and multiple-measurement Random-HMM.

5.1. Experimental Environment

An experiment for target scattering was conducted in a water tank with size of 35 m (length) $\times$ 20 m (width) $\times$ 9 m (depth). A simple illustration of the water tank is shown in Figure 7a, and its details are provided comprehensively in [27]. A 1 s long linear frequency-modulated pulse signal with a bandwidth between 0.5 and 25 kHz from a transducer impinged on the cylindrical shell target, and the scattered signal from the target was measured using two hydrophones at different water depths (referred to as R1 and R2). After applying a matched filter to the measured signals (pulse compression), a specific time period of 1.5 ms, including the returns from the target (intense specular echo and two subsequent weak elastic surface waves) was selected, as illustrated in Figure 7b, and the corresponding observation size T was 150, with a sampling frequency of 100 kHz.

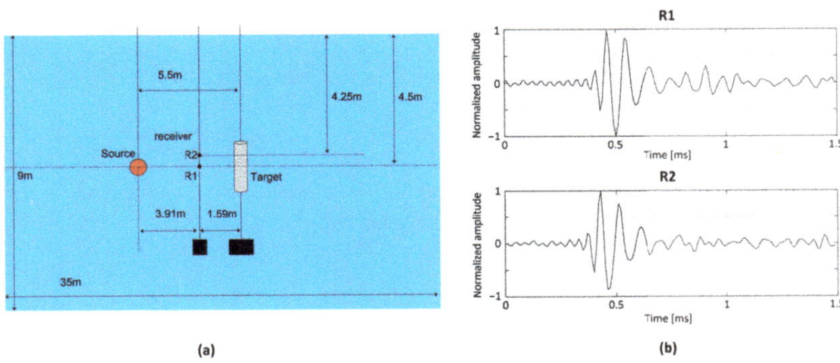

Figure 7. (**a**) Experimental environment; transmission signals are scattered by cylindrical shell and are received by two receivers at different water depths. (**b**) Portion of acoustic signals at two receivers after pulse compression (R1 and R2), which include intensive specular echoes and weak elastic surface waves.

The specular echoes from two measurements exhibited similar amplitudes and arrival times (approximately 0.5 ms) and were insensitive to the depth difference. The elastic waves, which exhibited distinct circumferential paths on the cylinder surface, were not consistent with the measurements in terms of amplitudes and time delays. A slight gap existed between the first elastic wave (approximately 0.8 ms) and the specular echo in the R1 measurement. On the other hand, the first elastic wave (approximately 0.6 ms) was immediately behind the specular echo and exhibited a small amplitude in the R2 measurement. The arrival times were confirmed by comparing the measured data with the simulated data based on the same environment [27]. The second elastic waves in the R1 and R2 measurements arrived at approximately 1.3 and 1.5 ms, respectively. The detection performance of the schemes was analyzed in terms of the identification of weak elastic waves, as will be described in the following subsection.

5.2. Detection Results of GA-HMM for Measured Acoustic Signals

Figure 8 shows the detection results from the EM-VA, the multiple-measurement Random-HMM, and the multiple-measurement GA-HMM. While the EM-VA was applied to the measurements individually, the Random-HMM and GA-HMM detected the SOIs after the shared parameters over the measurements were estimated. In the EM-VA, k means and pre-grouping were applied to evaluate the initial values, and obtain a consistent detection result for the same data. However, although the GA-HMM is less affected by the random initial value used in GA, the detection of the GA- and Random-HMM varies depending on the trials, even if the same data are used. Hence, the Random–HMM and GA-HMM were applied to the measurements repeatedly, and the average of 100 detection results were used. The value for the state of a certain sample in the measurements was between 0 and 1, and a sample with a higher (or lower) number was likely to be the signal (or noise). In this study, samples exhibiting values exceeding 0.7 and less than 0.3 were classified as "signal" and "noise", respectively. The remaining samples exhibiting values between 0.3 and 0.7 were neither "signal" nor "noise" and were referred to as "unclear samples".

The EM-VA detected the specular echoes without unclear samples owing to the consistent estimation, whereas it overlooked the weak elastic waves in both measurements. Furthermore, the elastic waves having similar magnitudes with noise made the Random-HMM using multiple measurements determine most samples as unclear samples, except for some samples within the specular echoes, and it was detrimental to identifying the SOIs. The GA-HMM using multiple measurements also suffered from detecting the weak elastic waves. Particularly, the second elastic wave in the R2 measurement was misidentified as "noise". In the water tank experiment, acoustic signals were measured by two hydrophones, and thus sparse measurements were used for the detection, which resulted in the diminished detection performance, compared to those using the synthetic data. The performance reduction could be mitigated by using additional measurements, which were unavailable in the current study. However, the multiple-measurement GA-HMM showed the best performance among the considered schemes. It significantly reduced the unclear samples and detected the specular echoes and elastic waves most confidently; even unclear samples locating between noise provided clues for the SOIs (e.g., the first elastic wave in R2 measurement).

Among the considered schemes, the GA-HMM exhibited the best signal detection and false alarm reduction in both the synthetic and measured data. Hence, the GA-HMM is applicable to sonar signal detection when ML-based schemes are unavailable because of inadequate training data.

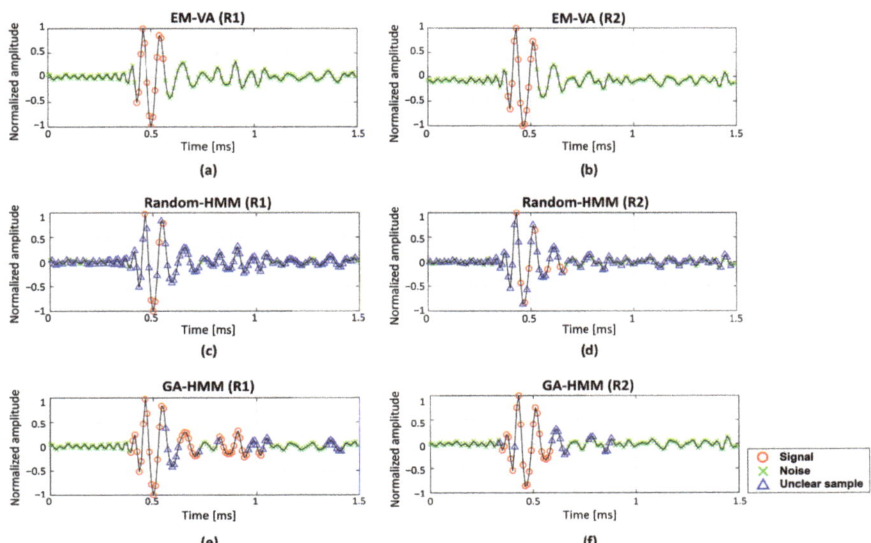

Figure 8. Detection results for signals measured by two receivers (R1 and R2). (**a**) Single-measurement detection result of EM-VA for R1; (**b**) single-measurement detection result of EM-VA for R2; (**c**) average detection result of multiple-measurement Random-HMM for R1; (**d**) average detection result of multiple-measurement Random-HMM for R2; (**e**) average detection result of multiple-measurement GA-HMM for R1; (**f**) average detection result of multiple-measurement GA-HMM for R2. Averaged values of the states from multiple-measurement Random-HMM and multiple-measurement GA-HMM are between 0 and 1. Samples exhibiting values exceeding 0.7 and less than 0.3 were classified as "signal" and "noise," respectively. The remaining samples exhibiting values between 0.3 and 0.7 were "unclear samples". Symbols of "o", "x", and "△" represent "signal", "noise", and "unclear samples," respectively.

6. Conclusions

We proposed a novel HMM-based detection method to accurately identify signals with a low FAR without requiring training data. However, the Baum–Welch algorithm for parameter estimation in the HMM is sensitive to the initial point and the problem of falling into the local optimum point often occur because of a random initial point. The GA provided a proper initial point for obtaining a global optimal point and determined the appropriate parameters by using the Baum–Welch algorithm with the initial point.

Furthermore, by using multiple measurements both in deriving the initial point with GA and updating the parameters with Baum–Welch algorithm, SOIs are detected more accurately and reliably; GA and multiple measurements improve the stability and accuracy of SOI detection, respectively. Thus, the multiple-measurement GA-HMM displayed superior performance in passive and active acoustic data, which are from simulation and real measurements, respectively. The detection results are compared with those from other detection schemes such as EM-VA and Random-HMM. Particularly, inconsistent and unclear detections from conventional HMM (single-measurement Random-HMM) are significantly alleviated by the multiple-measurement GA-HMM at the cost of computational complexity.

Author Contributions: Conceptualization, H.Y. and Y.C.; Data curation, H.Y.; Formal analysis, H.Y. and Y.C.; Funding acquisition, S.-H.B.; Methodology, H.Y.; Project administration, S.-H.B.; Software, H.Y.; Supervision, Y.C.; Validation, H.Y. and Y.C.; Writing—original draft, H.Y. and Y.C.; Writing—review & editing, H.Y., S.-H.B. and Y.C. All authors have read and agreed to the published version of the manuscript.

Funding: This study was supported by the research project funded by the Korea Research Institute of Ships and Ocean Engineering (KRISO, NTIS 1525012176).

Institutional Review Board Statement: Not applicable.

Informed Consent Statement: Not applicable.

Data Availability Statement: Not applicable.

Acknowledgments: Thanks to the Korea Research Institute of Ships and Ocean Engineering (KRISO) for providing the experimental data.

Conflicts of Interest: The authors declare no conflict of interest.

References

1. Kabalci, E.; Kabalci, Y. *From Smart Grid to Internet of Energy*; Academic Press: New York, NY, USA, 2019; pp. 909–248.
2. Abraham, D.A.; Willett, P.K. Active sonar detection in shallow water using the Page test. *IEEE J. Ocean. Eng.* **2002**, *27*, 35–46. [CrossRef]
3. Murphy, S.M.; Hines, P.C. Examining the robustness of automated aural classification of active sonar echoes. *J. Acoust. Soc. Am.* **2014**, *135*, 626–636. [CrossRef] [PubMed]
4. Zhou, X.; Yang, K. A denoising representation framework for underwater acoustic signal recognition. *J. Acoust. Soc. Am.* **2020**, *147*, EL377–EL383. [CrossRef] [PubMed]
5. Hu, G.; Wang, K.; Liu, L. Underwater acoustic target recognition based on depthwise separable convolution neural networks. *Sensors* **2021**, *21*, 1429. [CrossRef] [PubMed]
6. Tian, S.; Chen, D.; Wang, H.; Liu, J. Deep convolution stack for waveform in underwater acoustic target recognition. *Sci. Rep.* **2021**, *11*, 9614. [CrossRef] [PubMed]
7. Wang, M.; Qiu, B.; Zhu, Z.; Xue, H.; Zhou, C. Study on Active Tracking of Underwater Acoustic Target Based on Deep Convolution Neural Network. *Appl. Sci.* **2021**, *11*, 7530. [CrossRef]
8. Gales, M.; Young, S. *The Application of Hidden Markov Models in Speech Recognition*; Now Publishers Inc.: Hanover, MA, USA, 2008.
9. Coleman, J.; Coleman, J.S. *Introducing Speech and Language Processing*; Cambridge University Press: Cambridge, UK, 2005.
10. Jurafsky, D.; Martin, J.H. *Speech and Language Processing*, 3rd (draft) ed.; Stanford University: Stanford, CA, USA, 2020.
11. Rabiner, L.R. A tutorial on hidden Markov models and selected applications in speech recognition. *Proc. IEEE* **1989**, *77*, 257–286. [CrossRef]
12. Tugac, S.; Efe, M. Radar target detection using hidden Markov models. *Prog. Electromagn. Res.* **2012**, *44*, 241–259. [CrossRef]
13. Putland, R.L.; Ranjard, L.; Constantine, R.; Radford, C.A. A hidden Markov model approach to indicate Bryde's whale acoustics. *Ecol. Indic.* **2018**, *84*, 479–487. [CrossRef]
14. Diamant, R.; Kipnis, D.; Zorzi, M. A clustering approach for the detection of acoustic/seismic signals of unknown structure. *IEEE Trans. Geosci. Remote. Sens.* **2017**, *56*, 1017–1029. [CrossRef]
15. Oudelha, M.; Ainon, R.N. HMM parameters estimation using hybrid Baum-Welch genetic algorithm. In Proceedings of the 2010 International Symposium on Information Technology, Kuala Lumpur, Malaysia, 15–17 June 2010; IEEE: Piscataway, NJ, USA, 2010; Volume 2, pp. 542–545.
16. Koller, D.; Friedman, N. *Probabilistic Graphical Models: Principles and Techniques*; MIT Press: Cambridge, MA, USA, 2009.
17. Li, X.; Parizeau, M.; Plamondon, R. Training hidden markov models with multiple observations-a combinatorial method. *IEEE Trans. Pattern Anal. Mach. Intell.* **2000**, *22*, 371–377.
18. Viterbi, A. Error bounds for convolutional codes and an asymptotically optimum decoding algorithm. *IEEE Trans. Inf. Theory* **1967**, *13*, 260–269. [CrossRef]
19. Press, W.H.; Teukolsky, S.A.; Vetterling, W.T.; Flannery, B.P. *Numerical Recipes 3rd Edition: The Art of Scientific Computing*; Cambridge University Press: Cambridge, UK, 2007.
20. Zhang, X.; Wang, Y.; Zhao, Z. A hybrid speech recognition training method for hmm based on genetic algorithm and baum welch algorithm. In Proceedings of the Second International Conference on Innovative Computing, Informatio and Control (ICICIC 2007), Kumamoto, Japan, 5–7 September 2007; IEEE: Piscataway, NJ, USA, 2007; pp. 572–572.
21. Liu, T.; Lemeire, J.; Yang, L. Proper initialization of Hidden Markov models for industrial applications. In Proceedings of the 2014 IEEE China Summit & International Conference on Signal and Information Processing (ChinaSIP), Xi'an, China, 9–13 July 2014; IEEE: Piscataway, NJ, USA, 2014; pp. 490–494.
22. Huda, M.; Ghosh, R.; Yearwood, J. A variable initialization approach to the EM algorithm for better estimation of the parameters of hidden markov model based acoustic modeling of speech signals. In Proceedings of the Industrial Conference on Data Mining, Leipzig, Germany, 14–15 July 2006; Springer: Berlin, Germany, 2006; pp. 416–430.
23. Kipnis, D.; Diamant, R. A factor-graph clustering approach for detection of underwater acoustic signals. *IEEE Geosci. Remote. Sens. Lett.* **2018**, *16*, 702–706. [CrossRef]
24. Mathew, T.V. Genetic algorithm. Report Submitted at IIT Bombay. 2012. Avilable online: http://datajobstest.com/data-science-repo/Genetic-Algorithm-Guide-[Tom-Mathew].pdf (accessed on 22 October 2021)

25. Russell, S.J. *Artificial Intelligence a Modern Approach*; Pearson Education, Inc.: Upper Saddle River, NJ, USA, 2010.
26. Mansouri, T.; Sadeghimoghadam, M.; Sahebi, I.G. A New Algorithm for Hidden Markov Models Learning Problem. *arXiv* **2021**, arxiv:2102.07112.
27. Lee, K.; Choo, Y.S.; Choi, G.; Choo, Y.; Byun, S.H.; Kim, K. Near-field target strength of finite cylindrical shell in water. *Appl. Acoust.* **2021**, *182*, 108233. [CrossRef]

Article

The Design and Development of a Ship Trajectory Data Management and Analysis System Based on AIS

Chengxu Feng [1], Bing Fu [1], Yasong Luo [1,*] and Houpu Li [2]

[1] College of Weapon Engineering, Naval University of Engineering, Wuhan 430033, China; kerryfengcx@126.com (C.F.); 13476039901@139.com (B.F.)
[2] College of Electrical Engineering, Naval University of Engineering, Wuhan 430033, China; Lihoupu1985@126.com
* Correspondence: yours_baggio@sina.com; Tel.: +86-138-7116-8627

Abstract: To address the data storage, management, analysis, and mining of ship targets, the object oriented method was employed to design the overall structure and functional modules of a ship trajectory data management and analysis system (STDMAS). This paper elaborates the detailed design and technical information of the system's logical structure, module composition, physical deployment and main functional modules such as database management, trajectory analysis, trajectory mining and situation analysis. A ship identification method based on the motion features was put forward. With the method, ship trajectory was first partitioned into sub-trajectories in various behavioral patterns, and effective motion features were then extracted. Machine learning algorithms were utilized for training and testing to identify many types of ships. STDMAS implements such functions as database management, trajectory analysis, historical situation review, and ship identification and outlier detection based on trajectory classification. STDMAS can satisfy the practical needs for the data management, analysis, and mining of maritime targets because it is easy to apply, maintain and expand.

Keywords: AIS; ship trajectory; data analysis; system design; trajectory classification

Citation: Feng, C.; Fu, B.; Luo, Y.; Li, H. The Design and Development of a Ship Trajectory Data Management and Analysis System Based on AIS. *Sensors* **2022**, *22*, 310. https://doi.org/10.3390/s22010310

Academic Editors: Haixin Sun and Xuebo Zhang

Received: 25 October 2021
Accepted: 30 November 2021
Published: 31 December 2021

Publisher's Note: MDPI stays neutral with regard to jurisdictional claims in published maps and institutional affiliations.

1. Introduction

China faces an increasing demand for marine resources and ocean space along with economic and social development, bringing a variety of practical challenges to its maritime regulation and security. In the maritime regulation field, it has become more and more difficult to supervise and regulate the ships engaging in illegal fishing and maritime smuggling because these ships evade detection [1]. In the military and national security field, China's national defense and military security is severely threatened by military vessels in disguise, including survey ships and electronic reconnaissance ships that some countries sent to deliberately perform illegal activities such as seabed and hydrological surveying and mapping and military reconnaissance in China's offshore waters [2]. In the nontraditional maritime security field, China faces rampant criminal activities on its shipping routes across the South China Sea and the Strait of Malacca, etc., and the passing ships are also exposed to the serious threat posed by pirates, armed hijacking at sea, and maritime terrorism. Therefore, implementing an accurate classification of unknown ships the effective identification of suspected ships, and the timely detection of ships doing abnormal activity is of great significance to achieving effective maritime regulation and defending maritime security [3].

Apart from further strengthening the maritime military forces and increasing patrols and law enforcement efforts, "soft methods" must also be employed to cope with these challenges, e.g., improving the automatic identification system (AIS) receiving stations and radar surveillance stations [3]. The data analysis in AIS can be conducted to dig out some valuable information from the spatial and temporal ship trajectory data that are constantly accumulated. Computers are able to automatically learn the regular movement and behavioral patterns of ships [2]. In this way, the analysis can assist maritime supervisors and military commanders in rapidly locating any suspected and abnormal target based on the massive quantities of information on ship trajectory. This analysis is then combined with other detection means to comprehensively judge the actual identity of suspected and abnormal maritime targets and their purposes.

In recent years, attention has been focused on the classification and identification of moving targets using data mining technology and AIS trajectory data. Reference [1] studied temporal and spatial data mining algorithms, and illustrated the design and implementation of a moving object activity pattern recognition system. The system utilized methods as data cleaning, data interpolation, and compression to preliminarily process the original trajectory data. Subsequently, a trajectory clustering algorithm was designed to realize the identification and classification of similar motion modes. Nevertheless, it overlooked the influence of time on the behavioral characteristics of ships, so that it was not applicable to the classification and identification of long-duration ships. Reference [2] put forward a ship behavioral pattern recognition method based on machine learning. In this method, AIS trajectory data were segmented to generate sub-trajectories for dimension reduction and visualization of data. Subsequently, sub-trajectories were clustered using the spectral clustering algorithm to recognize the behavioral pattern of ships. In reference [3], a new port ship classification method based on behavioral clustering was proposed to first conduct behavioral clustering of ships in a port and then extract the clustering characteristics. The ships were then classified into different behavioral clusters based on the ship features. The proposed method helped resolve the classification and identification of ships in the port, but was not applicable for ships sailing at sea. Reference [4] presented a ship classification model based on a graph neural network (GNN). The trajectory features extracted with the model contained such temporal and spatial features as location, distance, and speed. Additionally, a topologic correlation network was constructed to effectively extract the spatial features for ship classification and identification. Nevertheless, it did not consider the intrinsic differences between different types of moving objects.

This paper takes the AIS as a fundamental platform to design and develop a ship trajectory data management and analysis system (STDMAS) with various functions following the trajectory data management and analysis method. A ship identification method based on the motion features was put forward. With the method, ship trajectory was first partitioned into sub-trajectories in various behavioral patterns, and effective motion features were then extracted. Machine learning algorithms were utilized for training and testing to identify many types of ships. By addressing the practical needs of ship management and regulation, STDMAS can help realize the effective management of ship trajectory data, so as to effectively support and safeguard China's maritime management and maritime security.

2. Demand and Functional Analysis

The main tasks of STDMAS include managing the storage of ship trajectory data from various sources to provide protection service for the historical data on spatial and temporal ship trajectory; assessing and analyzing the quality of trajectory data from various sources to provide the data support for evaluating the equipment performance based on the collected ship target trajectory data; and analyzing and mining the information on the hot spots and situation regularity in the waters around China to provide the technical support for the analysis and decision making in maritime supervision and management [5,6]. It can also analyze the navigational characteristics, pattern, and regularity of ships to overcome the impossible identification of many targets at sea and provide the basis for maritime administration and military commanders in the selection and tracking of key targets among many unknown ships [7].

STDMAS has five modules: database management module, trajectory analysis module, trajectory mining module, situation analysis module, and configuration management module. Among them, the database management module can be further divided into three sub-modules including original maritime information data management, trajectory data management, and geographical information data management. The trajectory analysis module is further divided into two sub-modules: situation review analysis and trajectory quality assessment [8]. The trajectory mining module is further divided into three sub-modules, including target identification, ship navigational pattern detection, and ship outlier detection. The situation analysis module is further divided into two sub-modules, including ship activity statistical analysis and ship activity hot spots. The system configuration management module is further segmented into two sub-modules, i.e., system configuration and user authority management. The main functional modules of STDMAS are presented in Figure 1.

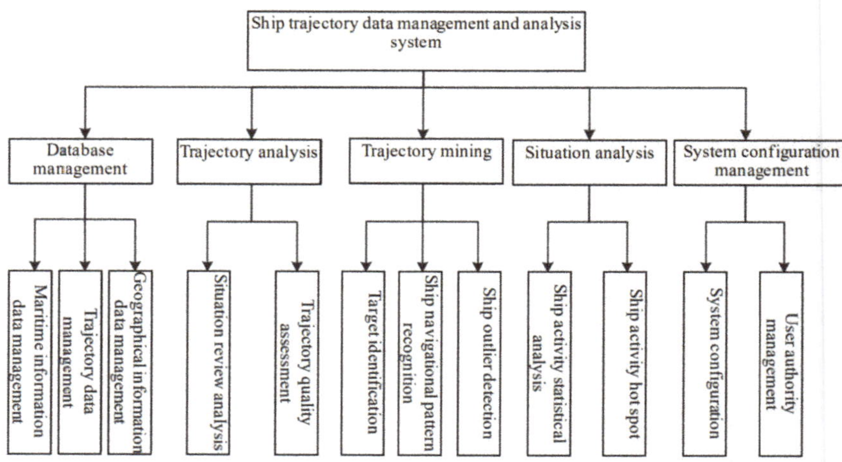

Figure 1. Main functional modules of STDMAS.

3. System Design

3.1. Overall Architecture

The logical structure of STDMAS is designed as shown in Figure 2. In general, the structure is divided into three levels, i.e., the data level, service level, and presentation level. The data level is mainly responsible for the storage, management, and maintenance of massive amounts of ship trajectory data, and it transmits the data to the service level. At the service level, the maritime data are sorted out and processed to remove invalid redundant data. Subsequently, a variety of algorithms are employed to obtain the regularity, pattern, and knowledge of ship navigation, which are directly provided to the presentation level or stored in the dataset. The presentation level allows for human-computer interaction so that a user request for computation service is sent to the service level to obtain the abstract data. After that, the abstract data are presented to the user in graphical form [9,10].

Data level: As the fundamental database of STDMAS, this level is mainly responsible for cleaning, filtering, and converting the massive multi-source heterogeneous data of maritime targets. Subsequently, it puts the maritime data into storage in a unified format and provides the corresponding interface for data management, maintenance, and query. For this level, attention is mainly paid to the marine geospatial representation model, trajectory data representation model, massive trajectory fast storage method, and fast temporal and spatial query method, etc.

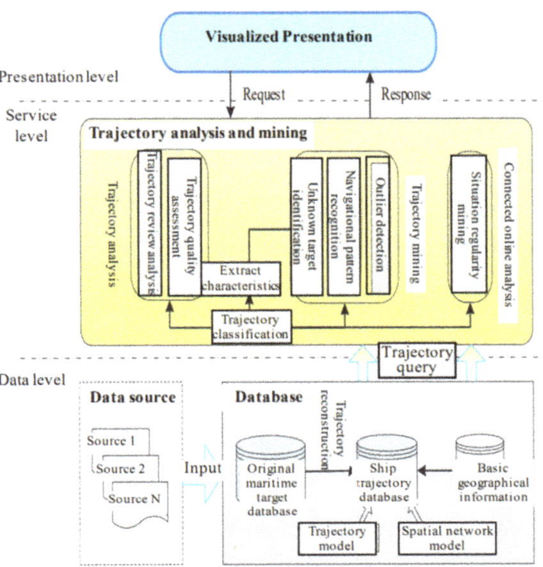

Figure 2. Logical structure diagram of STDMAS.

Service level: As the business center of STDMAS, this level is responsible for the central computation service of trajectory data. It involves three modules, i.e., trajectory analysis, trajectory mining, and online analysis processing. It implements such functions as ship historical trajectory review and replay, data quality assessment of various information sources, ship identification, ship navigational regularity mining, ship outlier detection, and situation regularity mining in the waters around China [11].

Presentation level: This level is responsible for presenting the human-computer interaction operations and the results of data analysis. After borrowing the data presentation methods from other similar platforms and soliciting opinions from maritime information supervision users, the GIS display technique and graphic user interface technique are comprehensively applied to present the results of trajectory data computation and analysis to users in the form of maps, diagrams, and data reports.

3.2. Technical Route

For the design and development of STDMAS, the technical route is determined according to the requirements for system construction and functions. It also ensures that the system is easy to expand and maintain.

(1) The object-oriented method is employed to design the system. The unified modeling language UML is taken as the analysis tool to enhance the reuse rate of code.
(2) The system adopts the C/S architecture based on the Visual Studio 2013.NET development environment. It is developed with the C# programming language that is absolutely object-oriented.
(3) The GIS spatial display module is developed using SharpMap. As a completely open source and lightweight GIS rendering tool, SharpMap offers a variety of underlying function interfaces.
(4) The database is the open source object-relational database PostgreSQL with the PostGIS spatial module plug-in. Additionally, Npgsql dynamic link library is employed to connect the system with the database [12,13].

3.3. Physical Deployment

The physical deployment of STDMAS, as shown in Figure 3, involves a data storage cluster, a computation service center, clients, and a network.

(1)   The data storage cluster is the foundation of the entire system. The data from different information sources have different security levels. Therefore, physical separation is ensured in the storage of the maritime data from different sources to guarantee the absolute security and control of the data. For the efficient, stable, and expandable management of data storage, IP-SAN two-layer data storage architecture is employed and the data storage disk array supports convenient future expansion. The storage capacity is 100 T.

(2)   The computation service center is mainly composed of a switch, a router, a computing server, and a management server. The management server is responsible for system configuration and user authority management. The computing server is connected by high speed Ethernet to the data storage cluster. It allows querying and sorting of the maritime data and other operations through the storage server in the data storage cluster. In this way, it can fulfill such functions as maritime and air situation data loading management, storage management, data query, system review analysis, trajectory mining, and situation analysis. Through the client, a user can send a data analysis request to the computation service center. The computing server can then extract the data from the data storage cluster for computation, and transmit the computation results to the client [14].

(3)   A client is a computer for any maritime data management user or common user in the local area network. Maritime data users can send a service request through a client to the computation service center, and obtain the corresponding trajectory data, graphic reports, or other views.

(4)   The network in the system is divided into three parts, i.e., the part in the data storage cluster, the part in the computation service center, and the part for interaction between the computation service center and the client user. After comprehensively considering the needs for functions and the construction cost, optical fiber connection was provided in the data storage cluster (between the storage server and the storage disk array), whereas the Gigabit Ethernet was used in the computation service center [15]. A 100 m local area network is employed to connect the computation service center with the clients.

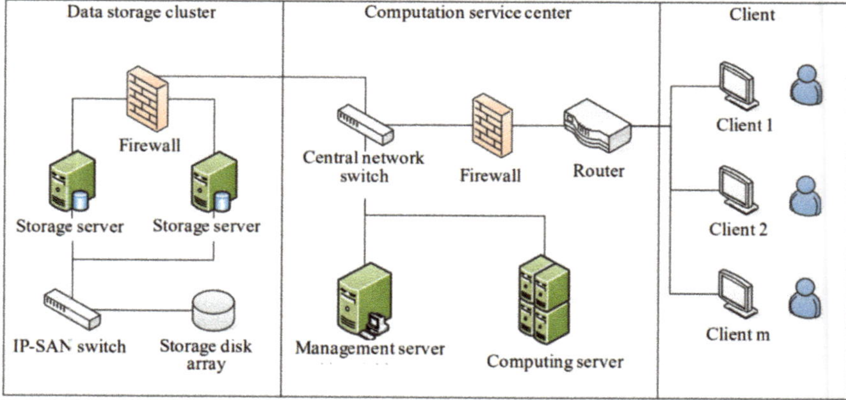

Figure 3. Physical deployment diagram of STDMAS.

4. Design of Functional Modules

4.1. Database Management Module

The database management module provides the basic data for the upper-layer services such as trajectory data analysis and mining [16]. The class diagram for the UML design of the module is presented in Figure 4. In the diagram, the GUI class of STDMAS represents the overall graphical user interface of the system. It can be used to call the graphical user interface of the database management module, i.e., the DatabaseManagerGUI class. The DatabaseManagerGUI class can call the functional sub-modules of the database management module, i.e., original maritime data management (OriginalDataManager class), trajectory data management (TrajDataManager class), and geographical information data management (GISDataManager class) [17].

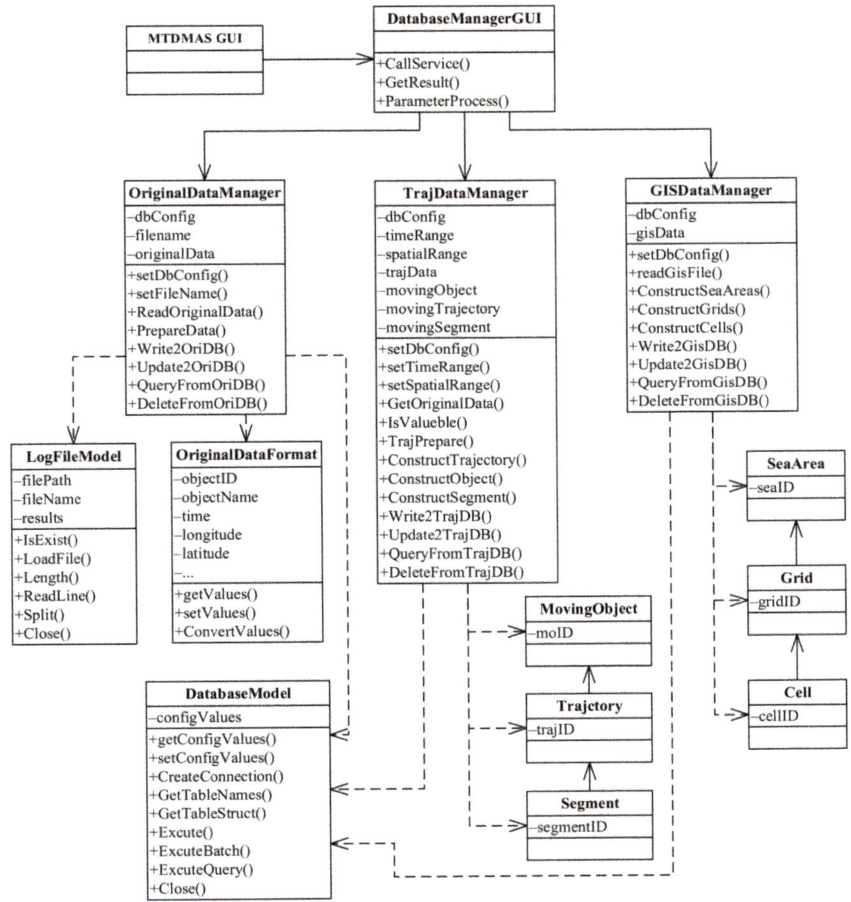

Figure 4. Class diagram of the database management module.

The OriginalDataManger class implements the storage, query, deletion, and other management services of original maritime information data. The class mainly contains the object dbConfig indicating the database connection configuration, the object fileName indicating the file name of original maritime information data files, and the object originalData indicating the storage of original data. Among the methods used in the class, ReadOriginalData() is responsible for reading the original maritime data of different formats from various sources and depends on the LogFileModel class representing the reading of files; PrepareData() is used for the cleaning, format conversion, and other operations of the original maritime data after being read and relies on the OriginalDataFormat class for the unified format; Write2OriDB() implements the storage of the original maritime data and depends on the DatabaseModel class for realizing the basic database services. Moreover, the OriginalDataManger class can also, upon the user's request, call the methods Update2OriDB(), QueryFromOriDB(), and DeleteFromOriDB() for the update, query, and deletion in the original maritime database.

The TrajDataManager class implements the storage, query, deletion, and other management services of trajectory data. The class mainly consists of the object dbConfig indicating the database connection configuration, the objects timeRange and spatialRange indicating the time and spatial range, respectively, selected by a user, the object trajData saving the constructed trajectory data, and other objects such as movingObject, movingTrajectory, and movingSegment. Among them, the objects movingObject, movingTrajectory, and movingSegment have their type of data depending on the MovingObject, Trajectory and Segment classes for the trajectory data model. Among the methods for the TrajDataManager class, GetOriginalData() is used to extract all the original data in the time and spatial ranges given by the user from the original maritime database; TrajPrepare() allows reconstructing the trajectory with all the trajectory data of a target, and IsValueble() is then employed to judge the validity of the trajectory, remove the invalid points, and smooth the trajectory; ConstructObject(), ConstructTrajectory(), and ConstructSegment() are used to construct the trajectory data model designed in this paper; Write2TrajDB(), Update2TrajDB(), QueryFromTrajDB(), and DeleteFromTrajDB() can store, update, query, and delete the data in the trajectory database upon a user's request. Like those in the OriginalDataManger class, these methods also rely on the DatabaseModel class [18].

The GISDataManager class implements the storage, query, deletion, and other management services of geographical information data. The class is mainly composed of the object dbConfig indicating the database connection configuration, and the object gisData indicating the storage of geographical information data. In the class, readGisFile() is used to read the geographical information data, and the methods ConstructSeaAreas(), ConstructGrids(), and ConstructCells() are responsible for converting the geographical information data into the spatial grid model. The three methods rely on the SeaArea class, the Grid class, and the Cell class, respectively. The methods Write2GisDB(), Update2GisDB(), QueryFromGisDB(), and DeleteFromGisDB() can implement the storage, update, query, and deletion in the geographical information database upon a user's request. Similarly, these methods are also dependent on the DatabaseModel model [19].

4.2. Trajectory Analysis Module

The class diagram of the trajectory analysis module is shown in Figure 5. In the diagram, the TrajAnalysisGUI class represents the graphical user interface of the trajectory analysis module. It can also be called by virtue of the STDMAS GUI class. The graphical user interface (TrajAnalysisGUI) of the trajectory analysis module can be also used to call two functional sub-modules, i.e., situation review analysis (the SituationAnalysis class) and trajectory quality assessment (the TrajQualityAssessment class).

The SituationAnalysis class mainly implements such functions as querying all the maritime data in the time and spatial ranges given by a user, performing the review, replay, and analysis of maritime situation, creating the sequence of events for the sea and air situation in the corresponding process, and reproducing the maritime situation evolution process of missions and events in the corresponding time and spatial ranges. The SituationAnalysis class contains the object dbConfig indicating the database configuration, the objects spatialRange and timeRange indicating the time and spatial ranges selected by a user, the object simStep determining the speed of replay, and the object trajData for the storage and query of trajectory data. Among the main methods for the class, SelectTrajs() is used to check the trajectory in the trajectory database based on the time and spatial ranges given by a user, and depends on the DatabaseModel class; RunSituation() calculates the location of each target at each time of simulation by calling the simulation module (the SimulationModel class) and replays the maritime situation using the simulation step simStep through the geographical information display module (the GISDisplayModel class); TargetStats() is employed for the statistics of the information on various targets, e.g., quantity, and displays the information in the form of a report [20–22].

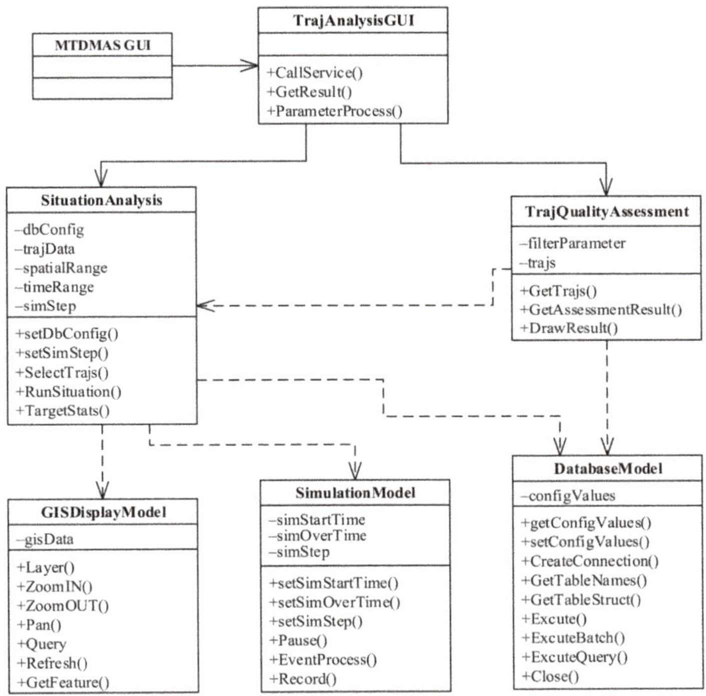

Figure 5. Class diagram of the trajectory analysis module.

The TrajQualityAssessment class is designed to analyze and assess the quality of the selected trajectories. In the class, the object filterParameter indicates the conditions for selecting trajectories, and the object trajs stores all the selected trajectories. The method GetTrajs() can extract the trajectory data from the database or from the trajectories in the replay and analysis module. It depends on the DatabaseModel class and the Situation-Analysis class. The method GetAssessment() is used to analyze and assess the quality of the selected trajectories. The method DrawResult() displays the results of analysis and assessment to the user in graphical form. The indicators of the trajectory quality assessment include target identification rate, average identification response time, trajectory outlier rate, continuous tracking rate, and fault tracking rate [23].

4.3. Trajectory Mining Module

The UML class diagram of the trajectory mining module is shown in Figure 6. In the diagram, the TrajMiningGUI class represents the graphical user interface of the trajectory mining module, and can be called by the STDMAS GUI class. The TargetIdentify class, the OutlierDetection class, and the PatternRecognition class are used to implement three functions of the trajectory mining, respectively, that is, identifying the unknown ships, detecting the ship outliers, and recognizing the navigational pattern of ships. The OutlierDetection class mainly involves the methods ConstructOutlierModel() and DetectOutlier(), which are used to construct and detect the outlier model, respectively. The ship outlier detection model in the STDMAS depends on the identification of unknown ships (the TargetIdentify class) and the recognition of ship navigational patterns (the PatternRecognition class) [24].

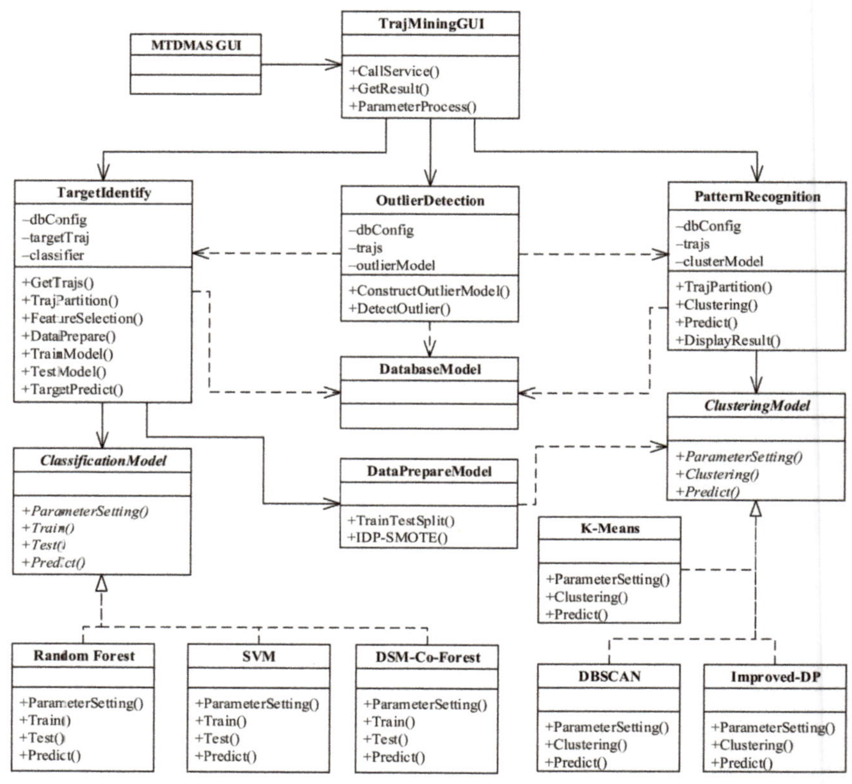

Figure 6. Class diagram of the trajectory mining module.

The TargetIdentify class can construct the ship trajectory classification model based on the parameters set by a user and then use the model to determine the category of any unknown ship. The class consists of the database configuration object dbConfig, the trajectory data set targetTraj, and the classifier, etc. In the TargetIdentify class, the method GetTrajs() is employed to extract the trajectory data needed from the trajectory database according to the user's request, and depends on the DatabaseModel class [25]. The method TrajPartition() can partition each ship trajectory using the proposed trajectory partitioning method based on the movement mode in this paper. Subsequently, the method FeatureSelection() is utilized to extract the characteristics of each trajectory in the way proposed in this paper. The method DataPrepare() aims to classify the characteristic data sets into the training set and test set and utilizes the algorithms including IDP-SMOTE to balance these data sets. The method depends on the DataPrepareModel class and indirectly on the clustering model, i.e., the ClusteringModel class [26]. TrainModel(), TestModel(), and TargetPredict() represent the training, test, and prediction of the classification model, respectively, and depend on the ClassificationModel class for the abstract classification model. It implements the RandomForest class for the random forest model, the SVM class for the support vector model, and the DSM-Co-Forest class for the semi-supervised learning model. Meanwhile, it can be further expanded based on the specific classification model.

The PatternRecognition class is used to construct the clustering model based on the trajectory selected by a user and display the trajectory clustering results in graphical form. The class consists of the objects dbConfig, trajs, and clusterModel [27]. The methods mainly include TrajPartition() for trajectory partitioning, Clustering() for trajectory clustering modeling, Predict() for cluster prediction, and DisplayResult() for displaying the trajectory clustering result. In the class, the construction of the cluster model depends on the abstract class, i.e., the ClusteringModel class, which implements the cluster models including the K-Means model, the DBSCAN, and the Improved-DP model.

4.4. Situation Analysis Module

The UML class diagram of the situation analysis module is shown in Figure 7. In the diagram, the TDWManagerGUI class represents the graphical user interface of the situation analysis module. It is also called through the STDMAS GUI class.

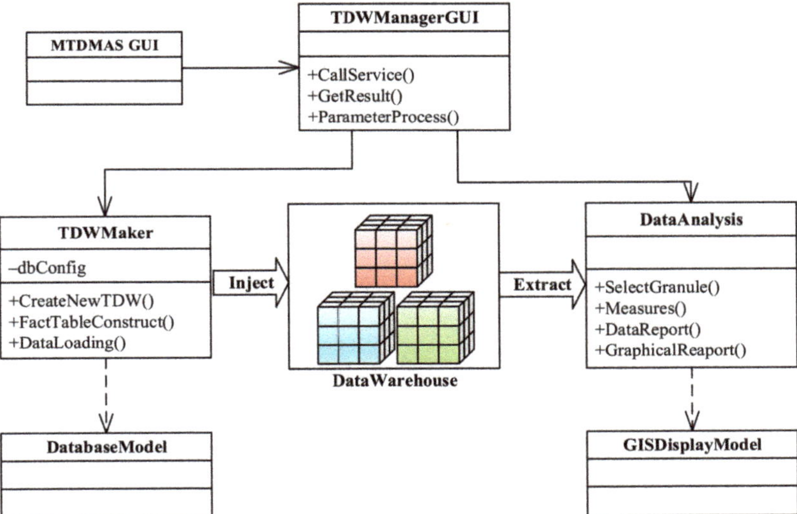

Figure 7. Class diagram of the situation analysis module.

The TDWMaker class is mainly used to construct the trajectory data warehouse. Among its methods, CreateNewTDW() can create the new data mart for a new subject; FactTableConstruct() can define the fact table and dimension table based on the configuration given by a user; DataLoading() can extract the processed data from the trajectory database in the extraction-transformation-loading (ETL) method and then input the data into the data warehouse, but it depends on the DatabaseModel class for implementation [28].

The DataAnalysis class is mainly intended to extract information from the trajectory data warehouse upon a user's request, and it generates the reports. Among its methods, SelectGranule() is used to select the granule of time, space, and other dimensions for data analysis; Measures() can further calculate and obtain the measured value of a subject (e.g., speed, distance, pass time, etc.) at each granule; DataReport() and GraphicalReport() can present the results of data analysis in the form of a data report or graphical report. However, the DataAnalysis class must rely on the GISDisplayModel class in the display of data analysis results, so as to more vividly present the changes of historical maritime situations.

5. Methods for Ship Trajectory Data Mining

Presently, trajectory mining technology offers four features in terms of mission and based on temporal and spatial characteristics, that is, trajectory pattern mining, trajectory clustering analysis, trajectory classification, and trajectory outlier detection [2]. Among them, the trajectory pattern mining feature discovers the valuable motion feature patterns for a single moving object or a group of moving objects, e.g., frequent pattern, periodic pattern, and adjoint pattern, so as to help people understand the motion regularity of the moving object and reasonably predict its future trend of motion. Trajectory clustering intends to devise a measure of similarity between trajectories and cluster the highly similar trajectories, in order to find out the representative path or common behavioral tendency of moving objects. Trajectory classification aims to predict the motion state of moving objects or their means of transport by extracting the features of trajectory or trajectory sections and then classifying them into different categories. Trajectory outlier detection can identify any suspected moving object or behavior based on the motion regularity or behavioral pattern of historical trajectories.

5.1. Ship Trajectory Partitioning

In the studies of trajectory motion mode, many scholars divide trajectory only into Stop Mode and Move Mode. However, ships at sea may take any turn for a specified reason or purpose unlike the moving objects on the ground, e.g., vehicles and pedestrians, which are restricted by road grids. In other words, turn is also an important motion mode of ship trajectory. On this basis, this paper partitions ship trajectory into three basic motion modes, that is, Stop Mode, Turn Mode, and Line Mode [7,8].

(1) Stop Mode: A ship trajectory contains a segment of continuous sub-trajectories in which the speed at all points is lower than the speed threshold. The distance between any two trajectory points must be lower than the distance threshold, and the duration is greater than the time threshold. At this time, the motion mode of the sub-trajectory is Stop Mode.

(2) Turn Mode: The turn threshold is set to Δ_θ. If the sum of trajectory direction changes within a range of time Δ_{t1} before and after a trajectory point p is greater than Δ_θ, the trajectory point p is a potential turn point. If the potential turn point is a segment of a continuous sub-trajectory, and the total time for such sub-trajectory exceeds the threshold Δ_{t2}, the motion mode of the sub-trajectory is Stop Mode.

(3) Line Mode: In a ship trajectory, the remaining sub-trajectory is in the Line Mode after taking out those in the Stop Mode and the Turn Mode. The three motion modes of a ship trajectory are illustrated in Figure 8.

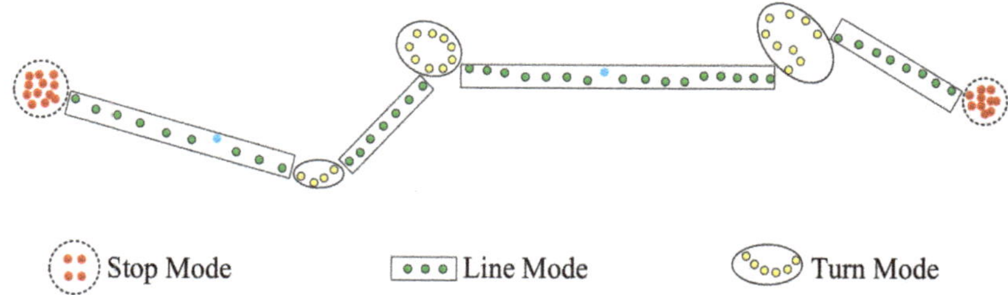

Figure 8. Three basic motion modes of a ship trajectory.

For further data mining and knowledge discovery, a ship trajectory can be partitioned in terms of these three basic motion modes in the following procedure:

Step 1: Partition a ship trajectory into several continuous sub-trajectory segments;

Step 2: Go through all the sub-trajectory segments. When the speed at most of the points in a sub-trajectory segment is less than the speed threshold δ_v, it is judged that the sub-trajectory is in the Stop Mode. Otherwise, it is in the Move Mode;

Step 3: Go through all the sub-trajectory segments in the Move Mode. When the sum of direction changes at the points of a sub-trajectory within the time threshold exceeds the turn threshold, it is judged that the sub-trajectory is in the Turn Mode. Otherwise, it is in the Line Mode;

Step 4: Use the outlier detection algorithm to rule out the outliers in the trajectory;

Step 5: Add the segmented ship sub-trajectories into the corresponding sets of Stop Mode, Turn Mode, and Line Mode.

5.2. Ship Trajectory Feature Extraction

For higher efficiency of the trajectory data mining algorithm, the trajectory features must be selected in terms of their contribution to ship identification. Based on the above ship trajectory partitioning algorithm, trajectory features are classified into four categories, that is, global features, stop features, line features, and turn features.

(1) Global features refer to the extracted general features of the entire trajectory and its sub-trajectories and reflect the features of the trajectory holistically. There were nine global features extracted in this paper, including total sailing time, total sailing distance, total sailing sinuosity, number of sub-trajectories in the Stop Mode, Line Mode, and Turn Mode, and their respective proportion of total trajectory time. Among these global features, trajectory sinuosity represents the ratio of the sailing distance between two trajectory points and the straight distance between them and is employed to indicate the curvature of the path. Total trajectory sinuosity is the ratio of the total sailing distance of a ship to the straight distance between its departure and destination. The total trajectory sinuosity is calculated by the following formula:

$$\text{sinuosity} = \frac{\sum_{i=1}^{n-1} \text{distance}(p(i), p(i+1))}{\text{distance}(p1, pn)} \tag{1}$$

(2) Stop features are the trajectory features extracted from the sub-trajectories in the Stop Mode. The stop features extracted in this paper include two parameters, that is, stop duration and stop range. Stop duration refers to the period of time from the start to the end of a sub-trajectory in the Stop Mode. Stop range is represented by the area of circle of uncertainty for all trajectory points available in a sub-trajectory in the Stop Mode. It is assumed that the stop sub-trajectory is extracted as StopT = {$p1, p2, \ldots \ldots, pn$}, the central point of stop $C(\overline{x}, \overline{y})$ is as follows:

Let the matrix stop error radius R_e be the maximum distance between all trajectory points in the trajectory segment and the stop central point, there is:

$$R_e = \max\{\text{distance}(p_i, \text{C})\}^N_{i=0} \tag{2}$$

Then the stop range *Sa* is:

$$S_a = \pi \cdot R_e^2 \tag{3}$$

(3) Line features are the trajectory features extracted from the sub-trajectories in the Line Mode. The line features extracted in this paper involve three parameters, that is, speed, acceleration, and heading. The "global features" for these parameters are calculated with seven statistical quantities including mean, median, standard deviation, average of three largest numbers, coefficient of variation, skewness, and kurtosis. Among them, skewness is a coefficient for measuring the deviation of data distribution from the symmetrical center and is normally represented by the ratio of three-order center distance to the third power of standard deviation. Kurtosis is a coefficient reflecting the aggregation level of data at the center and normally denoted by the ratio of four-order center distance to the fourth power of standard deviation. Skewness S_k and kurtosis K_u are calculated as follows:

$$S_k = \frac{\sum (x_i - \mu)^3}{N \cdot \sigma^3} \tag{4}$$

$$K_u = \frac{\sum (x_i - \mu)^4}{N \cdot \sigma^4} \tag{5}$$

(4) Turn features are the trajectory features extracted from the sub-trajectories in the Turn Mode. In this paper, the extracted turn features involve three parameters in total, that is, angular speed, turn speed, and turn angle. Angular speed is the ratio of the direction difference between two trajectory points and the time. Turn speed is the ratio of the distance between two trajectory points and the time. Turn angle is the difference between the destination direction and the departure direction of all sub-trajectories in the Turn Mode.

5.3. Ship Trajectory Classification Based on Motion Features

Based on the ship trajectory partitioning and trajectory feature extraction, the classification algorithm in the field of machine learning is employed to obtain a ship trajectory classification model through training. The feature extraction method proposed in this paper is taken to construct the overall framework of a ship trajectory classification model as shown in Figure 9.

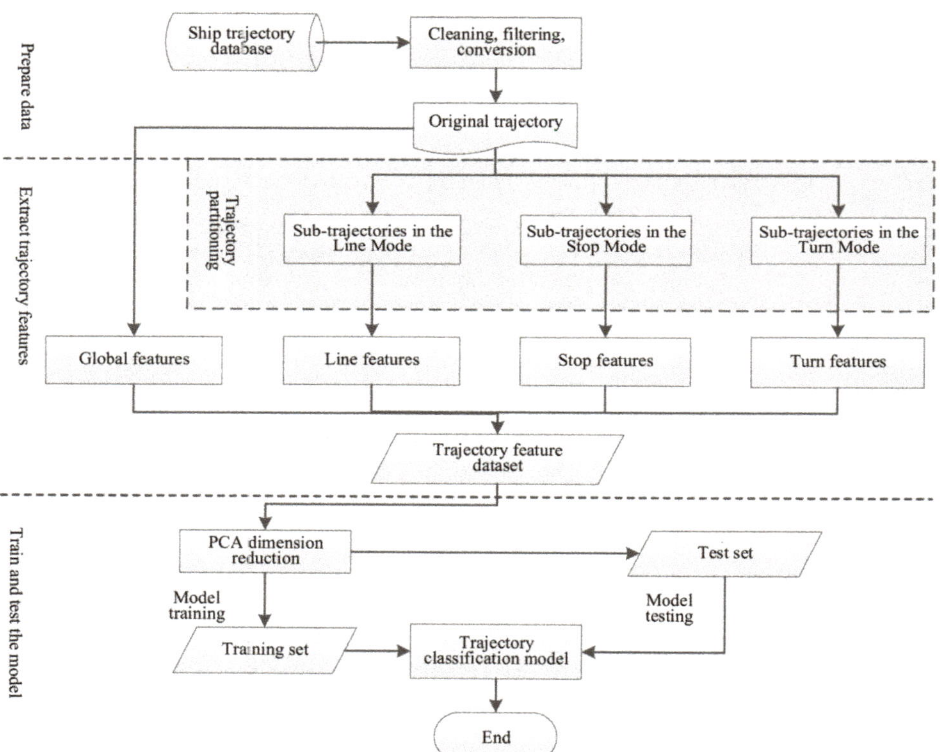

Figure 9. Overall framework of ship trajectory classification model.

In this paper, the real AIS data of ships were taken from the ship historical temporal and spatial trajectory library to extract 1000 effective trajectories of fishing boats and cargo ships (500 each) through data cleaning, filtering, conversion, and preliminary processing, for training and testing of the ship classification model [13]. The spatial distribution of the selected ship historical trajectories (50 items) is shown in Figure 10.

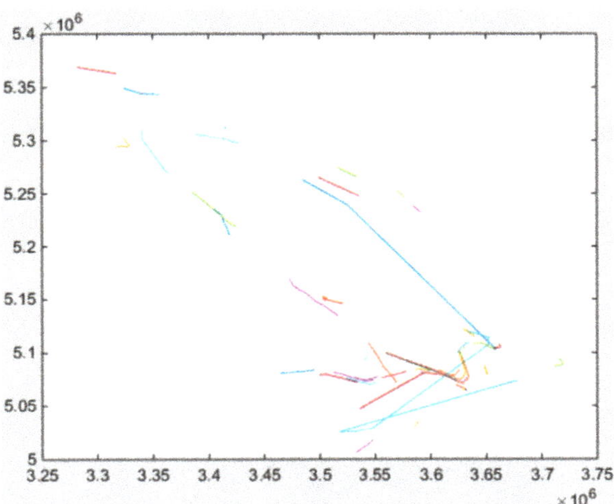

Figure 10. Spatial distribution of ship historical trajectories (50 items).

Based on the proposed ship trajectory feature extraction method, 158 trajectory features were extracted from the historical trajectories of fishing boats and cargo ships, including global features, stop features, line features, and turn features of each trajectory. Subsequently, the data sets of trajectory features for fishing boats and cargo ships were created together with the statistics and analysis of features.

As shown in Figure 11, the cargo ship has a longer and more centrally distributed sailing time and a significantly higher proportion of Line Mode than the fishing boat. However, the fishing boat has a larger turn count and line count (i.e., the number of sub-trajectories in the Turn Mode and Line Mode) and a higher proportion of Turn Mode than the cargo ship. The motion pattern of trajectory is not considered for the global features of trajectory in the calculation of total sailing distance and total sailing sinuosity. This may be the reason for some remaining outliers or outlier sub-trajectories. After normalization, these two features are dramatically reduced and not effectively reflected in the box plot.

5.4. Ship Classification Model Training and Testing

To verify the effectiveness of the proposed ship trajectory feature extraction method, some popular single classifier learning algorithms in the machine learning field were selected in this paper for training and testing the ship classification model, including Decision Tree (DT), Naïve Bayers (NB), Logistic Regression (LR), Artificial Neural Networks (ANN), and Support Vector Machine (SVM) [21]. The test was implemented with the Python programming language and the scientific computing environment Anaconda. The classification algorithm used the standard model given in the machine learning kit scikit-learn. The default parameters were employed. The classification model was constructed and tested in the following procedure:

(1) Feature selection and dimension reduction:

The principal component analysis (PCA) method [22] was adopted to select the features from the extracted trajectory features for dimension reduction. The threshold for retaining principal components was 95%. After dimension reduction, the number of extracted features is as shown in Table 1.

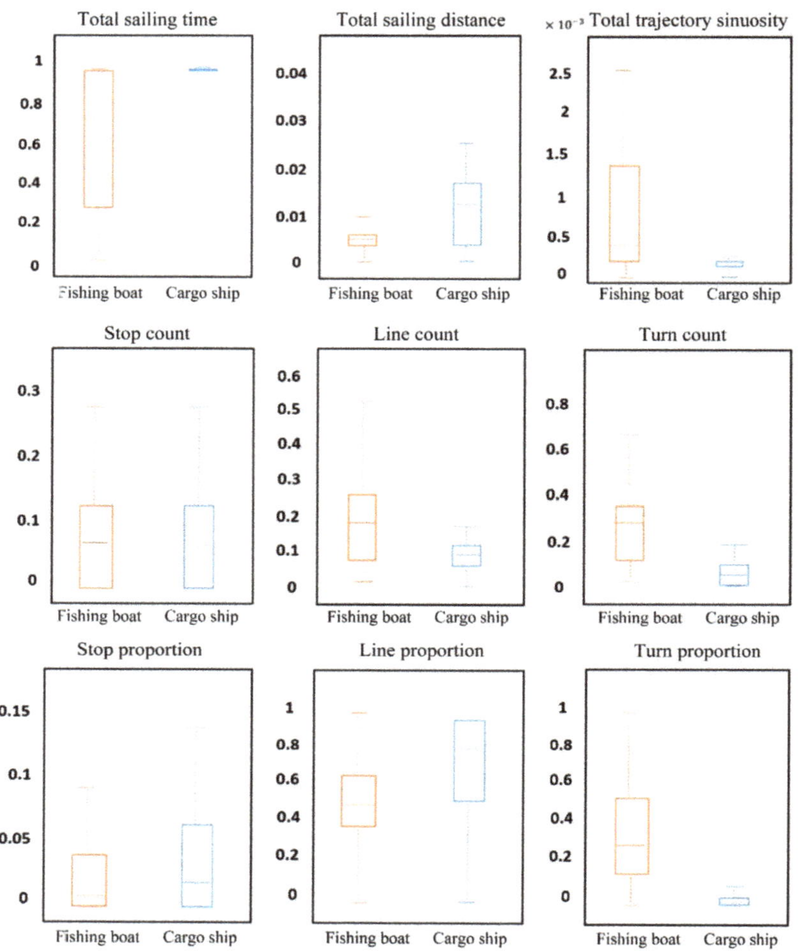

Figure 11. Comparison of global features of fishing boat (blue) and cargo ship (red) trajectories.

Table 1. Selected features using PCA.

Number of original features	158
Number of features after dimension reduction	24

(2) Classification of training set and test set:

The feature data set was partitioned with hold-out. Before each test, the feature data set was randomly divided into two mutually exclusive sets, that is, 75% training set and 25% test set.

(3) Model training and evaluation:

Two indicators were selected to evaluate the results of model prediction, that is, accuracy and area under curve (AUC) [22]. For this binary problem, the model predicted results and actual results of the test set form a confusion matrix presented in Table 2.

Table 2. Confusion matrix.

Actual Class	Predicted Result	
	Positive Class	Negative Class
Positive class	True positive (*TP*)	False Negative (*FN*)
Negative class	False Positive (*FP*)	True Negative (*TN*)

Accuracy indicates the proportion of all predicted results as follows:

$$Accuracy = \frac{TP + TN}{TP + FN + FP + FN} \tag{6}$$

The receiver operating characteristic curve (ROC) can be used to reflect the performance of model classification. The true positive rate (*TPR*) and false positive rate (*FPR*) are as follows:

$$TPR = \frac{TP}{TP + FN} \tag{7}$$

$$FPR = \frac{FP}{TN + FP} \tag{8}$$

Based on the predicted results of classifiers, samples were sequenced and correspondingly predicted as the positive class. Their *FPR* and *TPR* were calculated, respectively, and then used to obtain the ROC curve with *FPR* as the horizontal axis and *TPR* as the longitudinal axis, as shown in Figure 12. Normally, the closer the ROC curve gets to the upper left corner, the better performance of a classifier. AUC is the area encircled under the ROC curve. It is a quantitative description in place of the ROC curve. AUC ranges between 0 and 1. The larger the AUC, the better performance of a classifier.

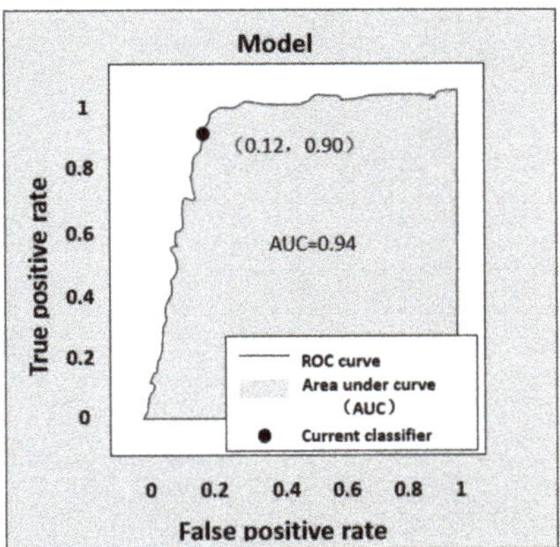

Figure 12. Diagram of ROC curve and AUC.

The average accuracy and AUC values of the classification model are indicated in Table 3. The specific distribution is given in Figure 13. Figure 13a shows the accuracy distribution of the classification model from 100 independent tests. Figure 13b provides the AUC distribution of the model.

Table 3. Comparison of classification performance of classifiers.

Classification Model	Evaluation Indicator	Result
Decision tree (DT)	Accuracy	0.8560
	AUC	0.8558
Naïve Bayers (NB)	Accuracy	0.8443
	AUC	0.8435
Logistic Regression (LR)	Accuracy	0.8837
	AUC	0.8833
Artificial Neural Networks (ANN)	Accuracy	0.8260
	AUC	0.8159
Support Vector Machine (SVM)	Accuracy	0.8456
	AUC	0.8451
K-nearest Neighbor (KNN)	Accuracy	0.8379
	AUC	0.8377

The analysis reveals that the prediction performance of the trained model was dramatically improved when the trajectory was reasonably partitioned and then trajectory features were extracted from sub-trajectories. In the meantime, features were extracted with the proposed ship trajectory partitioning method in terms of motion mode to further enhance the prediction performance of classifiers.

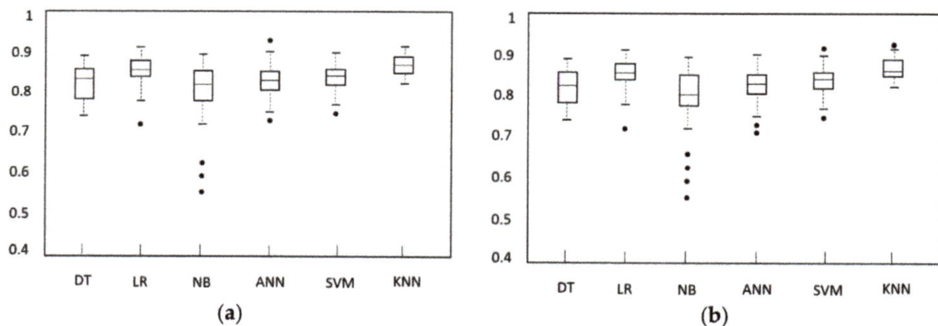

Figure 13. Comparison of the distribution of evaluation indicators of each classifier. (**a**) Comparison of classification accuracy of each classifier; (**b**) Comparison of AUC scores of each classifier.

6. System Implementation

Presently, the ship trajectory management and analysis system (STDMAS) has implemented such functions as identification of unknown ships and ship outlier detection in the database management module, trajectory analysis model, and trajectory mining module [29]. In this section, the functional modules of STDMAS are illustrated with some images including the main interfaces of system management configuration, database management, trajectory analysis, trajectory mining, and situation analysis.

6.1. Database Management Module

With the database management of the system, a user can read the original files of original maritime data, trajectory data, and GIS data, and can also convert, write, check, update, and delete the data from the database, as shown in Figure 14. When reconstructing the trajectory data, the parameters including time threshold and distance threshold can be set to filter the outliers in the original data. Additionally, interpolation or filter algorithms may be chosen to process the data as needed [30].

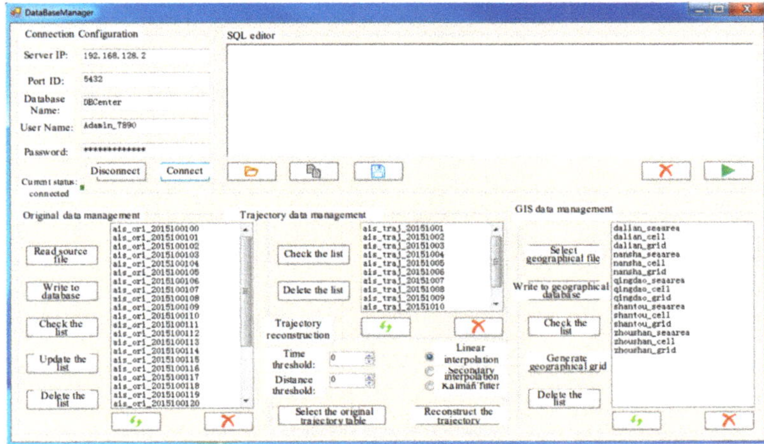

Figure 14. STDMAS database management interface.

6.2. Trajectory Analysis Module

This module can implement the fast query of radar detection or AIS target original data and trajectory target within the spatial and time ranges and analyze the variation tendency of the spatial curve, sampling time interval, and distance interval of trajectories [6], as shown in Figure 15. It can also calculate the outlier rate, error rate, and inferior rate of the radar detection data, so as to indirectly reflect the target detection performance of radar. Moreover, the trajectory analysis module of STDMAS is integrated with the review and replay feature of historical trajectories to review and analyze the historical situation of military exercises, major missions, or other special maritime activities. The functions of the module are as shown in Figure 16. The interface integrates the common measuring tools for distance, area, and angle, etc., in the GIS software to facilitate the user's analysis and calculation. A user may select and control the retention length of target wake and the replay speed of historical situation in this interface, so as to implement the detailed review and analysis of historical maritime situations [31].

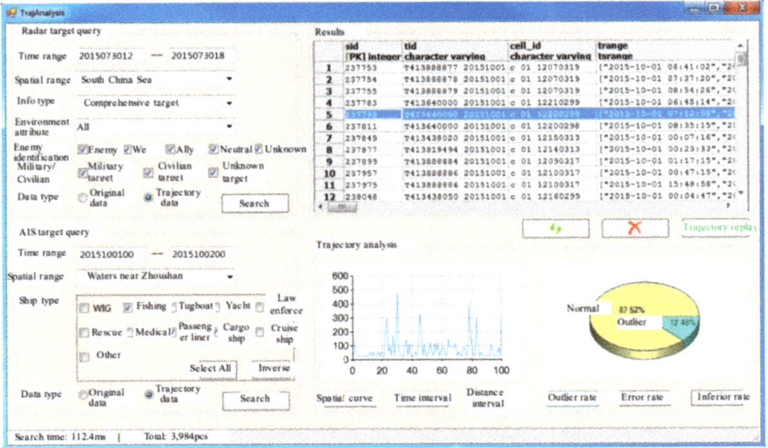

Figure 15. STDMAS trajectory analysis interface.

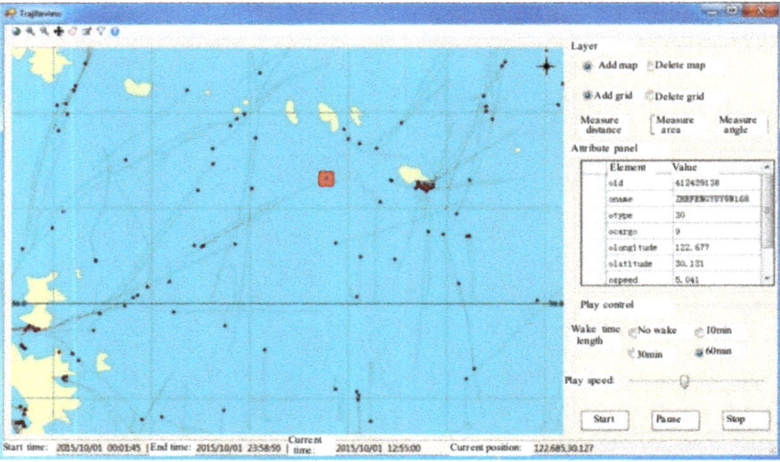

Figure 16. STDMAS historical trajectory review interface.

6.3. Trajectory Mining Module

Based on the abovementioned trajectory data mining method, the trajectory mining module of STDMAS can implement the identification of unknown targets and the outlier detection of targets. However, both functions depend on the training and implementation of the trajectory classification model [32].

Figure 17 presents the unknown ship identification function of the STDMAS trajectory mining module. The "trajectory data loading" panel in the upper left of the interface is used to check the ship trajectory within the given time or space ranges. All the searched trajectories are displayed in the GIS display module in the middle of the interface. The "ship type identification" panel in the lower left of the interface can load the trained trajectory classification model and identify the selected unknown ship trajectory. In the figure, an unknown ship trajectory with the target No. 900,411,284 is selected, and its ship identification and prediction results include fishing: 90.3%, tugboat: 1.7%, passenger liner: 6.8%, and cargo ship: 1.2%. Hence, the target is identified as a fishing boat in the system.

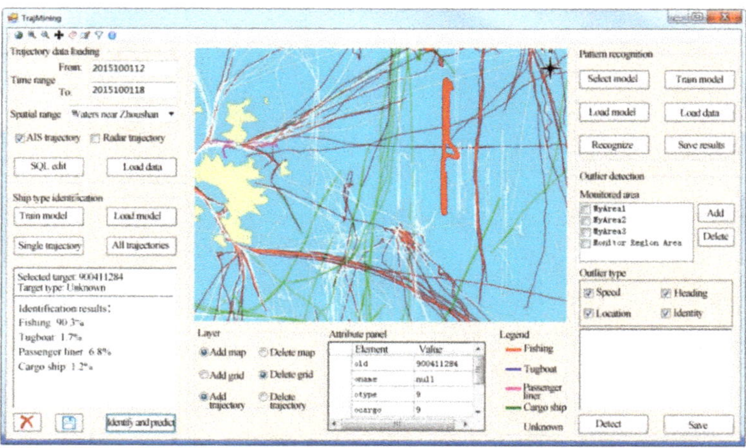

Figure 17. STDMAS unknown ship identification interface.

Figure 18 presents the outlier detection function of the STDMAS trajectory mining module. The buttons "Add" and "Delete" in the "outlier detection" panel in the right section of the interface are used to implement the setting and management of the monitored area. The tabs "Monitored Area" and "Outlier Type" are used to select the area to be monitored (for outlier detection) and the type of outlier. The text box in the lower right of the interface displays the system's target identity outlier detection results in the "Monitor Region Area".

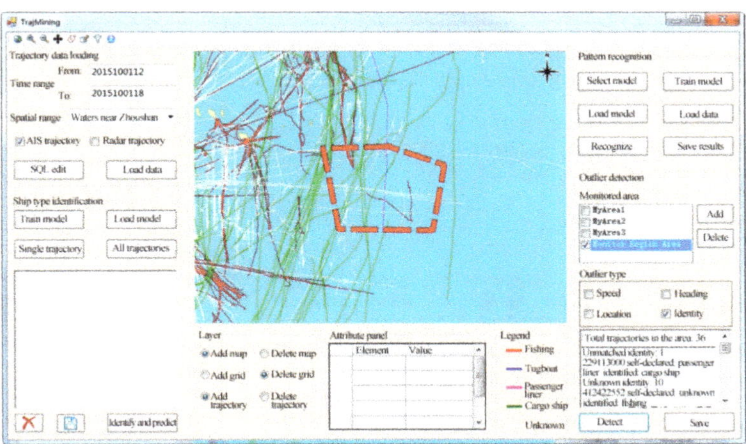

Figure 18. STDMAS ship identity outlier detection interface.

7. Conclusions

To address the data storage, management, analysis and mining, of maritime targets, an object-oriented method was adopted to design the overall structure and functional modules of the ship trajectory management and analysis system (STDMAS). This paper elaborates the design and technical details of the STDMAS functional modules including logical structure, module composition, physical deployment, database management, trajectory analysis, trajectory mining, and situation analysis. A ship identification method based on motion features was put forward. With the proposed method, ship trajectory was first partitioned into sub-trajectories in various behavioral patterns, and effective motion features were then extracted. Machine learning algorithms were utilized for training and testing to identify many types of ships. The functional modules implemented for the system include database management, trajectory analysis, historical situation review, ship identification, and outlier detection based on trajectory classification. STDMAS can satisfy the practical needs for the data management, analysis, and mining of maritime targets because it is easy to apply, maintain, and expand. Efforts will be made to integrate such functions as trajectory cluster analysis and mining and trajectory situation analysis in STDMAS. Moreover, STDMAS will be connected to the real-time maritime target data receiving system, and then tested to further provide the data support and computation service for research and application personnel in the management, analysis, and mining of maritime target data. Additionally, the system can process a limited size of AIS data at present, so that big data algorithm and cloud computing architecture may be employed to improve the efficiency of mass data processing algorithms in future research.

Author Contributions: The authors' individual contributions are summarized below: Conceptualization, C.F. and Y.L.; data curation, C.F.; formal analysis, B.F.; funding acquisition, Y.L. and H.L.; methodology, B.F.; software, C.F.; visualization, B.F.; writing—original draft, C.F.; writing—review and editing, H.L. All authors have read and agreed to the published version of the manuscript.

Funding: This research was funded by National Science Foundation for Outstanding Young Scholars (grant number 42122025) and Natural Science Foundation for Distinguished Young Scholars of Hubei Province of China (grant number 2019CFA086).

Institutional Review Board Statement: Not applicable.

Informed Consent Statement: Not applicable.

Data Availability Statement: Not applicable.

Conflicts of Interest: The authors declare no conflict of interest.

References

1. Lei, P.R. A Framework for Anomaly Detection in Maritime Trajectory Behavior. *Knowl. Inf. Syst.* **2016**, *47*, 189–214. [CrossRef]
2. Dina, F.; Sherin, M.; Nagwa, B. The Spatiotemporal Data Fusion (STDF) Approach: IoT-Based Data Fusion Using Data Analysis. *Sensors* **2021**, *21*, 7035.
3. Chen, X.; Liu, Y.; Achuthan, K.; Zhang, X. A ship movement classification based on Automatic Identification System (AIS) data using Convolution Neural Network. *Ocean Eng.* **2020**, *218*, 108182. [CrossRef]
4. Sheng, K.; Liu, Z.; Zhou, D. Research on Ship Classification Based on Trajectory Features. *J. Navig.* **2018**, *71*, 100–116. [CrossRef]
5. Ye, X.; Du, J.; Gong, X.; Zhao, Y.; Shamal, A. SparseTrajAnalytics: An Interactive Visual Analytics System for Sparse Trajectory Data. *J. Geovis. Spat. Anal.* **2021**, *3*, 5. [CrossRef]
6. Alsahfi, T.; Almotairi, M.; Elmasri, R. A Survey on Trajectory Data Warehouse. *Spat. Inf. Res.* **2019**, *28*, 53–66. [CrossRef]
7. Nardini, F.M.; Orlando, S.; Perego, R.; Raffaetà, A.; Renso, C.; Silvestri, C. Analysing Trajectories of Mobile Users: From Data Warehouses to Recommender Systems. In *A Comprehensive Guide through the Italian Database Research over the Last 25 Years*; Springer: Berlin, Germany, 2018; pp. 407–421.
8. Etemad, M.; Júnior, A.S.; Hoseyni, A.; Rose, J.; Matwin, S. A Trajectory Segmentation Algorithm Based on Interpolation-Based Change Detection Strategies. EDBT/ICDT Workshops. 2019. Available online: http://ceur-ws.org/Vol-2322/BMDA_4.pdf (accessed on 31 January 2020).
9. Leonardi, L.; Marketos, G.; Frentzos, E.; Giatrakos, N.; Orlando, S.; Pelekis, N.; Raffaetà, A.; Roncato, A.; Silvestri, C.; Theodoridis, Y. T-warehouse: Visual Olap Analysis on Trajectory Data. In Proceedings of the 2010 IEEE 26th International Conference on Data Engineering (ICDE 2010), Long Beach, CA, USA, 1–6 March 2010; pp. 1141–1144.
10. Alarabi, L.; Mokbel, M.F.; Musleh, M. St-hadoop: A Mapreduce Framework for Spatio-Temporal Data. *GeoInformatica* **2018**, *22*, 785–813. [CrossRef]
11. Zheng, Y. Trajectory Data Mining: An Overview. *ACM Trans. Intell. Syst. Technol.* **2015**, *6*, 1–41. [CrossRef]
12. Feng, Z.; Zhu, Y. A Survey on Trajectory Data Mining: Techniques and Applications. *IEEE Access* **2017**, *4*, 2056–2067. [CrossRef]
13. Mazimpaka, J.D.; Timpf, S. Trajectory Data Mining: A Review of Methods and Applications. *J. Spat. Inf. Sci.* **2016**, *13*, 61–99. [CrossRef]
14. Mirge, V.; Verma, K.; Gupta, S. Dense Traffic Flow Patterns Mining in Bi-Directional Road Networks using Density Based Trajectory Clustering. *Adv. Data Anal. Classif.* **2017**, *11*, 547–561. [CrossRef]
15. Zheng, Y. Urban Computing: Enabling Urban Intelligence with Big Data. *Front. Comput. Sci.* **2017**, *11*, 1–3. [CrossRef]
16. Wang, H.; Zheng, K.; Zhou, X.; Sadiq, S. Sharkdb: An In-Memory Storage System for Massive Trajectory Data. In Proceedings of the 2015 ACM SIGMOD International Conference on Management of Data, Melbourne, Australia, 31 May–4 June 2015; pp. 1099–1104.
17. Sistla, A.; Wolfson, O.; Chamberlain, S. Modeling and Querying Moving Objects. In Proceedings 13th International Conference on Data Engineering, Birmingham, UK, 7–11 April 1997; Volume 15, pp. 165–190.
18. Güting, R.H.; Behr, T.; Düntgen, C. SECONDO: A Platform for Moving Objects Database Research and for Publishing and Integrating Research Implementations. *Bull. Technol. Comm. Data Eng.* **2010**, *33*, 56–63.
19. Praing, R.; Schneider, M. Modeling Historical and Future Movements of Spatio-Temporal Objects in Moving Objects Databases. In Proceedings of the Sixteenth ACM Conference on Conference on Information and Knowledge Management, Lisbon, Portugal, 6–10 November 2007; pp. 183–192.
20. Ding, Z.; Yang, B.; Güting, R.H. Network-Matched Trajectory-Based Moving-Object Database: Models and Applications. *IEEE Trans. Intell. Transp. Syst.* **2015**, *16*, 1918–1928. [CrossRef]
21. Wu, Q.; Huang, J.; Luo, J. An All-Time-Domain Moving Object Data Model, Location Updating Strategy, and Position Estimation. *Int. J. Distrib. Sens. Netw.* **2015**, *2015*, 463749. [CrossRef]
22. Huang, C.; Jin, P.; Wang, H. IndoorSTG: A Flexible Tool to Generate Trajectory Data for Indoor Moving Objects. In Proceedings of the IEEE International Conference on Mobile Data Management, Milan, Italy, 3–6 June 2013; pp. 341–343.
23. Hajari, H.; Hakimpour, F. A Spatial Data Model for Moving Objects Database. *arXiv* **2014**, arXiv:1403.3304. [CrossRef]
24. Kucuk, A.; Hamdi, S.; Aydin, B. Pg-Trajectory: A PostgreSQL/PostGIS Based Data Model for Spatiotemporal Trajectories. In Proceedings of the IEEE International Conferences on Big Data and Cloud Computing, Atlanta, GA, USA, 8–10 October 2016; pp. 81–88.
25. Leonardi, L.; Orlando, S.; Raffae, T. A General Framework for Trajectory Data Warehousing and Visual OLAP. *Geoinformatica* **2014**, *18*, 273–312. [CrossRef]
26. Braz, F.J. Trajectory Data Warehouses: Proposal of Design and Application to Exploit Data. In Proceedings of the IX Brazilian Symposium on Geoinformatics, Campos do Jordão, Brazil, 25–28 November 2007; pp. 61–72.

27. Zhang, X.; Ying, W.; Yang, P.; Sun, M. Parameter estimation of underwater impulsive noise with the Class B model. *IET Radar Sonar Navig.* **2020**, *14*, 1055–1060. [CrossRef]
28. Zhang, L.; Qiang, M.; Yang, G. Mobility Transportation Mode Detection Based on Trajectory Segment. *J. Comput. Inf. Syst.* **2013**, *9*, 3279–3286.
29. Khan, S.; Hayat, M.; Bennamoun, M. Cost-Sensitive Learning of Deep Feature Representations from Imbalanced Data. *IEEE Trans. Neural Netw. Learn. Syst.* **2018**, *29*, 3573–3587.
30. AL-Dohuki, S.; Kamw, F.; Zhao, Y.; Ye, X.; Yang, J. An open source trajanalytics software for modeling, transformation and visualization of urban trajectory data. In Proceedings of the 2019 IEEE Intelligent Transportation Systems Conference (ITSC), Auckland, New Zealand, 27–30 October 2019; pp. 150–155. [CrossRef]
31. Shang, Z.; Li, G.; Bao, Z. DITA: Distributed in-memory trajectory analytics. In Proceedings of the 2018 International Conference on Management of Data, Houston, TX, USA, 10–15 June 2018; pp. 725–740.
32. Almeida, D.; Baptista, C.; Andrade, F. A Survey on Big Data for Trajectory Analytics. *Int. J. Geo-Inf.* **2020**, *9*, 88.

 *sensors*

 MDPI

Article

Multi-Stage Feature Extraction and Classification for Ship-Radiated Noise

Hamada Esmaiel [1,2], **Dongri Xie** [3], **Zeyad A. H. Qasem** [1], **Haixin Sun** [1,*], **Jie Qi** [4] **and Junfeng Wang** [5]

1   Department of Information and Communication, School of Informatics, Xiamen University, Xiamen 316005, China; h.esmaiel@aswu.edu.eg (H.E.); zeyadqasem@stu.xmu.edu.cn (Z.A.H.Q.)
2   Electrical Engineering Department, Faculty of Engineering, Aswan University, Aswan 81542, Egypt
3   China Electronics Technology Avionics Co., Ltd., Chengdu 610100, China; 23320171153152@stu.xmu.edu.cn
4   School of Electronic Science and Engineering, Xiamen University, Xiamen 361005, China; qijie@xmu.edu.cn
5   Department of Information and Communication Engineering, School of Electrical and Electronic Engineering, Tianjin University of Technology, Tianjin 300383, China; great_seal@163.com
*   Correspondence: hxsun@xmu.edu.cn; Tel.: +86-159-5925-0539

Abstract: Due to the complexity and unique features of the hydroacoustic channel, ship-radiated noise (SRN) detected using a passive sonar tends mostly to distort. SRN feature extraction has been proposed to improve the detected passive sonar signal. Unfortunately, the current methods used in SRN feature extraction have many shortcomings. Considering this, in this paper we propose a new multi-stage feature extraction approach to enhance the current SRN feature extractions based on enhanced variational mode decomposition (EVMD), weighted permutation entropy (WPE), local tangent space alignment (LTSA), and particle swarm optimization-based support vector machine (PSO-SVM). In the proposed method, first, we enhance the decomposition operation of the conventional VMD by decomposing the SRN signal into a finite group of intrinsic mode functions (IMFs) and then calculate the WPE of each IMF. Then, the high-dimensional features obtained are reduced to two-dimensional ones by using the LTSA method. Finally, the feature vectors are fed into the PSO-SVM multi-class classifier to realize the classification of different types of SRN sample. The simulation and experimental results demonstrate that the recognition rate of the proposed method overcomes the conventional SRN feature extraction methods, and it has a recognition rate of up to 96.6667%.

Keywords: ship-radiated noise; variational mode decomposition; weighted permutation entropy; local tangent space alignment

Citation: Esmaiel, H.; Xie, D.; Qasem, Z.A.H.; Sun, H.; Qi, J.; Wang, J. Multi-Stage Feature Extraction and Classification for Ship-Radiated Noise. *Sensors* **2022**, *22*, 112. https://doi.org/10.3390/s22010112

Academic Editor: Andrea Trucco

Received: 2 November 2021
Accepted: 22 December 2021
Published: 24 December 2021

Publisher's Note: MDPI stays neutral with regard to jurisdictional claims in published maps and institutional affiliations.

1. Introduction

Ships are playing an increasingly important role in many military and civilian applications. For example, in military field applications, an effective prediction for enemy ships helps us to take the correct action and activate our countermeasure to avoid enemy attacks and defeat them. For civilian applications, a logical comprehensive analysis of different port noise, including ship-radiated noise (SRN) can help researchers support the reproduction of marine life [1]. For improving the passive sonar operation in ship applications, SNR feature extraction has been proposed [1]. However, marine environmental diversity provides a rich noise environment, and that increases the difficulty of extracting features reflecting the intrinsic characteristics of the ships [1]. In recent years, studies of the SRN feature extraction have increased. Unfortunately, the current SRN feature extraction schemes have many drawbacks. For example, Fourier transform (FT) [1] is only useful in only estimating the signal spectral information, but it is unsuccessful at time-varying representation. To address this drawback, short-time Fourier transform (STFT) has been proposed to indicate the time-varying signal traits. However, a fixed STFT window width makes STFT unable to consider good representation for the time domain and frequency domain at the same time [1]. To overcome the STFT drawbacks wavelet transform (WT) has been proposed by using the unfixed length of the window) [2]. The time-frequency decomposition is improved based on WT, but the wavelet basis function and decomposition layers are required to be set in advance, which practically limits the further application for WT in practice.

Empirical mode decomposition (EMD) has been proposed in [3], to decompose the signal into a group of limited intrinsic mode functions (IMFs), the core of the Hilbert transform (HT). The main purpose of the HT is to snaffle the component of the instantaneous frequency. Therefore, EMD can carefully and rapidly describe the instantaneous frequency of multi-components, hence EMD is more suitable for analyzing non-linear and non-stationary signals. However, the main EMD drawback is the mode mixing problem and many researchers have tried hard to address this issue. To this end, the ensemble empirical mode decomposition (EEMD) [4] has been offered as an amended EMD method [4]. EEMD has been proposed to solve the mode mixing problem by adding Gaussian white noise to the construed signal, and at the end averaging the obtained multiple decompositions results to obtain the IMFs. However, EEMD introduces and has additional problems, firstly, as, in the EMD, the EEMD decomposition results include residual components. Secondly, due to the randomness of Gaussian white noise, the outcome of EEMD is diverse between each decomposition time. Hence, the EEMD lacks a solid mathematical foundation to be distributed and widely accepted. Unlike the EMD and EEMD, variational mode decomposition (VMD) has been proposed in [5], which assumes that each mode is concerning a central frequency with restricted bandwidth. Thus, to obtain the center frequency and bandwidth of each component, VMD constantly searches for the modes and center frequency of each mode by using an alternating direction method of a multiplier, thereby solving the variational problem. In recent years, many works tried to extend the current EMD, EEMD, and VMD methods and applied these schemes in the fields of biomedical engineering [6–9], mechanical fault diagnosis [10–12], and acoustic signal processing [13,14]. In [13], the VMD is firstly performed to decompose the SRN signal, and the permutation entropy (PE) of each IMF with the highest energy is then extracted, achieving a recognition rate of 94%. In [14,15] the VMD decomposition of the SRN signal was performed and the fluctuation-based dispersion entropy (FDE) of each IMF was studied; the obtained IMF with the smallest difference from the FDE of the prime signal was then chosen to describe the raw signal with a recognition rate of 97.5%. In the EMD-EIMF-PE method proposed in [16], the signal-dominant IMF by EMD was chosen based on the energy gauge, and its PE was regarded as the feature parameter effectively distributing the SRN. Although both [13,14] have high recognition rates, there are still flaws in these methods. The mode VMD number has specified the EMD results in [13,14], which will certainly influence the VMD decomposition accuracy. Since the pertinent studies failed to strictly obtain the VMD model number, this paper suggests a new enhanced VMD (EVMD) method with the mode number close to the variance of IMFs' center frequency.

Entropy is an important indicator to measure the uncertainty of time series and can consider the implicit system dynamics. When the system dynamics vary, the time series complexity will be varied as well. PE [15] has been used in mechanical fault diagnosis [16], agricultural commodity analysis [17], financial sequence analysis [18] due to its fast operation speed and excellent stability. Nevertheless, PE does not consider the case of neighboring vectors having the same ordinal patterns with different amplitudes, which will lead to the estimated value higher than the actual [19]. To this end, due to the preamble weights, the weighted permutation entropy (WPE) [20] is further critical to the amplitude-coded information in the signal and has outperformed the PE in combating the distortion caused by noise. To the best of the authors' knowledge, WPE has been widely used in uncertainty measurements in many fields [21,22], but is rarely used in the SRN feature extraction.

With current computer technology development and the fast growth of data, it is of great necessity to further process the original high-dimensional data. Feature extraction can be divided into two parts; extracting the feature vectors can reflect the essence of analyzing signals through suitable signal processing methods. The other is selecting an appropriate measure decreases algorithm to minimize the increases, and in that way support the confession execution. Up to now, researchers have suggested different dimension relief algorithms, such as principal component analysis (PCA) [23,24], independent component analysis (ICA) [24], and linear discriminant analysis (LDA) [25]. PCA recognizes measurement saving by realizing optimal variance without dropping the creative data. The PCA and ICA are linear unsupervised dimension reduction algorithms. LDA is also a linear projection method that achieves dimension reduction by making the most of the ratio of the discrete matrix among classes and the discrete matrix within the class. Due to the SRN non-linear characteristics, linear dimension reduction algorithms are scarce in removing the intrinsic features of SRN signals. As a non-linear manifold learning algorithm, local tangent space alignment (LTSA) has been extensively useful in dimension reduction, thanks to its fast process speed and selfishness to selected parameters [26–29]. To the best of our knowledge, there is no study combining EVMD, WPE, and LTSA to classify underwater acoustic targets. Each method has its pros and cons, motivating us to combine all methods to have the maximum benefits.

To that end, this paper puts forward a novel multistage feature extraction method proposing an EVMD method and combining it with the WPE, and LTSA for SRN samples classification. In this paper, the proposed EVMD method uses the variance of the IMFs' center frequency to calculate the mode number of VMD and enhance its operation. Next, the new EVMD algorithm is used to decompose the SRN signals plus calculate the WPE of each IMF. Then high-dimensional features are reduced to two-dimensional ones by the LTSA method. Finally, the feature vectors obtained are fed into the PSO-SVM multi-class classifier to recognize the different types of SRN samples.

The structure of the paper is presented as follows: the fundamental theories of the relevant algorithms are described in Section 2. In Section 3, the basic steps of the proposed method are presented. Section 4 applies the proposed method to the analysis of simulated signals. In Section 5, the proposed method is utilized for the feature extraction of SRN. Finally, the paper is concluded in Section 6.

2. Basic Theory

In this section, the theories of the related methods such as VMD, PE, WPE, and LTSA will be presented.

2.1. Variational Mode Decomposition (VMD)

The VMD defines the IMF in the function of the instantaneous amplitude $Am_k(t)$ and phase $Ph_k(t)$ as an amplitude-modulated-frequency-modulated (AM-FM) signal, given as below:

$$I_k(t) = Am_k(t)\cos(Ph_k(t)) \tag{1}$$

where the change of the $Am_k(t)$ and $\dot{Ph}_k(t)$ are slower than $Ph_k(t)$. Each $I_k(t)$ is compacted around a respective center frequency with limited bandwidth obtained by Gaussian smoothing demodulation. In the VMD algorithm, decomposing the raw signal $s(t)$ into a finite group of IMFs to find the variational problem can be expressed as follows:

$$\min_{\{I_k, f_k\}} \left\{ \sum_k \left\| \partial_t \left[\left(\delta(t) + \frac{j}{\pi t} \right) * I_k(t) \right] e^{-j2\pi f_k t} \right\|_2^2 \right\}, \qquad s.t. \sum_k I_k = s(t) \tag{2}$$

where ∂_t, $\delta(t)$ and f_k represent the partial derivative, impulse function, and center frequency of $I_k(t)$, respectively. The constrained variation problem in Equation (2) is addressed using the quadratic penalty term and the Lagrange multipliers below:

$$L(\{I_k\}, \{f_k\}, \lambda) = \alpha \sum_k \left\| \partial_t \left[\left(\partial_t + \frac{j}{\pi t} \right) * I_k(t) \right] e^{-j2\pi f_k t} \right\|_2^2 + \left\| s(t) - \sum_k I_k(t) \right\|_2^2 + \lambda(t), s(t) - \sum_k I_k(t) \tag{3}$$

where α and λ denote the penalty factor and Lagrange multiplier, respectively. I_k^{n+1}, f_k^{n+1} and λ^{n+1} are updated as follows:

$$\hat{I}_k^{n+1}(f) = \frac{\hat{s}(f) - \sum_{i \neq k} \hat{I}_i(f) + \frac{\hat{\lambda}(f)}{2}}{1 + 2\alpha(f - f_k)^2} \tag{4}$$

$$f_k^{n+1} = \frac{\int_0^\infty 2\pi f |\hat{I}_k(f)|^2 df}{\int_0^\infty |\hat{I}_k(f)|^2 df} \tag{5}$$

$$\hat{\lambda}^{n+1}(f) = \hat{\lambda}^n(f) + \varepsilon(\hat{s}(f) - \sum_k \hat{I}_k^{n+1}(f)) \tag{6}$$

where ε represents the update parameter. In this method, the stop condition is given by:

$$\sum_k ||\hat{I}_k^{n+1} - \hat{I}_k^n||_2^2 / ||\hat{I}_k^n||_2^2 < a \tag{7}$$

where a denotes the convergence accuracy. The VMD algorithm can be summarized in: (1) initialize $\{\hat{I}_k^1\}$, $\{f_k^1\}$, $\hat{\lambda}^1$ and $n = 0$; (2) update the values of $\{\hat{I}_k^{n+1}\}$, $\{f_k^{n+1}\}$ and $\hat{\lambda}^{n+1}$ based on Equations (4)–(6); (3) check the covariance condition based on Equation (7), and the details about the VMD algorithm are published in [5].

2.2. Permutation Entropy (PE)

PE [15] can not only characterize the randomness of the time series but also detect its dynamic changes. In addition, PE does not consider the amplitude value, but only compares the neighboring values, which makes its operation speed faster. For the given time series $x = \{x_j\}_{j=1}^N$, the PE can be reconstructed as:

$$X_i = \{x(i), x(i + \tau), \cdots, x(i + (m-1)\tau)\}, \ i = 1, 2, \cdots, N - (m-1)\tau \tag{8}$$

where m is the embedding dimension, τ is the time delay and $i = 1, 2, \cdots, N - (m-1)\tau$. The elements in X_i can be rearranged in increasing order as:

$$x(i + (j_1 - 1)\tau) \leq x(i + (j_2 - 1)\tau) \leq x(i + (j_m - 1)\tau) \tag{9}$$

If two of the rearranged elements are equal, then,

$$x(i + (j_1 - 1)\tau) = x(i + (j_2 - 1)\tau) \tag{10}$$

hence, the new order can be denoted as:

$$x(i + (j_1 - 1)\tau) \leq x(i + (j_2 - 1)\tau)(j_1 \leq j_2) \tag{11}$$

Therefore, the symbols group can be obtained as:

$$S(g) = (j_1, j_2, \cdots, j_m) \tag{12}$$

where $S(g)$ represents one of $m!$ symbol sequences in phase space, $g = 1, 2, \cdots, k$, and $k \leq m!$. If the probability distribution of the symbol sequence is $P_1, P_2 \cdots, P_k$, for convergence, the normalized PE is defined as follows:

$$H_p(m) = -(\ln(m!))^{-1} \sum_{g=1}^{k} P_g \ln(P_g) \tag{13}$$

From Equation (13), we can observe that the value of PE ranges from 0 to 1. H_p indicates the randomness of time sequence, a larger H_p value means higher complexity of the time series; a smaller H_p value means lower uncertainty of the time series.

2.3. Weighted Permutation Entropy (WPE)

In the PE the neighboring vectors having the same ordinal patterns but with different amplitude values are unreasonably ignored. The WPE [20] has been proposed to take such a situation into account and overcome the PE shortcomings. In the WPE, forgiven embedding dimension m and time delay τ, first, the weight w_i of neighbouring vectors X_i is calculated as:

$$w_i = \sum_{k=1}^{m}\left[x_{j+(k-1)\tau} - \overline{X}_j^{m,\tau}\right]^2 \tag{14}$$

$$\overline{X}_j^{m,\tau} = \frac{1}{m}\sum_{k=1}^{m} x_{j+(k-1)\tau} \tag{15}$$

Then, the weighted relative frequency is calculated as:

$$p_w\left(\pi_i^{m,\tau}\right) = \frac{\sum j \leq N^{1_{u:\text{type}(u)=\pi_i(\overline{X}_j^{m,\tau})}}}{\sum j \leq N^{1_{\text{type}(u)\in\Pi}}\ (\overline{X}_j^{m,\tau})w_j} \tag{16}$$

Finally, the WPE definition is described below:

$$H_w(m,\tau) = -\sum_{i:\pi_i^{m,\tau}\in\Pi} p_w\left(\pi_i^{m,\tau}\right)\ln\left(p_w\left(\pi_i^{m,\tau}\right)\right) \tag{17}$$

2.4. Local Tangent Space Alignment (LTSA)

Due to the advantages of insensitivity to parameter selection and fast operation, the local tangent space alignment (LTSA) method has been widely used in dimension reduction in multiple fields [29,30]. The basic idea of the LTSA algorithm is constructing the local tangent space by using the sample neighborhood and mapping the coordinates of the local tangent space corresponding to the global low-dimensional coordinates through the local radiological transformation matrix. Given the data $X = \{x_1, x_1, \cdots, x_m\} \subset R^{M \times N}$, the principle of LTSA can be briefly described as below:

(1) Determine the K nearest neighbors of x_i to form the set of X_i, and centralize $\hat{X}_i$,

$$X_i = [x_{i,1},\ x_{i,2}, \cdots, x_{ik}] \tag{18}$$

$$\hat{X}_i = X_i - \overline{x}_i l_k^T \tag{19}$$

where $\overline{x}_i = \frac{1}{k}\sum_{j=1}^{k} x_{ij}$ and l_k is a unit vector with dimension K.

(2) Calculate the eigenvalues and eigenvectors of a matrix $\hat{X}_i$ by singular value decomposition. The eigenvectors corresponding to the first d largest singular values are the tangent space H_i.

$$\theta_{ij} = H_i^T\left(x_{ij} - \overline{x}_i\right) \tag{20}$$

(3) Construct the transformation matrix $L_i = \theta_i^+$, to retain as much information as possible, and the following conditions must be met,

$$\min\mu(Y) = \min\sum_{i=1}^{M}\left|Y_i\left(I - \frac{1}{k}ll^T\right) - L_i\theta_i\right| \tag{21}$$

where θ_i^+ represents the generalized inverse matrix of θ_i, Y_i represents the set of nearest neighbors of Y after dimension reduction, that is, $Y_i = (y_{i1}, y_{i2}, \cdots, y_{ik})$.

(4) Solve the optimization problem of Equation (21) by calculating the eigenvalues and eigenvectors of the matrix, and then the embedding matrix Y can be obtained. Equation (21) can be equivalent to the following equation:

$$\min \mu(Y) = \min(YHW) = \min tr\left(YHW^T H^T Y^T\right), \tag{22}$$

$$\begin{cases} H = (H_1, H_2, \cdots, H_M) \\ W = diag\left(W_1, W_2, \cdots W_M\right) \\ W_i = \left(I - \frac{1}{k}ll^T\right)\left(I - \theta_i^+ \theta_i\right) \\ I = YY^T \end{cases} \tag{23}$$

The low-dimensional embedding matrix Y can be obtained by calculating the eigenvectors corresponding to the second to d-th smallest eigenvalues of the alignment matrix B and it can be calculated as:

$$B = HWW^T H^T \tag{24}$$

3. Proposed Feature Extraction Method Based on Enhanced Variational Mode Decomposition (EVMD), Weighted Permutation Entropy (WPE), and Local Tangent Space Alignment (LTSA)

In Section 2 the details of the basic theories of VMD, WPE, and LTSA are presented. However, the model number of VMD needs to be determined in advance, so this paper proposes the EVMD for SRN signal processing. A multi-stage feature extraction method fully inheriting the advantages of EVMD, WPE, and LTSA is proposed in this paper. The flowchart of the proposed method is shown in Figure 1. The main steps of the proposed method can be summarized as follows:

(1) Based on the center frequency of the intrinsic mode functions, the variational mode decomposition mode number K will be calculated firstly.

(2) The VMD algorithm based on the optimum mode number K obtained in the first step is applied to the decomposition of the SRN signals.

(3) The WPE of each IMF by EVMD is then calculated.

(4) The feature vectors obtained will be randomly divided into two groups, the first group represent the training data used in classification information while the second one is used for evaluating the testing data and measuring its ability for classification.

(5) Definitively, PSO-SVM multiclass classifier uses the analysis data that are recognized and detects the underwater acoustic object.

To extract features of SRN, this paper classifies three types of SRN using a combination of VMD, WPE, LTSA, and PSO-SVM multi-class classifier. In the proposed method, the VMD mode number range is first set according to the EEMD decomposition results, and the variance of the IMFs' center frequency after each decomposition is calculated. The mode number corresponding to the maximum variance is used as an optimum value for VMD. Then, VMD is performed on three SRN signal types. The WPE value of each IMF for each VMD decomposition is calculated. The high-dimensional features obtained are reduced to be two-dimensional features using LTSA. Finally, the obtained feature vectors are input to the PSO-SVM multi-class classifier to achieve classification and recognition of the samples.

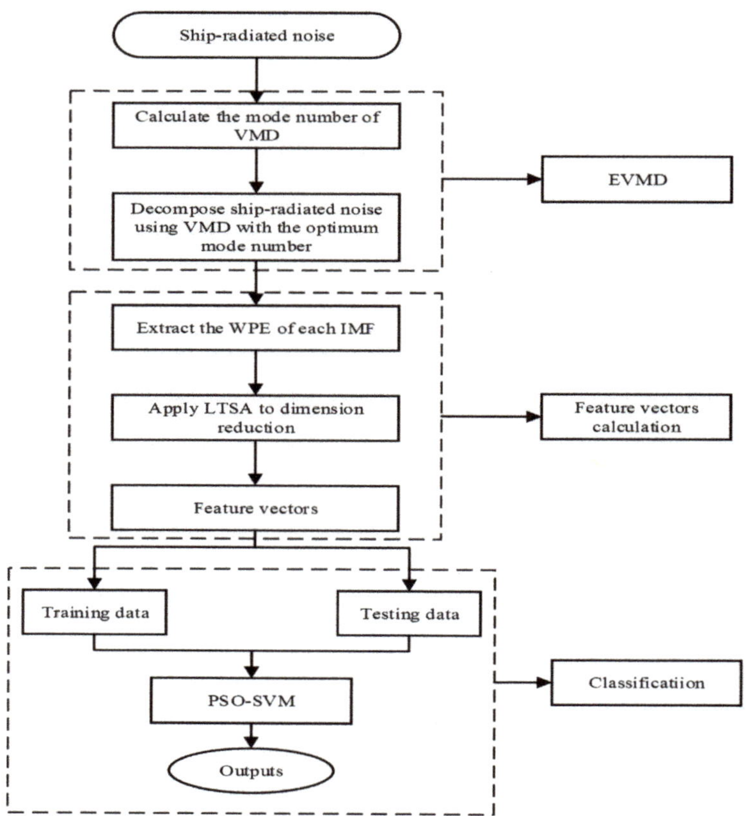

Figure 1. The flowchart of the proposed method.

4. Simulation Signals Analysis

4.1. Analysis of Simulated Signals Based on EVMD

In this paper, as the SRN signals contain the rich line spectral components, the simulated signals composed of the single-frequency components are introduced. In addition, the simulated signals are feasible as long as they satisfy the measurement signal conditions explained in [12], namely, their frequency interval should not be too small (like 1 Hz, 2 Hz, and 3 Hz). Following the same restriction conditions of [12] and verifying the effectiveness and feasibility of the proposed EVMD method. The simulated signals used in this paper are as follows:

$$\begin{cases} f_1(t) = \cos(10\pi t) \\ f_2(t) = \cos(60\pi t) \\ f_3(t) = \cos(110\pi t) \\ f(t) = f_1(t) + f_2(t) + f_3(t) + \eta \end{cases} \tag{25}$$

where the data length is set to be 5000 with a sampling frequency of 1 kHz and η denotes the Gaussian white noise with $\mathcal{CN}(0, 0.5)$.

Following the literature [5], before performing the VMD algorithm, the model number K needs to be adjusted, which serves as a main factor affecting the VMD performance. Other parameters are set as constants, namely the balancing parameter of the data-fidelity constraint is $\alpha = 2000$, the convergence tolerance level is $tol = 1 \times 10^{-7}$ and the update mode of the center frequency is $init = 0, 1, 2$ for center frequency iterated with 0, uniform distribution, or randomly. Too-large a K will lead to the occurrence of over-decomposition, which means undesirable spurious components will be generated during the decomposition process; too-small a K will cause under-decomposition, which discards some IMFs carrying useful information during the decomposition process. Hence, a properly chosen K value is crucial to the VMD method. Although the VMD method in [13,14] can successfully decompose the SRN signals to some extent, the method for determining the mode number of VMD has not been reasonably demonstrated, making it unacceptable scientifically. In general, the mode number of VMD will not be greater than the mode number of EMD and EEMD, so conventional EMD and EEMD methods are first employed to analyze the simulated signals above. Figure 2 shows the decomposition results and the time domain waveforms of the simulated signals.

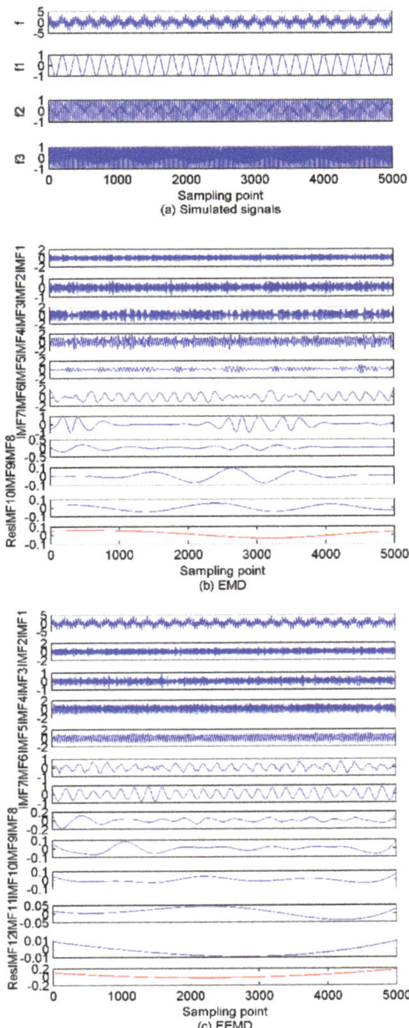

Figure 2. Modeled signals and decomposition results based on empirical mode decomposition (EMD) and ensemble empirical mode decomposition (EEMD) methods. (**a**) the initial modeled signals; (**b**) EMD procedure; (**c**) EEMD procedure.

Figure 2, show modeled indicator decomposed by using the EMD, one remaining part is obtained in addition to 10 IMFs. While in the EEMD decomposed method there are one remaining part and 10 IMFs, and they are separated. Hence, based on setting the variables for the mode number range to be $(2 \sim 12)$, and following calculation of every decomposition IMFs' center frequency, the good mode number K has the maximum variance which maximizes the IMFs center frequency difference. Figure 3, shows the IMFs' center frequency at different mode number K when the VMD decomposing method is used, and Figure 3 shows that $K = 9$ is the best choice.

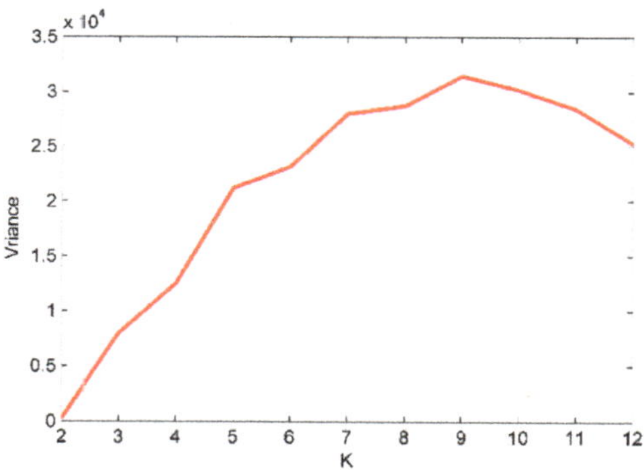

Figure 3. The variance curve of intrinsic mode functions (IMFs) center frequency with mode number K after simulated signals decomposition using variational mode decomposition (VMD).

When the mode number K is set to be more than the good, estimated value $K = 9$, the variance begins to reduce and that shows irrelevant variation among the IMFs' center frequency and the existence of the done decomposition. Figure 4, shows the decomposition results of the modeled signals based on the EVMD procedure. The IMFs' center frequency distribution at different mode numbers K are listed in Table 1. Table 1 shows how components are unseparated for $K = (2 \sim 6)$, and for $K \geq 7$ the modeled signal is separated. For $K = (10 \sim 12)$ further false components are created to recognize the split of the modeled signals.

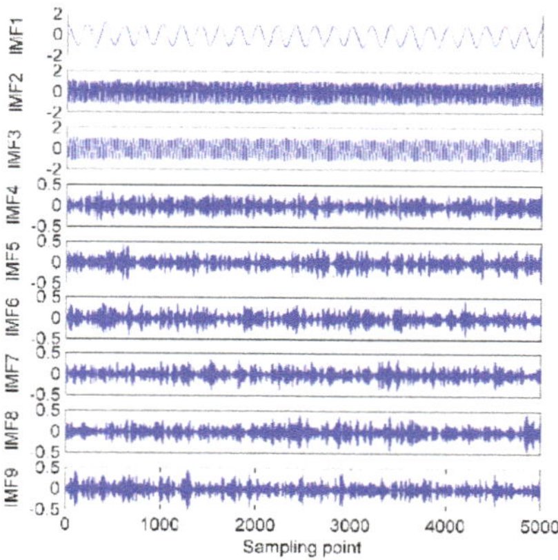

Figure 4. The EVMD decomposition results for the simulated signals.

Table 1. The mode number corresponding to the distribution of IMF's center frequency by VMD.

K	Center Frequency/Hz											
2	7.79	52.50										
3	7.77	52.50	310.22									
4	7.73	52.51	193.88	358.92								
5	7.71	52.52	178.13	306.10	429.27							
6	7.63	52.53	132.93	231.63	325.67	438.33						
7	5.04	55.14	30.02	178.15	281.03	365.89	468.10					
8	5.04	55.12	30.02	150.48	230.17	304.96	377.06	470.06				
9	5.04	55.11	30.02	133.53	202.41	280.99	349.96	413.83	476.09			
10	5.04	55.09	30.01	122.45	181.76	240.15	300.08	359.29	418.56	477.08		
11	5.03	55.09	30.01	118.11	173.42	227.27	276.65	321.81	369.88	424.40	478.29	
12	5.03	55.05	30.01	99.61	146.79	192.70	237.79	283.66	325.96	372.20	425.66	468.56

Based on the analysis above, the proposed EVMD method using the variance of the IMFs' center frequency is feasible in calculating the mode number of VMD. To further verify this method in the VMD algorithm, the VMD method in [4,20] is also introduced for comparison. The correlation coefficients between the corresponding IMF and simulated signals are calculated and the results are listed in Table 2. As shown in Table 2, the corresponding three components of the proposed EVMD have the highest correlation coefficients with the simulated signals, and its decomposition performance is significantly better than the conventional methods of the EMD, EEMD, and VMD. This confirms the validity and feasibility of the method proposed in this paper for determining the VMD mode number.

Table 2. Correlation coefficients between corresponding IMF with simulated signals.

	EMD	EEMD	VMD	EVMD
f_1	IMF6: 0.9244	IMF7: 0.9665	IMF1: 0.9938	IMF1: 0.9944
f_2	IMF4: 0.8959	IMF5: 0.9806	IMF3: 0.9914	IMF3: 0.9925
f_3	IMF3: 0.8875	IMF4: 0.9460	IMF2: 0.9872	IMF2: 0.9880

4.2. Analysis of the Properties Concerning Weighted Permutation Entropy (WPE) and Permutation Entropy (PE)

According to the basic principles of WPE and PE explained in Section 2, when the signal is mutated, it is difficult for PE to detect this state, while WPE should be more sensitive to this mutation. To validate the conjecture, we generate a standard Gaussian white noise series with a length of 5000. As the pulse series means a larger fluctuation, we add the pulse series to these Gaussian white noise series. As in [31], the time delay and the embedding dimension are set as 1 and 6, respectively. The PE and WPE are calculated using a window function with a length of 500 and a sliding step of 50. The time-domain waveforms of the Gaussian white noise series and the signal plus additive pulse series are shown in Figure 5. The results of the calculated entropy are shown in Figure 6.

As shown in Figure 6, the WPE values of two signals of the SRN are smaller than the corresponding PE ones. This indicates how Gaussian white noise has a higher complexity and contains more information. There is no difference in the PE of the two signals in the pulse region, which means the inability of PE to distinguish between these two signals. This can be attributed to the neglect of the amplitude difference between neighboring vectors having the same ordinal patterns in PE calculation. As a result, PE performs poorly in effectively detecting fluctuations caused by noise. In contrast, the WPE value of the signal after superimposed pulses decreases significantly in the pulse region, indicating that the WPE can effectively detect the amplitude-encoded information contained in the signal due to the introduction of weights, which outperforms the PE in noise detection and fluctuation observation.

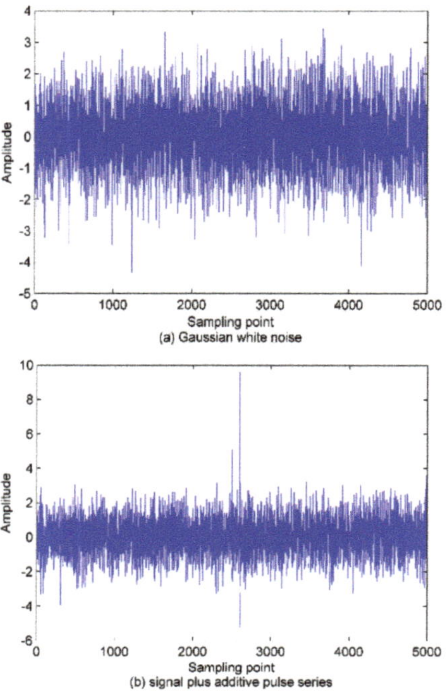

Figure 5. The time domain diagrams of Gaussian white noise and the signal plus additive pulse series.

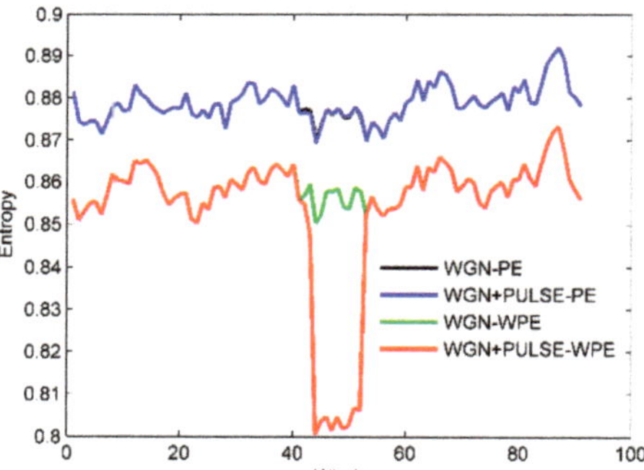

Figure 6. The curve of PE and WPE with window.

For further comparison between the performance of the PE and WPE, 50 1/f noise samples with a length of 500 are generated and the pulse series is superimposed on the raw 1/f noise series. For a fair comparison, PE and WPE calculations concern the two signals. The time-domain waveforms of the analyzed signals and the scatter plots of the calculation results are given in Figures 7 and 8, respectively.

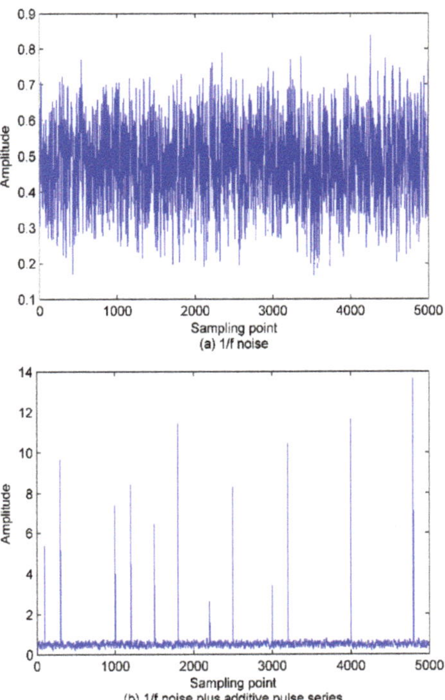

Figure 7. The time domain waveforms for; (**a**) 1/f noise (**b**) the signal plus additive pulse series.

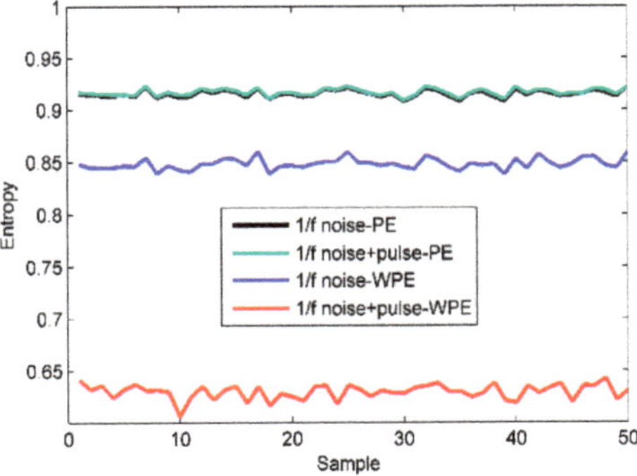

Figure 8. The scatter plots of calculation results for PE and WPE.

As shown in the results, thanks to the good anti-noise ability of WPE, making the estimated WPE value of the two signals is lower than the PE value. Also, sudden change detection is a hard task in the PE method as it neglects the amplitude information. In contrast, except for the significant difference in WPE values between the two signals, the WPE fluctuation trends are more dramatic than PE. Larger fluctuation means stronger discrimination ability. In short, the analysis of experimental results proves the advantages of WPE over PE.

5. Feature Extraction of Ship-Radiated Noise Based on the Proposed Method

5.1. Parameter Selection

The WPE calculation is dependent on the time delay τ and embedding dimension m. If m is too small, the reconstructed sequence will contain less state information, and the WPE algorithm cannot adequately detect the dynamic change of the time series. However, larger m indicates that the time series is homogenized by the reconstructed phase space and cannot detect the time series slight change. Also, if τ is too small, a strong correlation will occur between different delay vector elements, resulting in information redundancy. The phase space trajectory cannot be fully expanded when τ is too large. Therefore, such parameters should be adjusted first before WPE calculation.

To study the influence of m and τ on PE and WPE, the three types of SRN are randomly selected from the data set used in [32]. The samples were recorded on the Atlantic coast in north-western Spain $(42^\circ\ 14'\ \text{N},\ 008^\circ\ 43.4'\ \text{W})$ at a depth of 10 m. The sampling frequency is 52.734 kHz and the data length is set to be 5000. The three types of SRN signals are named class A, B, and C, respectively. Figure 9 shows the time-domain waveforms of the normalized signals.

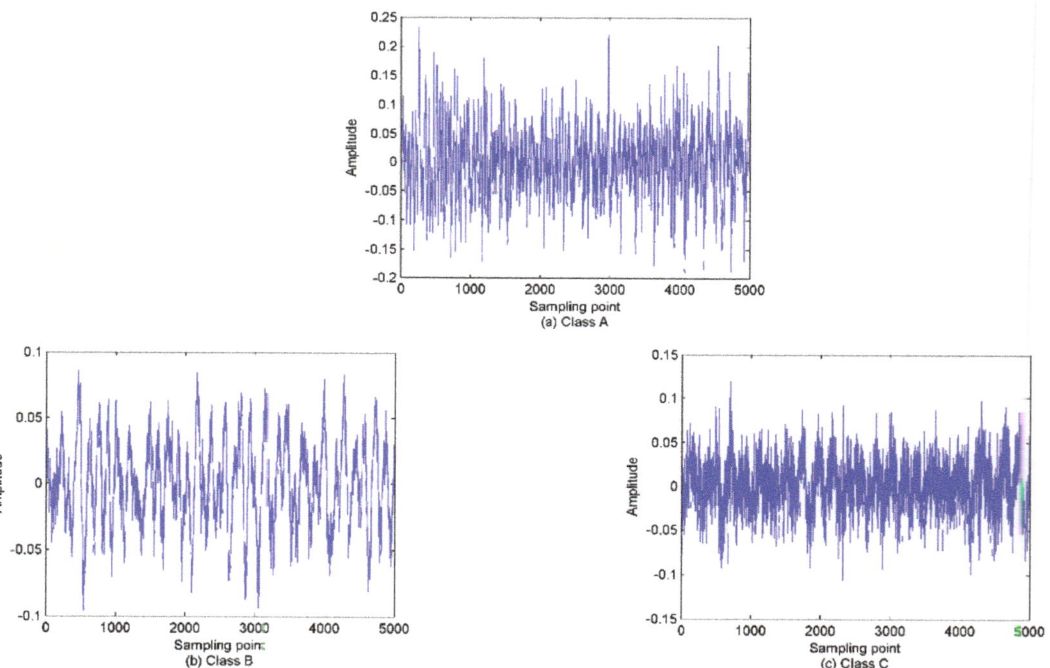

Figure 9. The time-domain waveforms for the three types of normalized ship$-$radiated noise (SRN).

As in [30], the PE and WPE of the three types of SRN signals are calculated under the condition m = 3, 4, 5, 6, 7, and the time delay ranges from 1 to 20. The results are shown in Figure 10. As seen in Figure 10, the class C SRN signal has the maximum entropy value and thus the highest complexity.

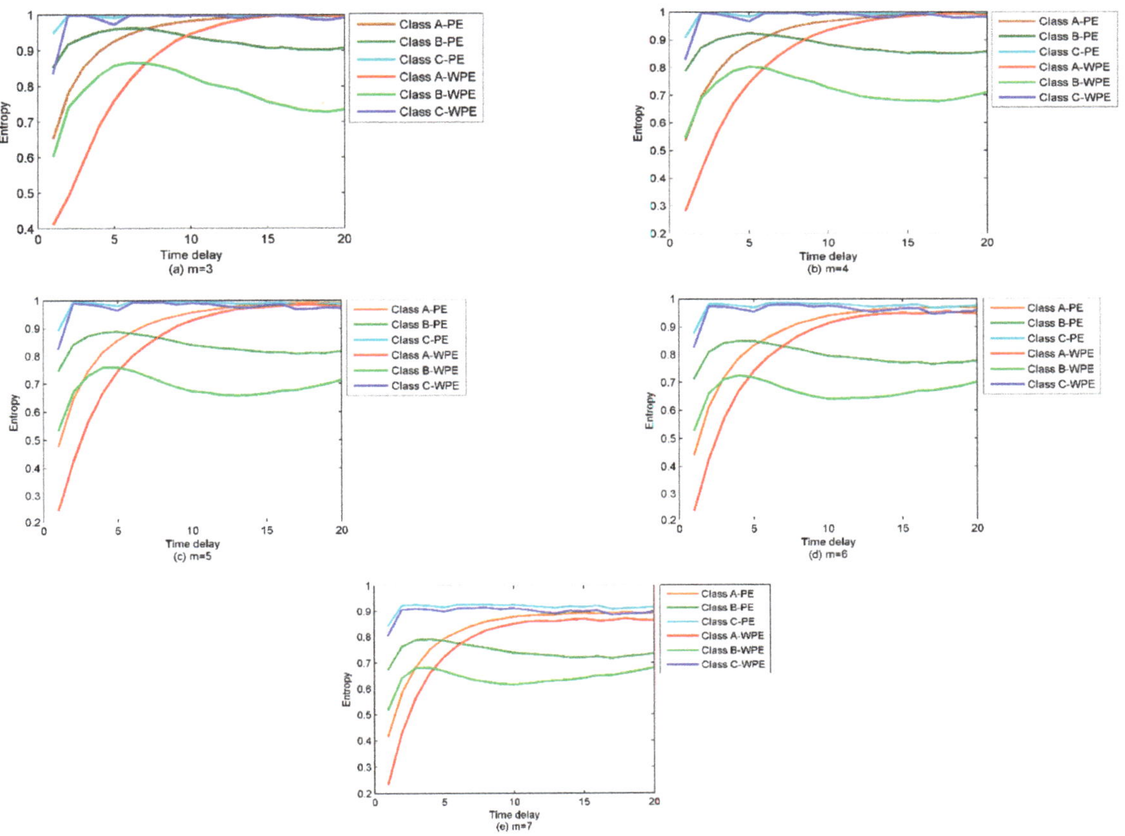

Figure 10. PE, WPE comparison of the three types of SRN under different embedding dimensions.

Thanks to the excellent anti-noise performance of WPE, the WPE value of the same time series is less than the PE. WPE is more sensitive to the time delay compared to the PE, as the PE fluctuates subtly with the time delay increasing. The WPE and PE begin to separate when $m = 6$. However, when m is increased to 7, the computational complexity will be increased without improving the accuracy of calculation results. Based on the results obtained in Figure 10, a significant difference occurs between the WPE and PE when the time delay is equal to 1. Therefore, in the proposed method when calculating the WPE and PE, the delay and embedding dimensions are set to be 1 and 6, respectively. These parameters are consistent with the recommendations given by [15,31]. The PE and WPE of the signals versus the time delay are shown in Figure 11. As shown in Figure 11, the time delay has a greater influence on PE and WPE in some ranges, while less in others as the embedding dimension increases. Compared to PE, the trends of WPE fluctuate relatively sharply, as the WPE is more sensitive to the pattern extracted from signals containing amplitude information.

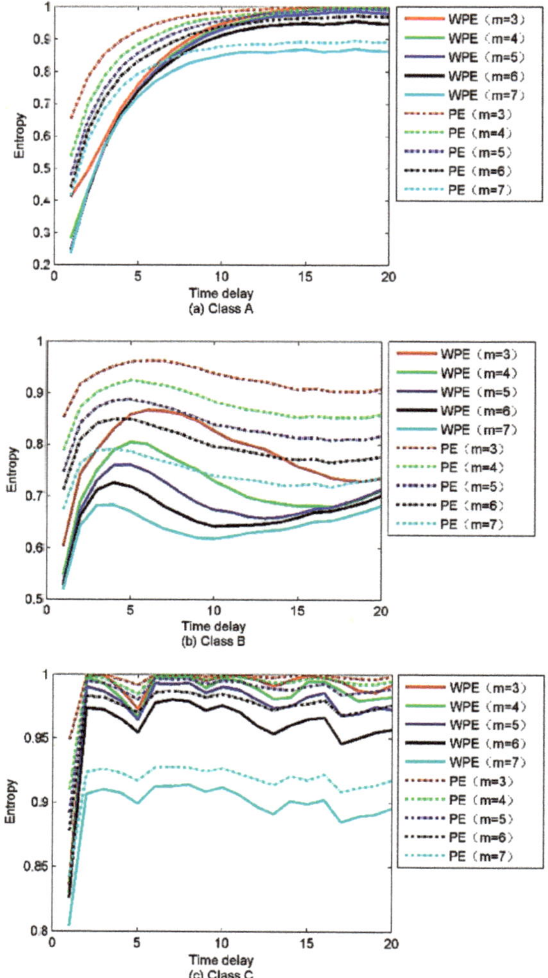

Figure 11. PE and WPE of the signals varying with the time delay.

5.2. Decomposition of Ship-Radiated Noise Using VMD

As described in Section 3, to calculate the VMD model number, the EEMD algorithm is first employed to decompose the SRN signals. The decomposition results are presented in Figure 12. Figure 12 shows that 12 IMFs and one residual component are obtained after the EEMD of each type of SRN signal. For the sake of observation, K is set as $(2 \sim 15)$ to calculate the variance of the IMFs' center frequency. The results are given in Figure 13. It can be observed from Figure 13 that the optimal K is 12, and when K is higher than 12, the variance starts to decrease sharply, implying the occurrence of over-decomposition. The decomposition results by EVMD are given in Figure 14.

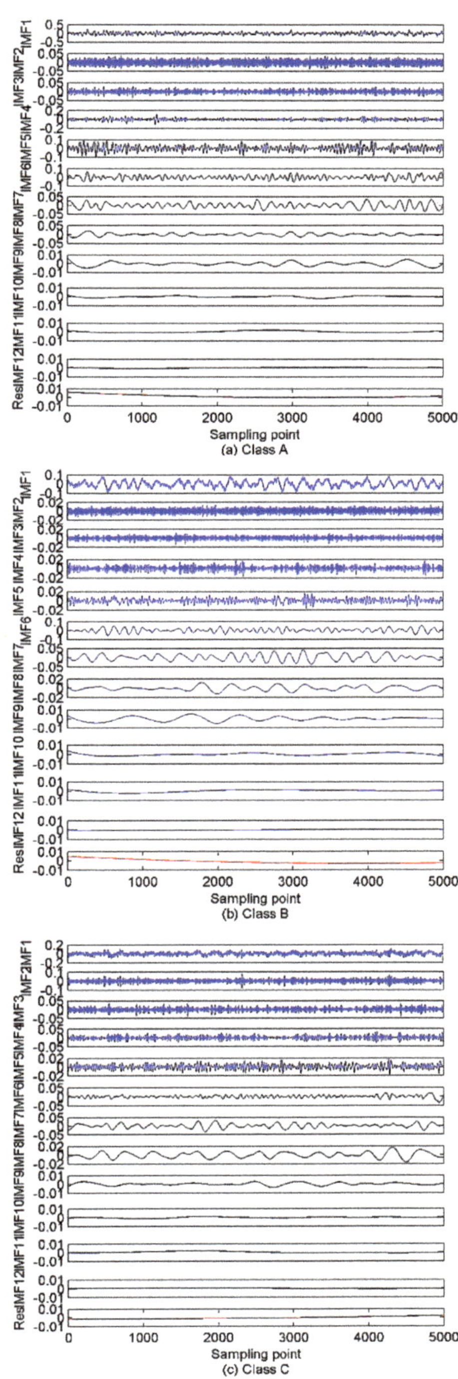

Figure 12. The EEMD results for SRN signals.

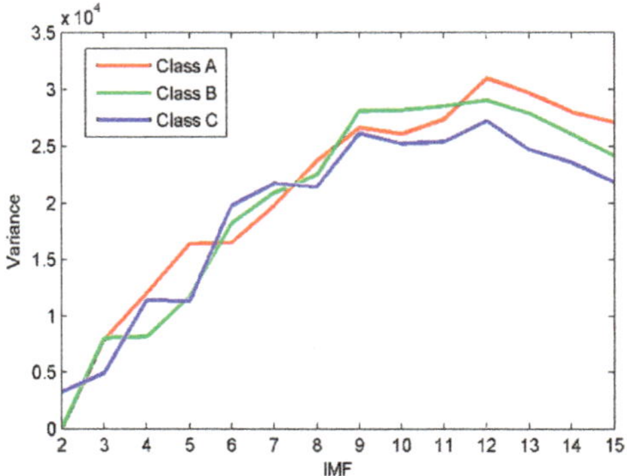

Figure 13. The results of variance analysis for the SRN signals.

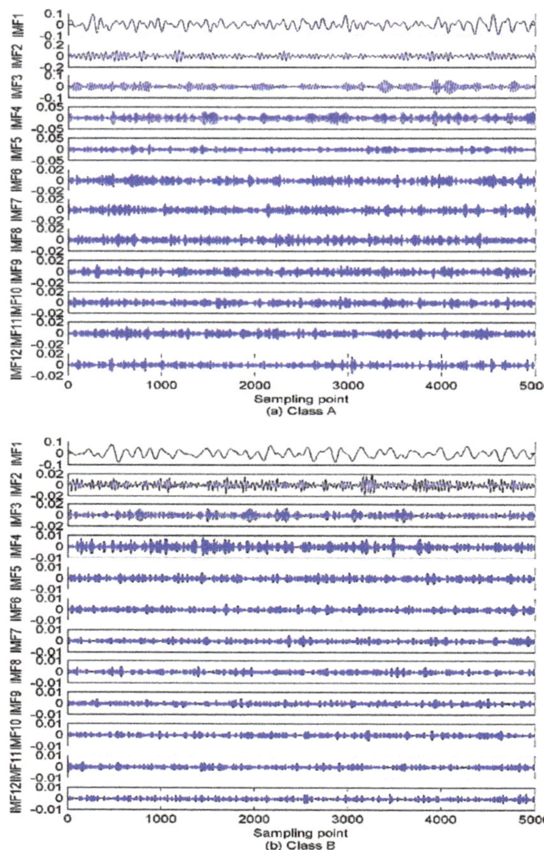

Figure 14. *Cont.*

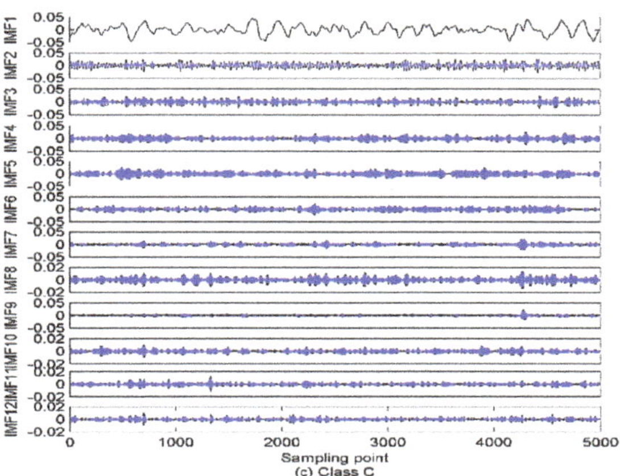

Figure 14. The EVMD results for SRN signals.

5.3. Classification of Ship-Radiated Noise (SRN)

In this section, the proposed method is applied to the SRN samples classification. 100 samples are randomly selected from each type of SRN sample and thus a total of 300 samples can be obtained. The EVMD is first performed to decompose these samples and the WPE of each IMF is calculated. Figure 15 shows the WPE mean and standard deviation.

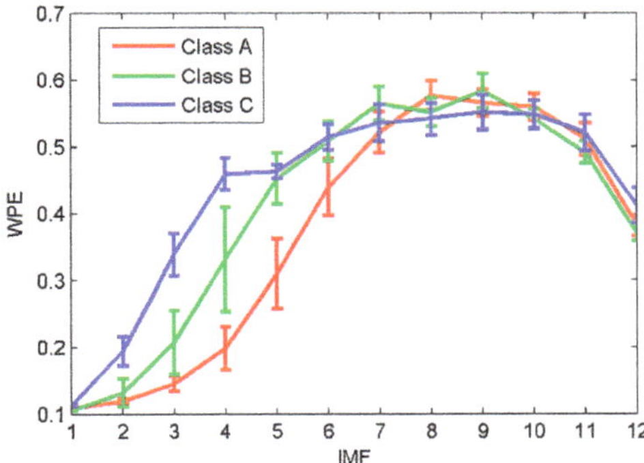

Figure 15. The mean and standard deviation of WPE.

Figure 15 shows, when the mode is IMF3 or IMF4, that the three types of samples can be distinguished, and the WPE values of class C samples are higher than that of the other two classes. This can be due to the dynamic behavior changes at different signals. Hence, WPE can effectively reflect the dynamic changes of the time series. However, when the mode is IMF2, the class A and B samples cannot be identified; when IMF5, only class A can be identified. In other modes, all three types of sample are indistinguishable. Therefore, not every IMF can fully characterize the raw SRN signal after the proposed EVMD decomposition. The results can be attributed to two points.

Firstly, due to the pollution of marine environmental noise, some IMFs belong to noise or noise-dominant components. Secondly, the occurrence of over-decomposition in the VMD algorithm allows different IMFs to share the same spectrum information. In this way, the dimension reduction algorithms are introduced to avoid dimensional disasters, thus preparing for the classification below. Next, the PCA, MDS, LLE, and LTSA methods are utilized for the low-dimensional feature extraction. The number of neighboring points and the target dimension are set as 15 and 2, respectively. The results are given in Figure 16. As shown in Figure 16, overall, most of the three types of samples can be distinguished by four algorithms despite the partially overlapping samples between classes A and B. From qualitative point of view, the combined EVMD-WPE-PCA algorithm has shown the wo performance with a small number of samples crossing between class A and B, and m samples overlapping between class B and C, while the performance of EVMD-WPE-M and EVMD-WPE-LLE has been significantly improved compared to EVMD-WPE-P with only a few samples of class A and B crossed. Despite the good performance of EVM WPE-PCA, EVMD-WPE-MDS, and EVMD-WPE-LLE in roughly identifying the three ty of samples, the degree of clustering within the class is still small, especially for clas In contrast, the proposed combined EVMD-WPE-LTSA method has the best cluster performance. The three kinds of SRN sample can be separated well, and the samp of classes B and C are better clustered. For a fair comparison, the scatter plots of W EMD-WPE-LTSA and EEMD-WPE-LTSA are represented in Figure 17.

As shown in Figure 17a, affected by the oceanic environment and the ambient no we cannot recognize the majority of the three forms of samples as they will be oppo together. Hence in such a case, the algorithms of the decomposition should be used. the EMD-WPE-LTSA shown in Figure 16b, a large proportion of samples of classe and B are overlapped, which cannot be separated well. The samples in the C categ are also intersected with the others. Hence, recognition of the three samples beco difficult. While concerning the EEMD-WPE-LTSA, the samples in categories A and B ca completely separated, but the degree of separation between categories B and C is still s From a qualitative point of view, there is no significant difference between EEMD-V LTSA and EVMD-WPE-LTSA, and EEMD-WPE-LTSA even looks more clustered withi class, but we cannot conclude on the merits of both algorithms. Data visualization is o qualitative tool rather than a quantitative one. In this situation, the PSO-SVM multi classifier is introduced to compare the algorithms above more accurately. Next, 60 san are randomly selected from each class to be used as a training classifier and the remai samples are left to be used in testing the performance. Also, both EMD-EIMF-PE [21] VMD-WPE-LTSA (the mode number of VMD is set as [4,20]) methods are considered. classification outputs are shown in Figure 18. Table 3 lists the classification accuracy computational time for SRN feature extraction under different algorithms.

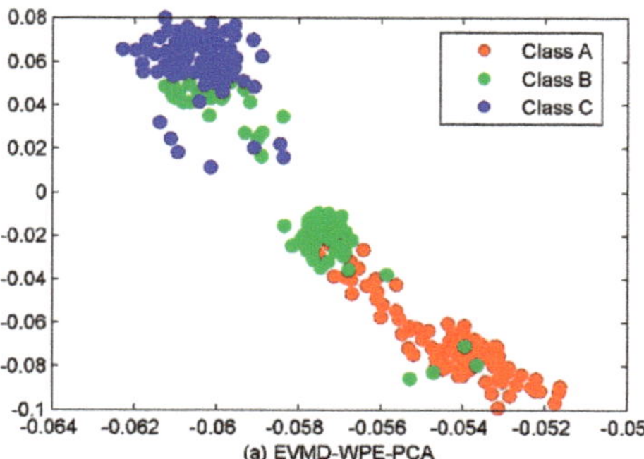

Figure 16. *Cont.*

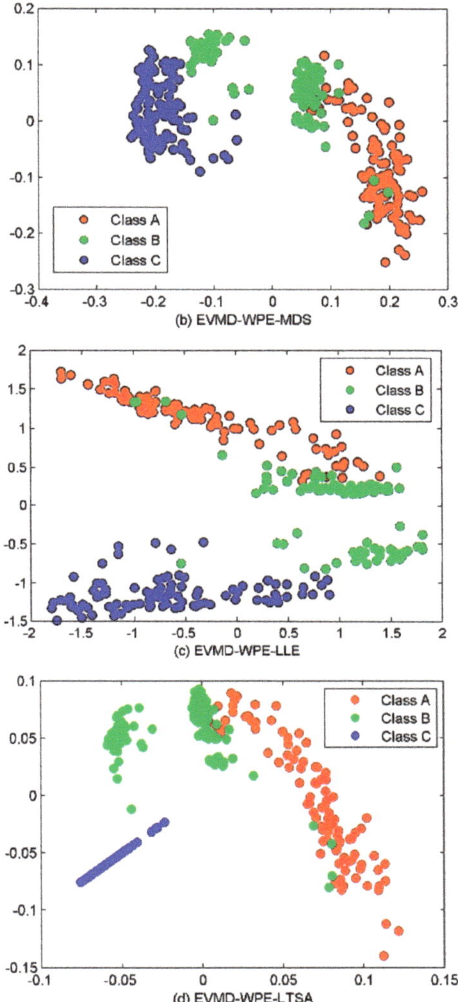

Figure 16. The clustering results of the four algorithms.

As shown in Figure 18 and Table 3, the direct calculation of the WPE for the samples fails to achieve the SRN samples identification as it has only a 62.5% recognition rate and that is far from the classification standard. The multistage classification techniques based on the proposed EVMD have the highest classification accuracy of 95.8333% and 96.6667% in EVMD-WPE-LLE and EVMD-WPE-LTSA, respectively. The proposed multistage classification techniques based on EVMD outperform the other conventional algorithms with the additional computational time cost. In general, computational complexity and classification accuracy are determined by the configuration of the hardware and algorithm design.

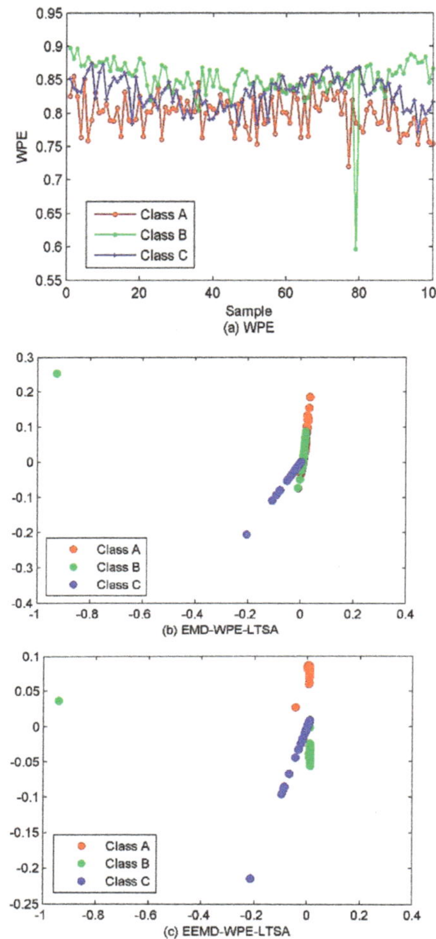

Figure 17. The scatter plots of the three algorithms.

Table 3. The classification accuracy and computational complexity under different algorithms.

Method	NUMBER OF MISCLASSIFIED SAMPLES			ACCURACY RATE (%)	COMPUTATIONAL TIMES (SECOND)
	CLASS A	CLASS B	CLASS C		
WPE	11	7	27	62.5	787.35041
EMD-WPE-LTSA	11	9	2	81.6667	6956.81681
EEMD-WPE-LTSA	0	12	0	90	23,643.2661
EVMD-WPE-PCA	2	5	5	90	22,254.0133
EVMD-WPE-MDS	4	2	0	95	22,255.2634
EVMD-WPE-LLE	1	4	0	95.8333	22,255.0276
EVMD-WPE-LTSA	3	0	1	96.6667	22,255.0041
EMD-EIMF-PE	5	0	8	89.1667	1342.94618
VMD-WPE-LTSA	3	5	0	93.3333	12,404.5999

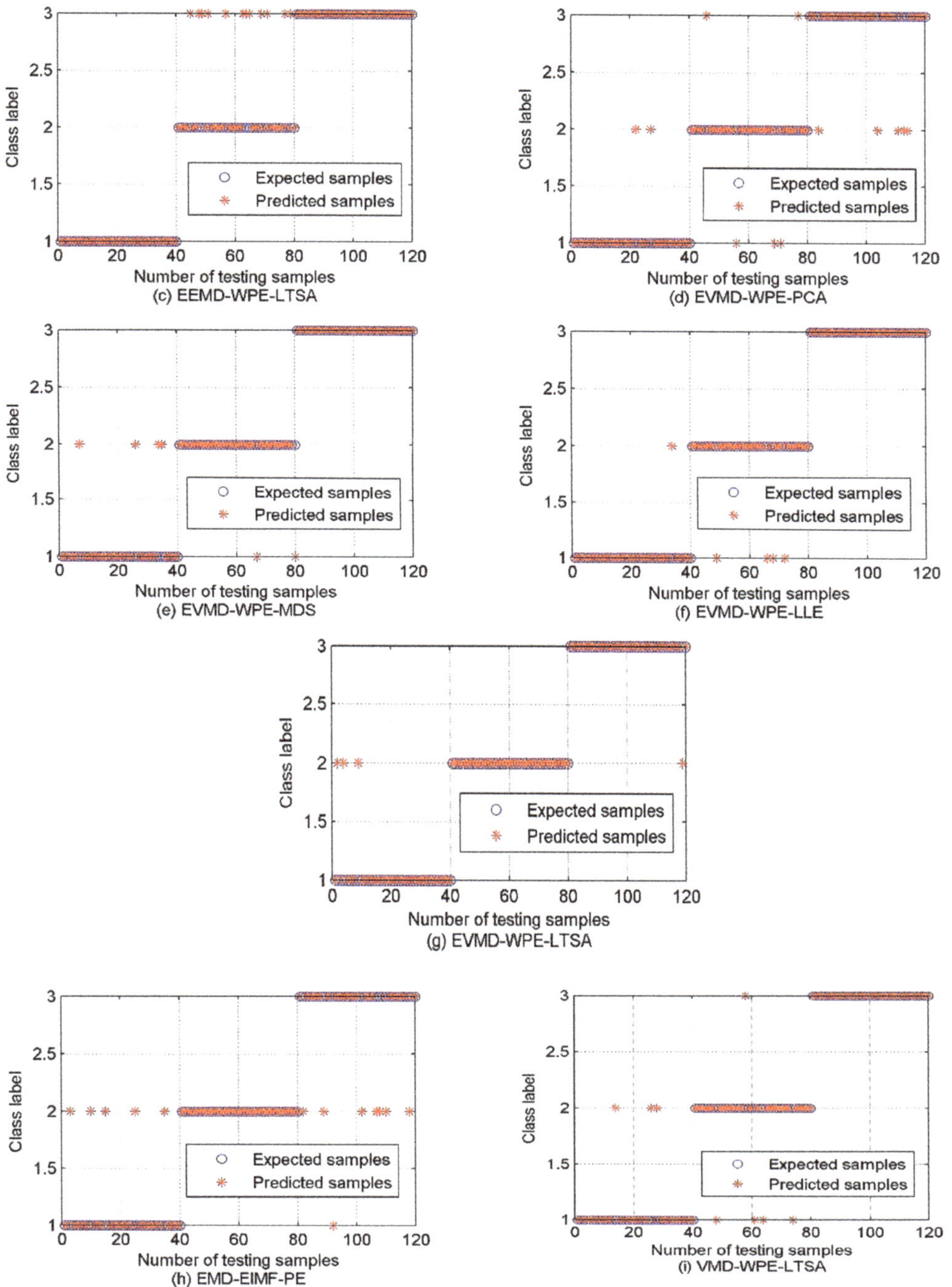

Figure 18. The outputs of classification under different algorithms.

6. Conclusions

A novel multi-stage feature extraction method for underwater acoustic signals is proposed in this paper based on combining the new EVMD method with WPE and LTSA. The main innovations and contributions of this work can be summarized as follows:

(1) The proposed EVMD method mode number was calculated based on the variance of IMFs' center frequency.
(2) WPE and LTSA were combined with the new EVMD and applied to underwater SRN signals for the first time.
(3) The proposed EVMD-WPE-LTSA was successfully applied, and extracted the features of the underwater acoustic signals. Compared with other algorithms, EVMD-WPE-LTSA had a higher recognition rate and better classification performance at the cost of computational time.

The EVMD algorithm proposed in this paper to overcome the shortage of the VMD performed accurately in the field of underwater acoustic communication. Nevertheless, this paper only addresses the VMD mode number. In future, the optimization between both the model number and the quadratic penalty term will be considered to obtain high decomposition accuracy for the EVMD. Also, we will try to reduce the computation complexity.

Author Contributions: Conceptualization, H.E. and D.X.; methodology, H.E. and D.X.; software, D.X.; validation, Z.A.H.Q. and H.S.; formal analysis, J.Q. and J.W.; investigation, J.Q. and J.W.; resources, H.S.; data curation, D.X.; writing—original draft preparation, H.E. and D.X.; writing—review and editing, H.E. and D.X.; visualization, Z.A.H.Q. and H.S.; supervision, J.Q. and J.W.; project administration, J.Q. and J.W.; funding acquisition, H.S. All authors have read and agreed to the published version of the manuscript.

Funding: This work is supported by the National Natural Science Foundation of China (61671394), The Fundamental Research Funds for the Central Universities (20720170044), and the Natural Science Foundation of Tianjin (16JCQNJC01100).

Data Availability Statement: The data used to support the findings of this study are available from the corresponding author upon request.

Conflicts of Interest: The authors declare no conflict of interest.

References

1. Xie, D.; Esmaiel, H.; Sun, H.; Qi, J.; Qasem, Z.A. Feature Extraction of Ship-Radiated Noise Based on Enhanced Variational Mode Decomposition, Normalized Correlation Coefficient and Permutation Entropy. *Entropy* **2020**, *22*, 468. [CrossRef] [PubMed]
2. Rao, R. Wavelet Transforms. In *Encyclopedia of Imaging Science and Technology*; Wiley: New York City, NY, USA, 2002.
3. Huang, N.E.; Shen, Z.; Long, S.R.; Wu, M.C.; Shih, H.H.; Zheng, Q.; Yen, N.-C.; Tung, C.C.; Liu, H.H. The empirical mode decomposition and the Hilbert spectrum for nonlinear and non-stationary time series analysis. *Proc. R. Soc. Lond. Ser. A Math. Phys. Eng. Sci.* **1998**, *454*, 903–995. [CrossRef]
4. Wu, Z.; Huang, N.E. Ensemble empirical mode decomposition: A noise-assisted data analysis method. *Adv. Adapt. Data Anal.* **2009**, *1*, 1–41. [CrossRef]
5. Dragomiretskiy, K.; Zosso, D. Variational mode decomposition. *IEEE Trans. Signal Process.* **2013**, *62*, 531–544. [CrossRef]
6. Zhang, X.; Wang, J.; Lin, L. Feature extraction of ship radited noises based on wavelet transform. *Acta Acust. Peking* **1997**, *22*, 139–144.
7. Shih, M.-T.; Doctor, F.; Fan, S.-Z.; Jen, K.-K.; Shieh, J.-S. Instantaneous 3D EEG signal analysis based on empirical mode decomposition and the hilbert–huang transform applied to depth of anaesthesia. *Entropy* **2015**, *17*, 928–949. [CrossRef]
8. Xie, P.; Yang, F.-M.; Li, X.-X.; Yang, Y.; Chen, X.-L.; Zhang, L.-T. Functional coupling analyses of electroencephalogram and electromyogram based on variational mode decomposition-transfer entropy. *Acta Phys. Sin.* **2016**, *65*, 118701.
9. Taran, S.; Bajaj, V. Clustering variational mode decomposition for identification of focal EEG signals. *IEEE Sens. Lett.* **2018**, *2*, 1–4. [CrossRef]
10. Hsieh, N.-K.; Lin, W.-Y.; Young, H.-T. High-speed spindle fault diagnosis with the empirical mode decomposition and multiscale entropy method. *Entropy* **2015**, *17*, 2170–2183. [CrossRef]
11. Zhang, X.; Liang, Y.; Zhou, J. A novel bearing fault diagnosis model integrated permutation entropy, ensemble empirical mode decomposition and optimized SVM. *Measurement* **2015**, *69*, 164–179. [CrossRef]
12. Lei, Y.; He, Z.; Zi, Y. Application of the EEMD method to rotor fault diagnosis of rotating machinery. *Mech. Syst. Signal Process.* **2009**, *23*, 1327–1338. [CrossRef]

13. Li, Y.; Li, Y.; Chen, X.; Yu, J. A novel feature extraction method for ship-radiated noise based on variational mode decomposition and multi-scale permutation entropy. *Entropy* **2017**, *19*, 342. [CrossRef]
14. Yang, H.; Zhao, K.; Li, G. A new ship-radiated noise feature extraction technique based on variational mode decomposition and fluctuation-based dispersion entropy. *Entropy* **2019**, *21*, 235. [CrossRef] [PubMed]
15. Bandt, C.; Pompe, B. Permutation entropy: A natural complexity measure for time series. *Phys. Rev. Lett.* **2002**, *88*, 174102. [CrossRef] [PubMed]
16. Tian, Y.; Wang, Z.; Lu, C. Self-adaptive bearing fault diagnosis based on permutation entropy and manifold-based dynamic time warping. *Mech. Syst. Signal Process.* **2019**, *114*, 658–673. [CrossRef]
17. de Araujo, F.H.A.; Bejan, L.; Rosso, O.A.; Stosic, T. Permutation entropy and statistical complexity analysis of Brazilian agricultural commodities. *Entropy* **2019**, *21*, 1220. [CrossRef]
18. Henry, M.; Judge, G. Permutation entropy and information recovery in nonlinear dynamic economic time series. *Econometrics* **2019**, *7*, 10. [CrossRef]
19. Li, R.; Wang, J. Interacting price model and fluctuation behavior analysis from Lempel–Ziv complexity and multi-scale weighted-permutation entropy. *Phys. Lett. A* **2016**, *380*, 117–129. [CrossRef]
20. Fadlallah, B.; Chen, B.; Keil, A.; Príncipe, J. Weighted-permutation entropy: A complexity measure for time series incorporating amplitude information. *Phys. Rev. E* **2013**, *87*, 022911. [CrossRef]
21. Wu, E.Q.; Zhu, L.-M.; Zhang, W.-M.; Deng, P.-Y.; Jia, B.; Chen, S.-D.; Ren, H.; Zhou, G.-R. Novel Nonlinear Approach for Real-Time Fatigue EEG Data: An Infinitely Warped Model of Weighted Permutation Entropy. *IEEE Trans. Intell. Transp. Syst.* **2019**, *21*, 2437–2448. [CrossRef]
22. Wu, P.; Guo, L.; Duan, Y.; Zhou, W.; He, G. Control loop performance monitoring based on weighted permutation entropy and control charts. *Can. J. Chem. Eng.* **2019**, *97*, 1488–1495. [CrossRef]
23. Wold, S.; Esbensen, K.; Geladi, P. Principal component analysis. *Chemom. Intell. Lab. Syst.* **1987**, *2*, 37–52. [CrossRef]
24. Reza, M.S.; Ma, J. ICA and PCA Integrated Feature Extraction for Classification. In Proceedings of the IEEE 13th International Conference on Signal Processing (ICSP), Chengdu, China, 6–10 November 2016.
25. Tharwat, A.; Gaber, T.; Ibrahim, A.; Hassanien, A.E. Linear discriminant analysis: A detailed tutorial. *AI Commun.* **2017**, *30*, 169–190. [CrossRef]
26. Zhang, Z.; Zha, H. *Principal Manifolds and Nonlinear Dimension Reduction via Local Tangent Space Alignment*; Tech Rep CSE-02-019; Department of Computer Science, Pennsylvania State University: University Park, PA, USA, 2002.
27. Ma, P.; Zhang, H.; Fan, W.; Wang, C. Early fault detection of bearings based on adaptive variational mode decomposition and local tangent space alignment. *Eng. Comput.* **2019**, *36*, 509–532. [CrossRef]
28. Li, S.; Qian, P.; Gu, H. Fault Detection of Discriminant Improved Local Tangent Space Alignment. In Proceedings of the International Conference on Intelligent Informatics and Biomedical Sciences (ICIIBMS), Shanghai, China, 21–24 November 2019.
29. Ye, Y.; Zhang, Y.; Wang, Q.; Wang, Z.; Teng, Z.; Zhang, H. Fault diagnosis of high-speed train suspension systems using multiscale permutation entropy and linear local tangent space alignment. *Mech. Syst. Signal Process.* **2020**, *138*, 106565. [CrossRef]
30. Zhang, N.; Sun, Y.; Zhang, Y.; Yang, P.; Lin, A.; Shang, P. Distinguishing Stock Indices and Detecting Economic Crises Based on Weighted Symbolic Permutation Entropy. *Fluct. Noise Lett.* **2019**, *18*, 1950026. [CrossRef]
31. Zheng, J.; Pan, H.; Yang, S.; Cheng, J. Generalized composite multiscale permutation entropy and Laplacian score based rolling bearing fault diagnosis. *Mech. Syst. Signal Process.* **2018**, *99*, 229–243. [CrossRef]
32. Santos-Domínguez, D.; Torres-Guijarro, S.; Cardenal-López, A.; Pena-Gimenez, A. ShipsEar: An underwater vessel noise database. *Appl. Acoust.* **2016**, *113*, 64–69. [CrossRef]

Review

Void Avoiding Opportunistic Routing Protocols for Underwater Wireless Sensor Networks: A Survey

Rogaia Mhemed [1,*], **William Phillips** [1], **Frank Comeau** [2] and **Nauman Aslam** [3]

[1] Department of Engineering Mathematics and Internetworking, Dalhousie University, Halifax, NS B3H 4R2, Canada
[2] Department of Engineering, St. Francis Xavier University, Antigonish, NS B2G 2W5, Canada
[3] School of Computing Engineering and Information Sciences, Northumbria University, Newcastle upon Tyne NE1 8ST, UK
* Correspondence: rogaia.mhemed@dal.ca

Abstract: One of the most challenging issues in the routing protocols for underwater wireless sensor networks (UWSNs) is the occurrence of void areas (communication void). That is, when void areas are present, the data packets could be trapped in a sensor node and cannot be sent further to reach the sink(s) due to the features of the UWSNs environment and/or the configuration of the network itself. Opportunistic routing (OR) is an innovative prototype in routing for UWSNs. In routing protocols employing the OR technique, the most suitable sensor node according to the criteria adopted by the protocol rules will be elected as a next-hop forwarder node to forward the data packets first. This routing method takes advantage of the broadcast nature of wireless sensor networks. OR has made a noticeable improvement in the sensor networks' performance in terms of efficiency, throughput, and reliability. Several routing protocols that utilize OR in UWSNs have been proposed to extend the lifetime of the network and maintain its connectivity by addressing void areas. In addition, a number of survey papers were presented in routing protocols with different points of approach. Our paper focuses on reviewing void avoiding OR protocols. In this paper, we briefly present the basic concept of OR and its building blocks. We also indicate the concept of the void area and list the reasons that could lead to its occurrence, as well as reviewing the state-of-the-art OR protocols proposed for this challenging area and presenting their strengths and weaknesses.

Keywords: void avoiding; opportunistic routing (OR); underwater wireless sensor networks (UWSNs); void area; routing

Citation: Mhemed, R.; Phillips, W.; Comeau, F.; Aslam, N. Void Avoiding Opportunistic Routing Protocols for Underwater Wireless Sensor Networks: A Survey. *Sensors* **2022**, *22*, 9525. https://doi.org/10.3390/s22239525

Academic Editors: Haixin Sun and Xuebo Zhang

Received: 18 November 2022
Accepted: 2 December 2022
Published: 6 December 2022

Publisher's Note: MDPI stays neutral with regard to jurisdictional claims in published maps and institutional affiliations.

1. Introduction

With a large area of the earth (more than 2/3) covered by water [1,2], investigating the underwater environment and exploiting the UWSNs in various areas of underwater studies have become imperative due to the increasing human requirements and needs. Applications in different human activities in the underwater environment have become very important and opened a new field for investigators interested in this area. Many researchers such as [3–12] have proposed solutions to fulfill human requirements and needs in such a harsh environment for industry (detecting chemical pollution, pipeline monitoring, biological phenomena, and seismic studies), government (military applications and maintaining the coast), and nature (hazard events, marine farms, ecological monitoring, and contamination studies). The underwater sensor design in such applications ranges from simple to complex [13]. However, this UWSN research area is very challenging, and most work that has been conducted using terrestrial wireless sensor networks (TWSNs) cannot be directly implemented into UWSNs because different communication channels are used and the characteristics of underwater environments are unique [3,14–16]. Moreover, underwater sensor nodes are expensive battery devices, and they need better protection of their hardware to resist the water characteristics [17–19].

One of the main tasks that faces researchers in the networks is determining how to route the information collected by sensor nodes to reach the sink(s) while minimizing routing costs (minimum delay, energy cost, number of hops, and shortest path) and ensuring network connectivity (void area problem). That is, to assure the network continues to function for as long as possible. OR is one of the routing techniques used to transmit data in UWSNs. It is an emerging routing technology that was proposed to overcome the drawback of unreliable transmission, especially in UWSNs. OR uses the broadcast nature of wireless communication to forward data packets to reach the sink(s) through one-hop or multi-hops. It addresses the major challenges of UWSNs, such as energy efficiency, void avoidance, reliability, and network stability. This OR approach takes into consideration the limited resources (battery and memory) of the underwater sensor nodes, and over the years, many forwarder methods in OR were proposed to prolong the network lifetime and increase the chance that every node has a direct or indirect link to the sink(s). Therefore, OR extends the lifetime of the UWSNs. However, based on previous research, it has been determined that there is still space for development in this area [20].

Many survey papers for UWSNs have been published, such as [21–26] which cover different directions of studies in the UWSN area, and [2,27–32] summarize existing UWSN routing protocols. However, in order to present a more specific survey paper from the perspective of routing, we aim through this survey paper to collect and present a comprehensive overview of the state-of-the-art of routing protocols for UWSNs that focuses on addressing the void area problem by utilizing OR. We also predict future trends and challenges that remain unexplored in order to bring these to researchers' attention.

The main contributions of this survey paper are:

- We explain how the void area problem in UWSNs can be addressed by utilizing the OR technique, and we review up-to-date OR protocols. To the best of our knowledge, our survey paper is the first one that reviews up-to-date void-avoidance OR protocols for UWSNs.
- We classify up-to-date void avoiding OR protocols for UWSNs based on their most important characteristics and features.
- We identify some of the open issues and challenges in UWSNs, which can assist the designers of OR protocols for UWSNs.

The rest of this paper is structured as follows: Section 2 presents the routing protocols in general. It includes the main challenges facing UWSN routing protocol designers, the void area problem, and the reasons that may lead to its existence in the network architecture. It also includes the concept of OR, the main components, and the classification of the OR protocols based on their construction blocks. A review of the state-of-the-art of routing protocols for UWSNs related to our paper is presented in Section 3. The comparison study between the reviewed protocols, including their architectural features, benefits, and drawbacks, is presented in Section 4. Future challenges to be faced in this area are reported in Section 5. Finally, we conclude this paper in Section 6.

2. Routing Protocols

In general, the underwater sensor nodes are deployed in the area of interest following one of the underwater network architectures (i.e., 1, 2, 3, or 4 dimensional UWSNs, which are presented in many papers such as [16,26,32–35]). The deployed underwater sensor nodes must be organized in such a way that they cover the entire area of interest in order to gather the data whenever an event occurs. Routing protocols are responsible for discovering and maintaining transmission routes. Thus, a route between sensor nodes and the sink needs to be established for effective and reliable data transmission. Routing is the backbone of any network. The sensor nodes can communicate with the sink(s) either by: (1) direct link, where the data packets can be sent directly from the source node towards the sink(s). or (2) through a multi-hop path where the data packets are forwarded by the relay nodes until they reach the sink(s). However, multi-hop communication suffers from the complexity of establishing a route, which has effects on network performance such as capacity, reliability, and efficiency.

2.1. Main Challenges Facing UWSN Routing Protocol Designers

In this context, to better design an OR routing protocol for UWSN, a number of challenges that UWSN routing protocol designers encounter are listed and briefly discussed below, as in [36]:

1. Limited bandwidth and data rate: UWSNs suffer from limited available bandwidth (i.e., acoustic waves use a frequency between a few Hz and tens of kHz) and low data rate (i.e., the transmission rate hardly exceeds 100 kbps). The limited accessible acoustic bandwidth depends on the communication range and acoustic frequency.
2. High propagation delay: The UWSNs use an acoustic channel for communication between the underwater sensor nodes themselves and with the sink(s). In the acoustic channel, the propagation speed is five orders of magnitude lower than in the radio channel. This high propagation delay (0.67 s/km) can significantly decrease the throughput of the network.
3. High noise and interference: Basically, there are two kinds of noises, man-made and natural. These noises are caused by water currents, machines, marine mammals, and shipping. The noise under water is much more serious than in the terrestrial environment. The interference is essentially caused by the surface, the bottom, or animals and contamination reflections.
4. High bit error: Due to the shadow zones caused by animals, water currents, and human-made noise, the acoustic channel suffers from a high bit error rate and temporary losses of connectivity.
5. Limited resources: In UWSNs, sensor nodes are constricted resource devices (i.e., they have limited energy and memory). Therefore, after deploying the sensors in an underwater environment, it becomes difficult and costly to replace or recharge the node batteries due to the harsh underwater environment. Moreover, underwater sensor nodes are vulnerable to deterioration and damage due to corrosion and pollution.
6. Topology changes: Due to the flow of water, the underwater sensor nodes cannot stay in one location; instead, they move randomly, which gives UWSNs a mobile or changeable topology.

2.2. Void Area Problem in UWSNs

Sensor nodes drift at different depths in the commonly deployed three-dimensional UWSN architecture to make it easier to identify or monitor a certain phenomenon. Multi-hop routing protocols depend on the relay nodes with positive advancement to transmit the data collected from the phenomenon from the source node to reach the target location on the water surface (sink(s)), as illustrated in Figure 1. In this figure, data can be transmitted through the path with dotted arrows to reach the middle sink or through the path with solid arrows to reach the sink on the right side. One of the very critical issues that face data transmission, particularly with UWSNs, is known as the void area problem as it appears in Figure 1. The data will be stuck in the relay node in the dotted path since the upper hemisphere area of the relay node is empty and there is no other node closer to the sink than this relay node. Numerous researchers have recently become more interested in this issue; however, much more research is required before it can be fully addressed. In this paper, we will follow the same UWSN routing protocol classification, location based and location-free based categories, as presented in previous works [15,25,29,31,37] in order to give a clear understanding of void area characterization. In location-based routing protocols, the detected data is transmitted from the source through the relay nodes with a shorter Euclidean distance to the destination on the sea surface. The upward region of the node's sphere is referred to as the void area if, during this procedure, any node that is holding sensed information could not locate a relay node in its communication range with a shorter distance to the destination to transmit the information to it. In this case, the node is known as a void node. In the location-free based routing protocols, the sensed data is forwarded by sensor nodes using their depths until it is delivered to the sea surface. The sensed data is transmitted from the source through low-depth relay nodes to the surface. In this category of routing protocols, a node holding information is referred to as a void node and the space above it as a void region if it is unable to connect with a node that is in its transmission range but has a lesser depth than itself.

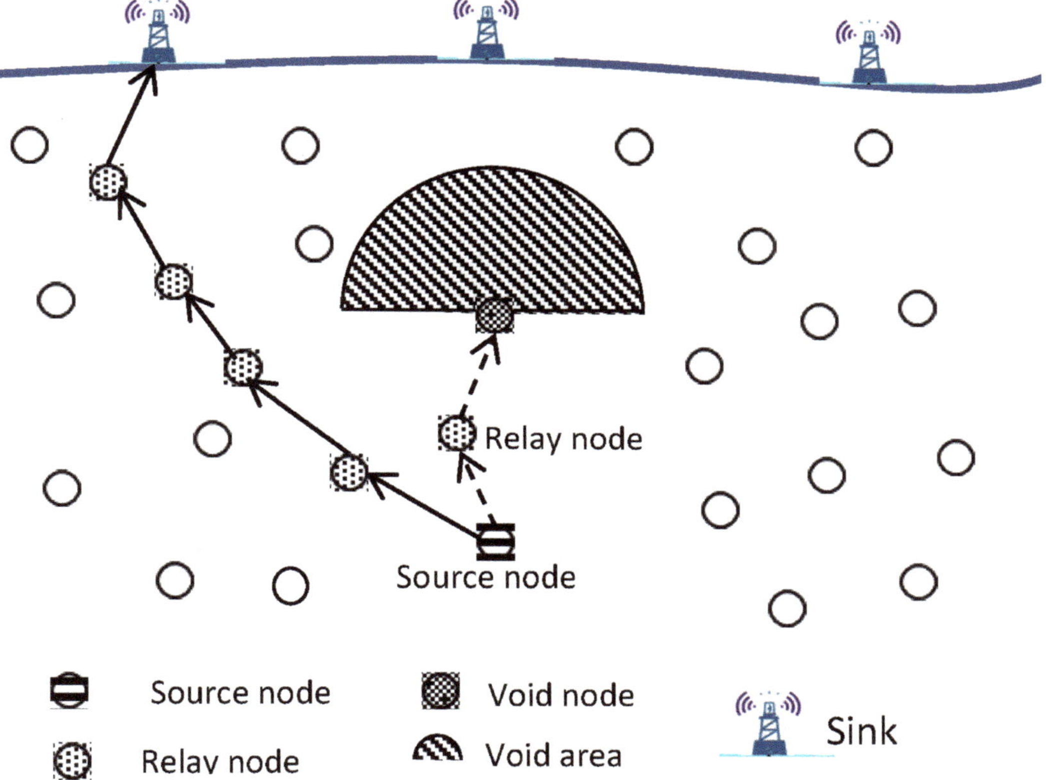

Figure 1. Void area in UWSN architecture.

Therefore, a communication void or void area between underwater nodes exists when the area has an absence of nodes. The void area is one of the essential problems to investigate in the UWSN field. It can prevent communication between two or more network sensor nodes, which in turn can lead to a topological partition that results in decreased network connectivity and increased packet loss, which lowers the performance of the entire network.

Through our research and review of the literature, we have come to conclude that any network architecture can experience the void area phenomenon due to one or more of the following causes [36]:

1. Sparse topology deployment: Because underwater sensor nodes are expensive, it may not be possible to deploy enough of them to cover the area of interest. These sparse networks are prone to empty areas.

2. Underwater sensors failure: Owing to the harshness and peculiarities of the underwater environment, it is more likely that the sensors would malfunction due to corrosion and fouling, which could result in a void area issue.

3. Movement of the underwater sensor nodes: Water currents cause the underwater sensor nodes to move in both horizontal and vertical directions. Both node placements and the network topology will change because of this relocation, and a void area might be created.
4. Temporary obstacles: Many living organisms are found in the underwater environment. The movement of organisms could interfere with the underwater sensor nodes' ability to communicate. In addition, ships, boats, and other water-surface machinery might block the communication link between the network devices. As a result, void areas could be created.
5. The acoustic channel characteristics: The underwater environment characteristics have an impact on the acoustic communication channel, changing the signal's quality and strength at different water depths due to disturbed pressure, temperature, and salinity at varying water levels.

2.3. Opportunistic Routing

At the beginning of this section, we will give the definitions of some terminologies that are used in this paper to make it simple and clear.

Neighbor nodes/neighboring nodes of node (i): are a set of nodes, which are in the transmission range of the node (i).

The qualified set: is a subset of the node's neighbor, which meet the rules adopted by the author(s) to design an efficient routing protocol.

Next-hop forwarder of node (i): the nodes within the transmission range of the node (i) and at the same time have depths less than the depth of node (i).

Beacon messages: Status messages that contain the status information of the nodes, which are used to exchange this information between the neighbors.

P_{holder}: is the node carrying the data packet in the current round. It could be the source node, which originally generated the data pocket, or a relay node.

Relay nodes (i): any node located between source node and sink(s) that carries on forwarding the data packet starting from the source hop by hop until it reaches the sink(s).

OR is a promising technique that was proposed for overcoming acoustic signal fading, high bit errors and losses due to shadow zones, limited bandwidth, high power consumption, and signal spreading [38]. The main concept of OR is to use the broadcast nature of wireless networks, which allows multiple nodes to overhear the transmissions made by any in-range sensor node. Therefore, various underwater OR protocols have been suggested in order to enhance communication in underwater networks.

In OR protocols, a subset of a node's neighbors will be selected as next-hop forwarder set candidates. These nodes collaborate in a coordinated manner to continue forwarding the packet along toward the destination (sink) by using a prioritized technique according to the rules implemented by the protocol [1,14,39]. The forwarding candidate set selection and the coordination manner between these forwarding candidates to deliver the packet are the two main parts of OR construction. This OR approach is preferable to the traditional multi-hop routing approach, in which only a single node is selected to act as a next-hop forwarder, to increase the probability of delivering the packet [1,25,30]. This can be illustrated through the following example:

Let us assume that the delivery probability of each link (which presented by the arrow in Figure 2 is p and ($0 < p \leq 1$).

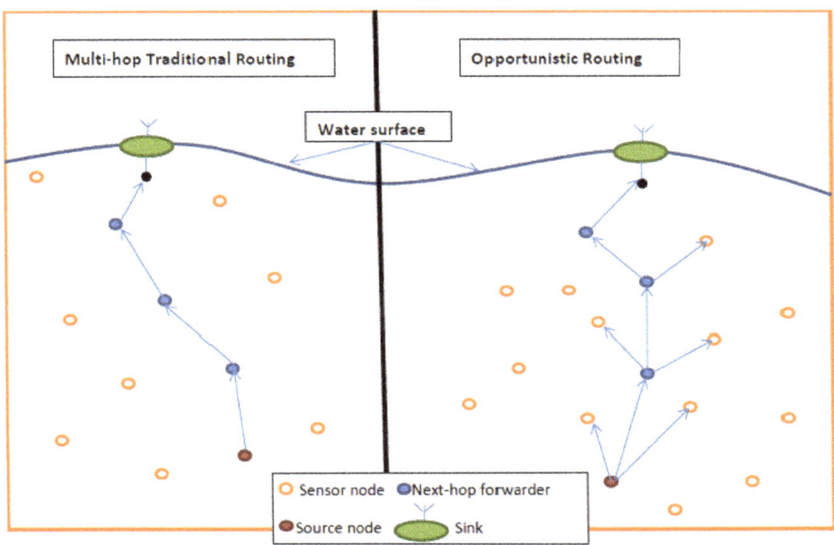

Figure 2. Multi-hop traditional routing vs. OR.

In the traditional multi-hop routing approach, the delivery probability, D_{Prob}, from the source node to the sink using h hops can be presented mathematically as.

$$D_{Prob} = p^h \tag{1}$$

In contrast, if all the relay nodes can transmit the packet by using the OR approach, the probability of delivering the packet to the sink is increased, as explained in [40]. For OR with m possible relay nodes in each hop, as shown in Figure 2, we can express the D_{Prob} mathematically as

$$D_{Prob} = \left(1 - (1 - p)^m\right)^h \tag{2}$$

where h is the number of hops between the node that originally generated the packet and the final sink, and m is the number of relay nodes in each hop.

Consider the following numerical example: assume $p = 0.8$, $m = 3$, and $h = 4$. By using Equation (1), the delivery probability is 0.4096 for the traditional routing, while by using Equation (2), we get a delivery probability of 0.9684 for the OR routing. Figure 2 below illustrates both routing protocols.

Hence, by taking into account the advantage of the broadcast nature of the wireless transmission medium [41] and using the OR forwarding technique, it has become possible to mitigate the effects of the underwater environment and its characteristics on the acoustic communication channel and improve the efficiency of the underwater acoustic physical links [1,25,42]. That is, the OR technique has been proposed to enhance network performance by reducing high bit errors and losses caused by limited bandwidth, high power consumption, and signal spreading [38]. Moreover, using OR reduces packet retransmission; retransmission will only take place when none of the next-hop forwarder set candidates receive that packet. Taking into account OR features, a number of OR protocols for UWSNs have been developed in recent years. These OR protocols utilize multicast mode, in which a single source node transmits its data to multiple nodes by utilizing more than one link at the same time to form the next forwarder candidate set.

2.3.1. OR Construction Blocks

The OR protocol technique is essentially constructed on two important building blocks, as illustrated with their classifications in Figure 3. These building blocks are candidate forwarding set selection and candidate set coordination [1,21,25].

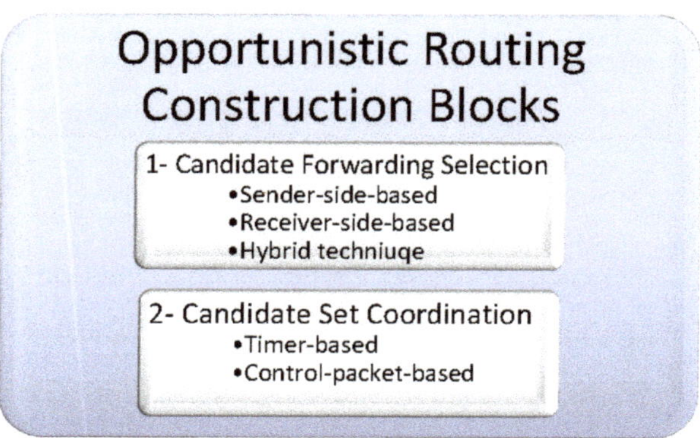

Figure 3. Opportunistic routing building blocks.

Candidate Forwarding Selection

The first building block in OR protocol design is the candidate forwarding set selection process. Selecting a subset of nodes from the source's neighboring nodes to be the qualified set to carry on the packet and continue the forwarding procedure is the responsibility of this process. More generally, based on the next-hop forwarder node-selecting technique, the candidate forwarding set selection procedures can be classified into the three following categories [1,25]:

1. Sender-side-based candidate set selection: in this category, beacon messages between the nodes in the networks are used to exchange the information of the sensor nodes and make it available within their neighborhood. The current forwarder node, which has a data packet to transmit, will use this information to facilitate its mission to determine its next-hop forwarder candidate set.
2. Receiver-side-based candidate set selection: in contrast to the first category, this one requires the neighbors to check the data packet's header when they receive a data packet from the sender in order to identify which received nodes are eligible to be candidate nodes and which ones are not. In this category, each neighboring node is responsible for determining whether it will be included in the list of the potential next-hop forwarder candidate set or not.
3. Hybrid candidate set selection: In this category, the next-hop forwarder candidate set is determined cooperatively by the current forwarder node, which has the data packet to transmit, and its neighbor nodes by exchanging their information.

Candidate Set Coordination

The coordination phase is the second and most significant building block in constructing an OR protocol. In order to continue forwarding the data packet until it reaches its destination, the nodes in the next-hop forwarder candidate set must cooperate in a coordinated manner. According to the protocol's regulations, the node with the highest priority (i.e., the most suitable node) will transmit the packet in this case, deferring transmission to other candidates with lesser priorities. If the node with the higher priority cannot finish its transmission, the node with the next higher priority will begin its transmission, and so on, until the packet reaches its destination.

By functioning in a coordinated manner, this building process supports enhancing the network's throughput and the routing protocol's accuracy by preventing packet duplication.

Packet duplication causes unnecessary and redundant transmissions, wasting the node's energy. Additionally, the overall collision rate can be decreased.

The coordination procedures between the candidate nodes can be divided into the two following categories [1,25]:

1. Timer-based candidate set coordination: Each candidate node in this process has a holding time based on its priority. As a result, the candidate keeps the source's received data packet in their possession for a while (the holding period). The remaining candidates will suppress their transmission if the highest-priority node successfully transmits the packet and if they get an indication during the waiting period. If not, the packet will begin to be forwarded by the node with the next highest priority when its holding time expires, and so on.

2. Control packet-based candidate set coordination: The candidate nodes in this approach communicate with one another by exchanging control packets. Therefore, a candidate node responds to a packet with a brief control message. This control packet transmission is used to notify the currently active forwarder node that the packet has been successfully received. It also notifies the other low priority candidate nodes to pause their transmissions.

2.3.2. OR Classification

The existing OR protocols in UWSNs can be classified based on their positioning information into two main classifications: geography-based and pressure-based. In the first category (geographic-based), selecting the forwarding set candidates and making the forwarding packet decisions in OR requires information about the geographic location of sensor nodes. While in the pressure-based category the depth information of nodes is needed to select the next forwarding set candidates and make the forwarding packets decisions. This classification with the state-of-the-art reviewed protocols can be seen in Figure 4.

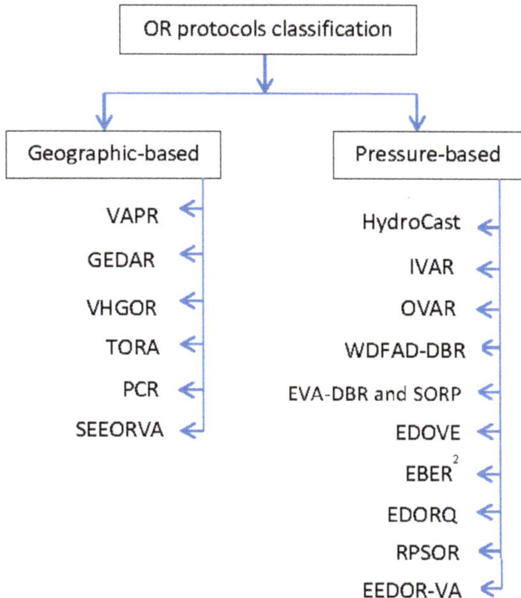

Figure 4. Classification of OR protocols for UWSNs based on position information.

3. Review on Opportunistic Routing Void Avoidance Protocols for UWSNs

Only a few protocols have been proposed to deal with the void communication area problem in UWSNs using the opportunistic routing technique. In this section, we will give a quick review of all the existing protocols.

HydroCast

Authors in [43] presented a hydraulic pressure routing for underwater sensor networks protocol (HydroCast). HydroCast forms a cluster of nodes by using only the local knowledge of the topology, excluding hidden terminals among them, while also maximiing the expected packet advance (EPA) of this cluster. When adopting the time of arrival technique, which is frequently used in UWSNs, the current forwarder node in HydroCast can define the pairwise distances and two-hop connections for the nearby nodes in order to determine its forwarding set. Additionally, the forwarding set candidates are prioritised using a distance-based timer approach. To help organise the transmission and reduce collisions, when nodes in the forwarding set receive a data packet from a recent forwarder node, they set their timers in order, starting with the node with the longest distance.

HydroCast also proposes a *Local Lower-Depth-First Recovery* approach and *2-D Void Floor Surface Flooding for Recovery Path Search* for a recovery mode. Where each void node (i.e., local minimum node as used in the paper) seeks out its neighbors to find a node with a lesser depth than itself, this lesser depth node could be another void node with a new recovery path or a sensor node in a position that helps to resume the greedy forwarding techniques. Figure 5 shows the recovery path in the HydroCast protocol.

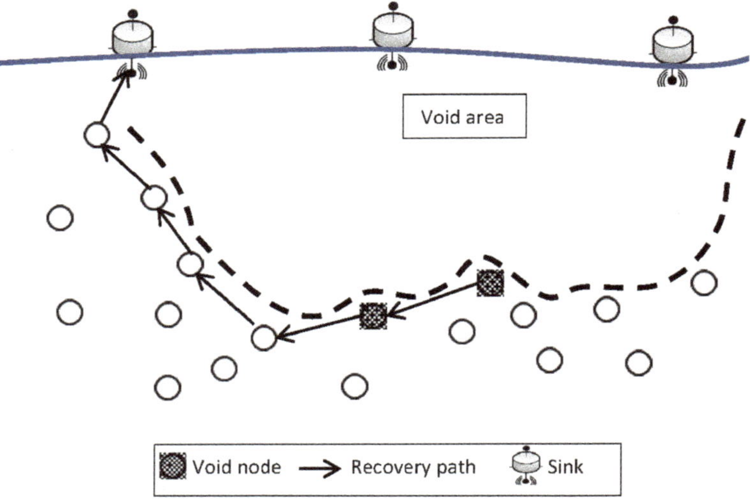

Figure 5. HydroCast void handling technique.

In the 3D network topology, nodes experiencing a void area employ a costly flooding technique to learn which nodes are best suited to resume greedy forwarding or identify alternative routes to better forwarding channels. However, it is difficult to estimate the limited 3D flooding probability value because the flooding could involve every sensor node and affect the entire network topology. They suggest 2D flooding on the surface of the void floor to get around this restriction and increase the effectiveness of the procedure. This flood will include the best possible collection of nodes. As a result, nodes on the surface will monitor their void floor surface status using their local connectivity information and forward packets accordingly, whereas nodes that are not on the surface but are controlled by surface neighbors will not forward packets.

HydroCast addresses the void area issue using an OR approach, which also successfully enables increasing the packet delivery ratio with small end-to-end delays since a subset of the neighboring nodes simultaneously receive the data packet appropriately. However, at the same time, as a result of using opportunistic routing, the HydroCast protocol suffers from redundant packet transmission, where a data packet may be delivered to the sink multiple times, causing the depletion of network resources. In addition, in terms of energy efficiency, implementing the recovery mode results in additional energy costs. Moreover, there is no evidence provided about the energy consumed by the pressure sensor in order to find its depth.

VAPR

Void-Aware Pressure Routing (VAPR) [44] is an anycast soft-state routing protocol. It was proposed in order to address the void node issue in UWSNs. VAPR is built up of two stages: the enhanced beaconing stage and the opportunistic data forwarding stage. Instead of falling into a void area and then implementing a recovery mode, VAPR takes advantage of the geographic routing and employs the regular beaconing messages method, which includes some useful local information about the node, in the forwarding set selection stage.

In VAPR, any node that receives a beaconing message from a neighbor updates its neighboring table and examines its depth with the received depth information. The node then makes its own routing decision by removing void nodes (dead ends or local maxima) from its forwarding sets and chooses the overall best route to the destination as shown in Figure 6; this will help avoid the packet from falling into a void area in the network. In fact, implementing the VAPR protocol will prevent data packets from being stuck in a node because the protocol relies on the surface station and the beaconing message sent from it to the sensor nodes below as well as the stored information in the nodes.

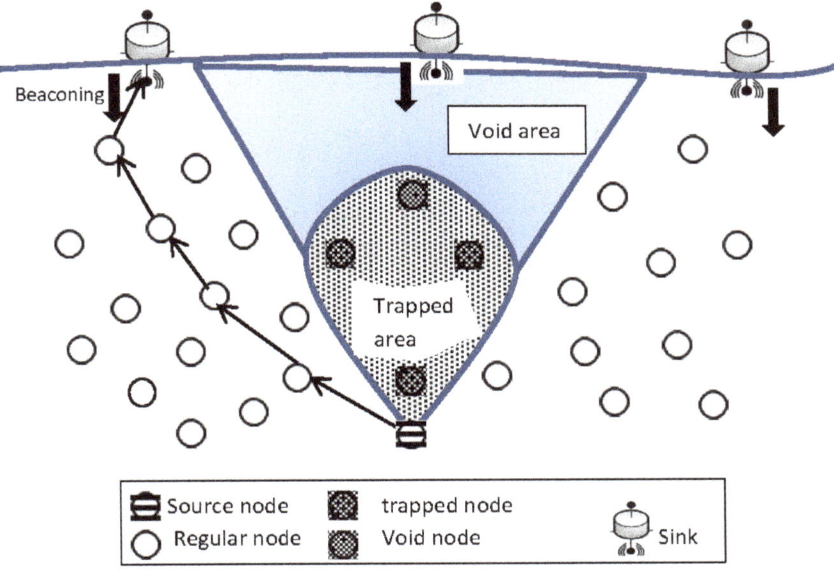

Figure 6. VAPR voids and trapped areas.

GEDAR

In [45], the proposed protocol, geographic and opportunistic routing with depth adjustment-based topology control for communication recovery (GEDAR), utilizes the greedy forwarding technique by knowing the position information of each current forwarding node, its neighbors, and the known sink. GEDAR follows the sender-side OR category, where the forwarding set candidates are determined in each hop by the sender node. Initially, GEDAR uses a greedy, opportunistic forwarding mode to route the packets. Once a node has gathered some data and needs to transmit these data to a sink(s) node, it includes IDs of its forwarding set candidates in the data packet header and broadcasts the packet to its neighbors. When a neighbor node receives the transmitted packet, it checks whether its ID is in the packet header or not. If it is not a forwarder candidate node, it just drops the packet. Otherwise, it calculates the holding time to decide when it can transmit the packet. This procedure will continue until the packet reaches the sink(s) on the water's surface. If the packet is trapped in a void node, the recovery mode is applied by GEDAR. In the recovery mode, when the packet gets stuck in a void node (node v in Figure 7), the protocol deals with the problem by taking advantage of a network topology control strategy where any node in a void area can move in a vertical direction (from D_1 to D_2) to adjust its depth. Then it bypasses the void area to be able to communicate with other nodes trying to resume the greedy forwarding. Therefore, the void node first discontinues sending the gathered packets and starts calculating a new depth that will allow it to continue its OR greedy forwarding to deliver the data packet to the next hop.

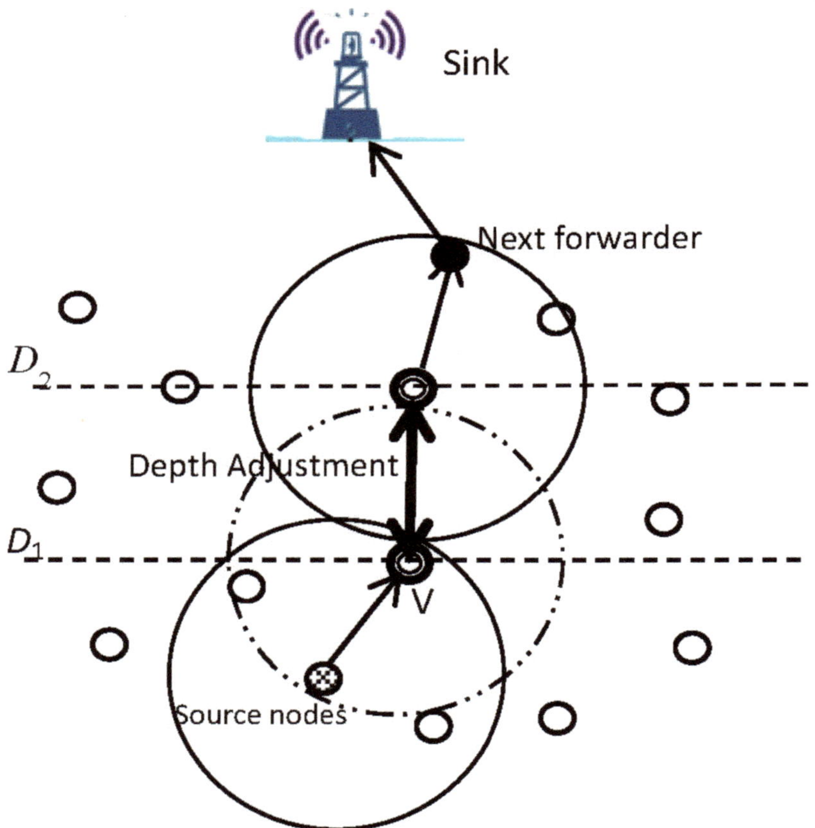

Figure 7. Depth Adjustment.

The recovery technique used by GEDAR helps by bypassing the void area, which as a result, improves the networks connectivity and increases the packet delivery ratio. On the other hand, in energy consumption terms, this Depth Adjustment technique implemented by GEDAR exhausts a very high amount of energy in physical movement to adjust the network topology, and this will make nodes exhaust their energy rapidly and reduce the network lifetime.

IVAR

An Inherently Void Avoidance Routing Protocol for Underwater Sensor Networks (IVAR) [46] is a receiver-based forwarding protocol, so the forwarding node does not need to store its neighbor's information. In IVAR, a hop-by-hop forwarding set selection technique is used to forward the data packets from the sensed node to the sink. Each packet holder uses local information about hop distance and packet advancement to determine its own forwarding set, and the nodes in these forwarding sets are arranged and given a priority depending on two metrics: their hop count as a first metric and their depth as a second one, to forward the packets. IVAR uses beaconing messages sent from the destination to the source; this helps the sensor nodes get the reachable information of the sink(s) and relay nodes. Therefore, the void nodes (yellow and red nodes), as Figure 8 shows, will be excluded from the forwarding set of the sensor node, and the route with a lower hop count will be chosen.

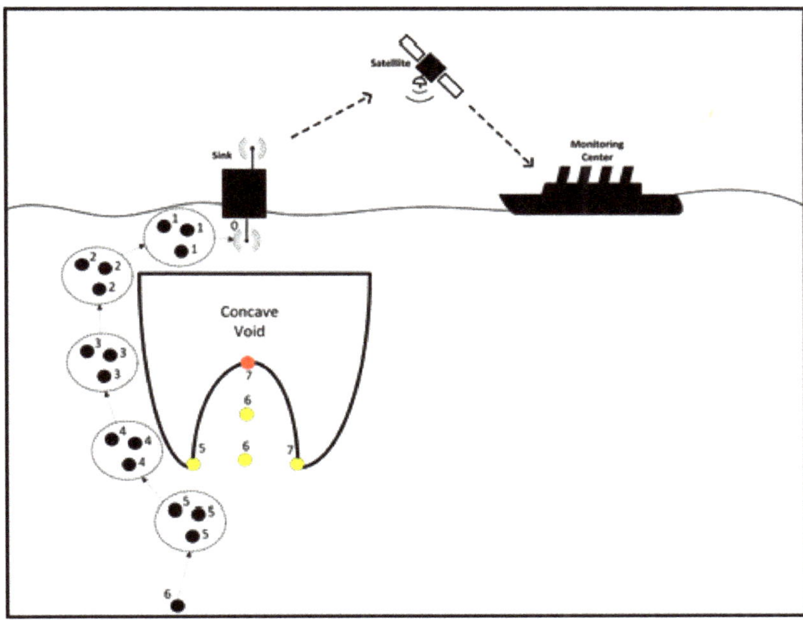

Figure 8. Void-handling technique [46].

Choosing a route with a lower hop count manages the energy consumption and reduces the packet delivery time. Besides, using the node's depth assists in preventing packet duplication. On the other hand, because of the broadcast nature of the protocol and because the qualified forwarding nodes may be distributed around the forwarding node in various directions, the protocol cannot completely suppress route and transmission duplication. This limitation will cause the hidden terminal problem and, consequently, extra energy consumption. IVAR uses a periodic beacon by the sink to update the underwater nodes with their current position in the network. Therefore, all the routes from the sink to the sensor nodes will be established in advance, and all the routes directing the packets to void areas will be excluded. However, the beacon interval has to be chosen cleverly because it has a great effect on node information and communication efficiency, which consequently will impact network performance.

OVAR

The opportunistic void avoidance routing (OVAR) protocol [47] is a sender-side method and a soft-state routing protocol, that requires some local reachability information (e.g., hop count distance, forwarding direction, etc.) about one-hop neighbors to be held in every node. This provides a general observation of each node on the topology. OVAR was proposed to handle IVAR weaknesses (i.e., hidden terminal problems and duplicated packets of transmission). In the same way as IVAR, to handle the problem of void areas, OVAR implements the beaconing procedure and considers its benefits. Different from the receiver side IVAR protocol, in the sender side OVAR protocol, the one-hop neighboring information is held in the sensor node to establish an adjacency graph at each forwarding node. In terms of energy consumption management, OVAR deals with the number of nodes in the forwarding sets, where the size of the forwarding set can be adjusted based on the network density to save energy by reducing the energy consumed by a node when it receives a packet. Reducing the forwarding set size may reduce the delivery ratio and increase packet retransmission, which will lead to more energy consumption. In terms of void area, OVAR includes the high-depth nodes in the forwarding set, which may be inefficient in terms of reliability, energy consumption, and protocol latency. OVAR is slightly more complicated than IVAR, which is caused by the procedure OVAR implemented to eliminate the hidden nodes problem and its effects on the protocol's performance, in addition to the trade-off procedure between energy consumption and reliability.

VHGOR

Void handling using geo-opportunistic routing in underwater wireless sensor networks (VHGOR) [48] adopts geography-based opportunistic routing (GOR) to forward data packets to reach the destination over multi-hops. It is a heuristic protocol implemented using two metrics to form optimal forwarder selection. OREPP metrics try to positively advance the data packets towards their destination. The first metric is opportunistic routing based expected packet progress (OREPP), which is calculated based on the difference between the geographic distance between the source and destination and the geographic distance between any node and the destination, residual energy, and packet delivery probability. The second metric is node closer to the destination (NCD); NCD can be defined as the best node with maximum OREPP to forward the current packet. VHGOR uses a greedy forwarding approach to advance the packet through each hop towards the destination, and if the packet becomes stuck in one of the forwarding nodes, then it switches to the void mode. VHGOR handles the void problem using the two following techniques:

1. Convex void handling: If the packet is stuck at the NCD node, VHGOR attempts to identify a different path to forward the same packet to the destination by removing the present forwarder and re-establishing the convex structure with remaining neighbors founded in the neighbor table (NT).
2. Concave void handling or recovery mode: The void becomes concave when a packet gets trapped in a node without any neighbors with lower pressure levels, which means that its NT entry is empty. In order to reroute the packet along a different path to the destination, VHGOR manages the concave void by rerouting it down the recovery path, which runs from downward to upward. The previous sender chooses the subsequent NCD node from its NT to continue sending the same packet after receiving the packet from the concave empty node.

Figure 9 demonstrates the forwarding packet route and recovery mode that VHGOR has adopted. Node n_1 chooses node n_2 to be the next forwarding node, since node n_2 has the highest Expected Packet Progress (EPP) value in its neighbor table (direction number 1 in the figure). In the same manner, node n_2 chooses node n_3 as the next forwarding node and transmits the packet to it (direction number 2 in the figure). However, since node n_3 is a void node and has no nodes to forward to, node n_3 returns the message back to node n_2 (direction number 3 in the figure). The next node in Node n_2's neighbor table is subsequently chosen as the next forwarder (direction number 4 in the figure). Since D is inside the transmission range of node n_{10}, node n_{10} finally delivers the packet to D. In order to create the best forwarder from FCS, VHGOR takes into account residual energy, which helps cut down on energy consumption.

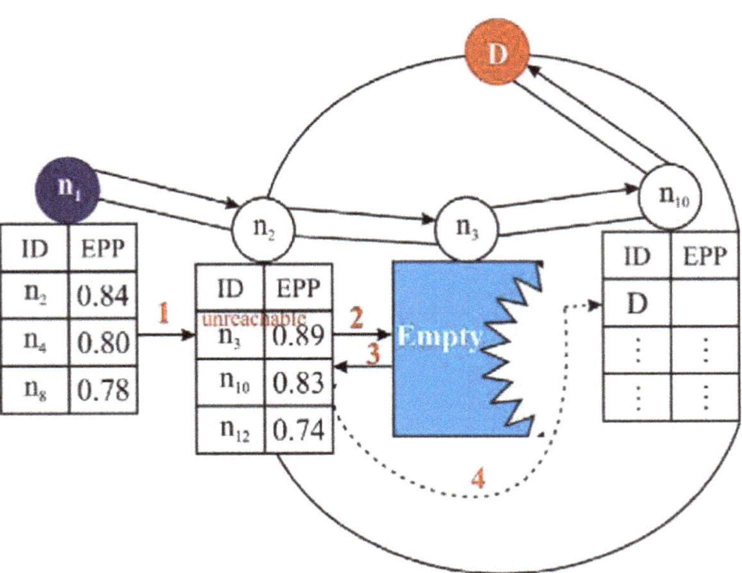

Figure 9. VHGOR recovery mode [48].

VHGOR considers the residual energy in forming the optimal forwarder from the forwarding candidate set (FCS), which assists in reducing the energy consumption. Besides, employing opportunistic forwarding works in improving the delivery ratio at the same time introduces end-to-end delay to some extent [48].

WDFAD-DBR

In [49], another pressure-based routing protocol was described in detail, namely the weighting depth and forwarding area division DBR routing protocol (WDFAD-DBR). To increase the reliability of the packet transmission and decrease the probability of the occurrence of a void area, WDFAD-DBR uses the weighting depth difference of two-hop nodes to construct its routing decision. As presented in Figure 10, node S is a source node, and the two forwarding candidate nodes with lesser depth are A and B. In the greedy protocol DBR, node A has a lesser depth than node B, giving A the priority to transmit first. Node B will suppress its transmission and drop the packet when it hears it from node A. However, a void area occurs since there are no nodes in node A's transmission area (S2) with less depth than node A to carry forward the packet. In contrast, WDFAD-DBR selects node B to forward the packet because it considers both depth differences, current depth difference (node B depth—source depth), and the difference depth of the expected next hop (node E depth—node B depth).

In WDFAD-DBR, the void nodes can remove themselves from the data packet routing to increase the opportunity for the other candidates in the forwarding set to forward the packet. In addition, to control the number of forwarding nodes, WDFAD-DBR divides the forwarding area into a constant primary forwarding area (the Reuleaux triangle) and two auxiliary forwarding areas, which might be extended or shrunk depending on node density and the quality of the channel. In terms of energy consumption, on one one hand, to help reduce the energy expenditure due to the duplicated packet transmission, the auxiliary forwarding area is divided into a number of smaller sub-areas, which helps save some energy. On the other hand, the periodic neighbor requests and the corresponding ACKs in a reply to each control packet exhaust the energy of the nodes and waste network resources. In order to bypass the void area, WDFAD-DBR successfully detects the void nodes and excludes them from the forwarding procedure. However, the protocol fails to detect the trapped nodes in advance. Moreover, when a fixed primary forwarding area is implemented by the protocol, the flexibility of the routing might be restricted in its ability to choose and adjust the forwarding nodes under various conditions.

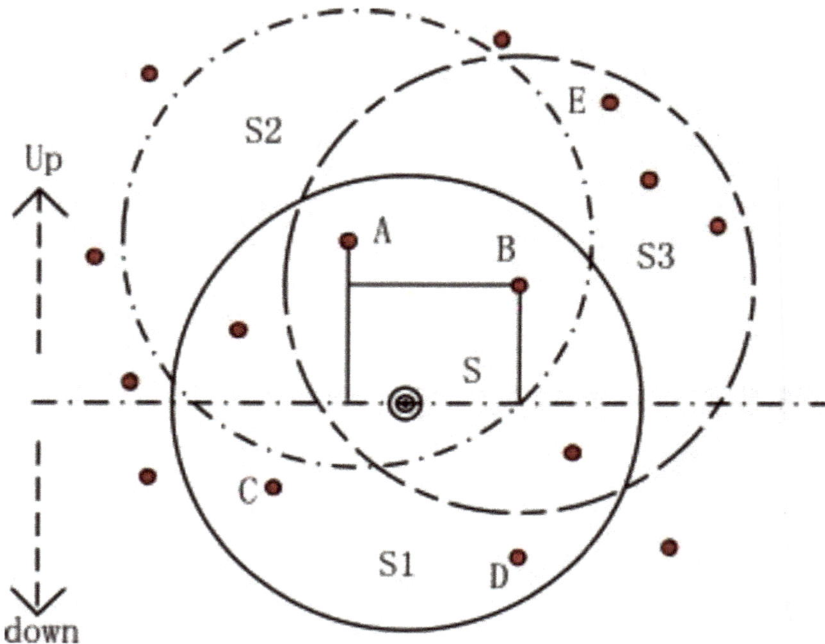

Figure 10. Void area problem [49].

EVA-DBR and SORP

In [50,51], the energy-efficient and void avoidance depth-based routing (EVA-DBR) protocol and A Stateless Opportunistic Routing Protocol for Underwater Sensor Networks (SORP) are proposed. SORP builds on the performance evaluation from [50], considering a realistic sensor mobility model, the shadow zone, variable propagation delays, and additional network parameters and results. EVA-BDR and SORP are routing protocols consisting of two phases: the updating phase and the routing phase. The protocols depend on the information broadcasted periodically in the updating phase from the neighbor nodes that are one-hop away from the source node for void detection and bypassing in the routing phase. Initially, all the nodes in the network are homogeneous. However, in the updating phase, the void and trapped nodes are detected over time by the broadcasted information from the neighboring nodes. In addition, through the updating phase, each regular node will choose its best candidate node in terms of the expected packet advancement (EPA) among the neighboring nodes with lesser depth to be used as a reference node in the opportunistic data forwarding [50,51]. In the routing phase, to increase the packet delivery probability in each data transmission operation, all the detected void and trapped nodes take themselves out of the forwarding set; this procedure will increase the opportunity for the other regular nodes in the forwarding set to forward the packet. In addition, the forwarding area can be resized depending on the density of the network, as presented in Figure 11, and all the qualified nodes will set their forwarding timer to forward the data packet. This forwarding time should guarantee a priority-based scheduling of the nodes in the forwarding set and should suppress the duplicate packets.

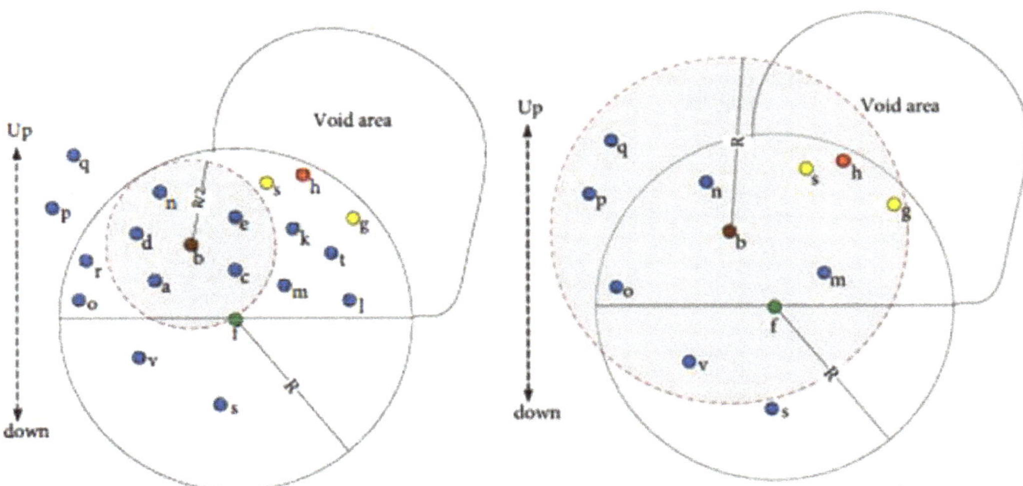

Figure 11. Resizing the forwarding area, sparse density on the right and dense density on the left [50].

In terms of energy consumption, since the nodes do not need to send an ACK to the node's neighbors as a reply to their control packets, the energy consumed per node will be somewhat reduced. In contrast, both protocols may allow the duplicated transmissions to increase the packet delivery probability in a sparse network in addition to periodic broadcasted information exhausting the node's battery and, as a result, decreasing the network lifetime as well as the node's life. And in terms of void avoidance, the state of excluded nodes from the forwarding set that announced themselves as void or trapped nodes may change during the transmission data packet or before the period of broadcasting information expires, which may effect the energy consumption and reliability of the network. Moreover, maintaining the neighboring table and the two-hop information will adversely affect the limited resources of the nodes (i.e., energy and memory).

EDOVE

This section reviews the energy and depth variance-based opportunistic void avoidance (EDOVE) protocol that was presented in [52]. EDOVE was proposed on the basis of the work presented in [24], called the WDFAD-DBR protocol. The protocol addresses the void area problem by selecting the forwarder candidates among the total distributed nodes that have i) a large residual energy and ii) several neighboring nodes within its transmission range (neighbors). Each node in the network architecture shares its information with its 1-hop neighbors using neighbor request and neighbor acknowledgment packets, and each node must keep its neighbor table updated in order to obtain this relevant node information when needed. Once a sender has a data packet to deliver, all of its neighbors will inevitably get it due to the broadcast nature of the protocol. From then, the packet must be transferred through one of these neighbors to the next hop or directly to the destination (sink(s)). In contrast to WDFAD-DBR, EDOVE uses the two-hop depth differences, the normalised residual energy of the node, the next hop depth difference to the source, and the depth difference variance between the neighbors to compute the holding time. This is because the receiving nodes have different residual energies, and EDOVE takes this diversity into account. The holding time parameters are shown in Figure 12.

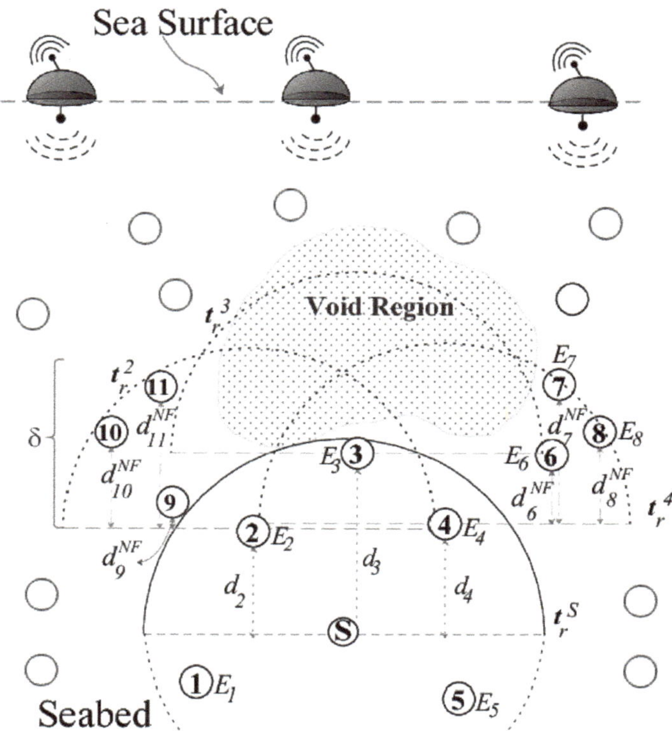

Figure 12. Holding time calculation parameters [52].

Finally, to choose the best forwarder node, EDOVE makes the decision by calculating the holding time and selecting the receiving node with the largest residual energy, the greatest depth difference to the source, the greatest depth difference to its neighbor, and many neighbors with a large variance in their depth differences. More factors are taken into account by the protocol, which increases energy efficiency, prevents packet collisions, and extends network lifetime. However, in dense networks or when the size of the network is increased, there are increases in the probability of duplicated packet transmission because the number of nodes with the same depth will increase, making their estimated holding times almost the same. This results in an increase in data packet traffic, which in turn increases energy consumption. Additionally, the protocol views the void area only as a series of energy holes, despite the fact that it serves a variety of purposes, as stated above.

TORA

The totally opportunistic routing algorithm (TORA) is proposed for UWSNs in [53]. TORA is a novel anycast, receiver-based opportunistic, and geographical routing protocol. It is suggested in order to prevent horizontal transmission, minimize end-to-end delay, address the issue of void nodes, and increase network performance and energy efficiency. The three steps of the proposed protocol's operation are node localization, candidate forwarder selection, and data transmission.

At the water surface, the multi-sink network architecture is installed, and ordinary nodes drift in different levels underwater, as shown in Figure 13. The ordinary nodes are divided into two types: 1) single transmission node (STN) that are in transmission range of surface sinks; and 2) double transmission node (DTN) that are not within transmission range of surface sinks, they estimate their position by communicating with STNs.

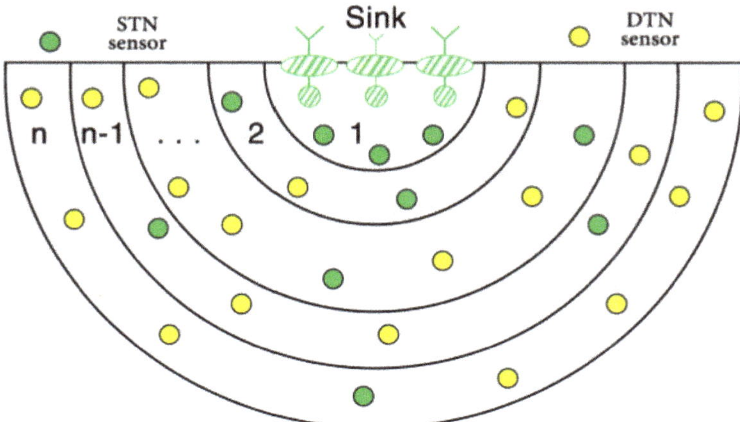

Figure 13. Layering structure in TORA at the node localization phase [53].

To locate nodes in the network, the time of arrival (TOA) and range are used. Sinks periodically send hello messages to help collect node information that will be used to determine ordinary node location in the localization phase. Next, based on the nodes' geographic coordinates and remaining energy, the best forwarding node that has a higher residual energy and is closer to one of the sinks will get a higher priority to relay the packet in the candidate forwarder selection phase. After that, the data transmission phase starts once a node has a data packet ready to send. This data packet should be delivered to one of the sinks in a multi-hop fashion through selected forwarding relay nodes. TORA utilizes 2-hop Ack to make sure that the packet has traveled for two hops and zero Acks to reduce end-to-end delay and retransmissions. As a conclusion, data is transmitted to the sink node using a combination of several short, active links.

EBER^2

In [54], an energy-efficient and reliable protocol called an energy balanced efficient and reliable routing protocol (EBER^2) has been proposed to address the void areas. EBER^2 adopts the potential forwarding nodes (PFN) concept to tackle WDFAD-DBR shortcomings. Since WDFAD-DBR experiences void area problems in some cases because it ignores taking into account the PFNs for the second hop, it suffers from high duplicate packets and collisions, which reduce protocol performance and efficiency. In EBER^2, the network architecture consists of three types of sensor nodes (sink nodes, anchored nodes, and relay nodes), as demonstrated in Figure 14.

The authors of EBER^2 take into account three factors as primary parameters for choosing the next forwarder in order to address the WDFAD-DBR weaknesses. The first parameter is the weighting depth difference of two hops; by choosing the next forwarder node based on the depths of the first two hops, the likelihood of a network void area problem is reduced. The second factor is the number of PFNs, which are nodes that are within the source node's upper hemisphere of its transmission range. A void node is one that has no PFNs; as a result, it is excluded from the upcoming forwarding set, which improves network stability. The residual energy is the third parameter, and it is used to provide PFNs with the same depth but varying holding durations in order to prevent duplicate packets. In addition to forming the next forwarder set and preventing void nodes from being chosen as candidates for the next forwarder, these three parameters also support energy efficiency, boost packet delivery ratio, and lengthen network lifetime by preventing duplicate packets and the ensuing collisions.

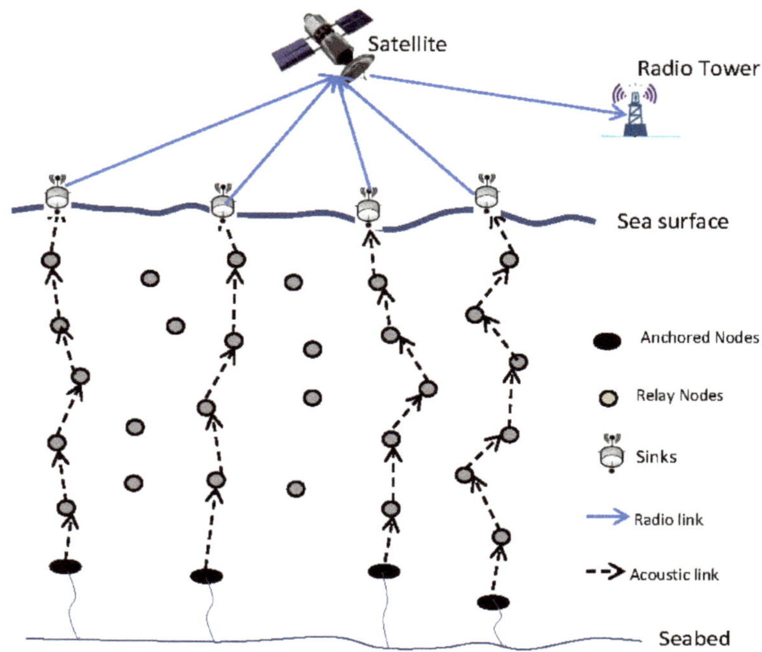

Figure 14. EBER^2 Network Topology.

To further assist these nodes in communicating with the embedded sinks and delivering data packets to them rather than travelling through a long path to reach the sinks on the surface, the EBER^2 protocol deploys two additional embedded sinks in the underwater area of interest that have high traffic density, as can be seen from Figure 15. In general, since the nodes placed in these dense and high traffic areas transfer the received packet to the closest embedded sink rather than transmitting further to the surface, this strategy enhances network packet delivery ratio while consuming less energy. Instead, the cost of communication rises because of the high-speed optical fiber links used to connect embedded sinks and on-surface sinks. Additionally, EBER^2 employs a transmission energy adaptation mechanism that enables nodes that are closer to sinks to reduce their transmission power level in accordance with their distance from that sink. This minimizes the void area created by the death of these nodes by preventing the nodes close to the sinks from rapidly exhausting their energy due to being involved in the majority of forwarding procedures.

EDORQ

The authors of [55] proposed a new receiver side-based routing protocol for UWSNs named Energy-efficient Depth-based Opportunistic Routing with Q-Learning (EDORQ). The EDORQ contains two phases: (1) the candidate set selection phase to choose a subset of neighbor nodes to carry on forwarding data packets until delivered to the destination; (2) the candidate set coordination phase, where the candidate nodes collaborate according to their priorities by applying the timer-based mechanism to suppress redundant forwarding. Moreover, the authors adopted the Q-learning technique to design the holding time of the candidate nodes. By defining a holding time for each candidate, the candidate node with the larger Q-value has a higher priority, a lower holding time, and will transmit the packet first.

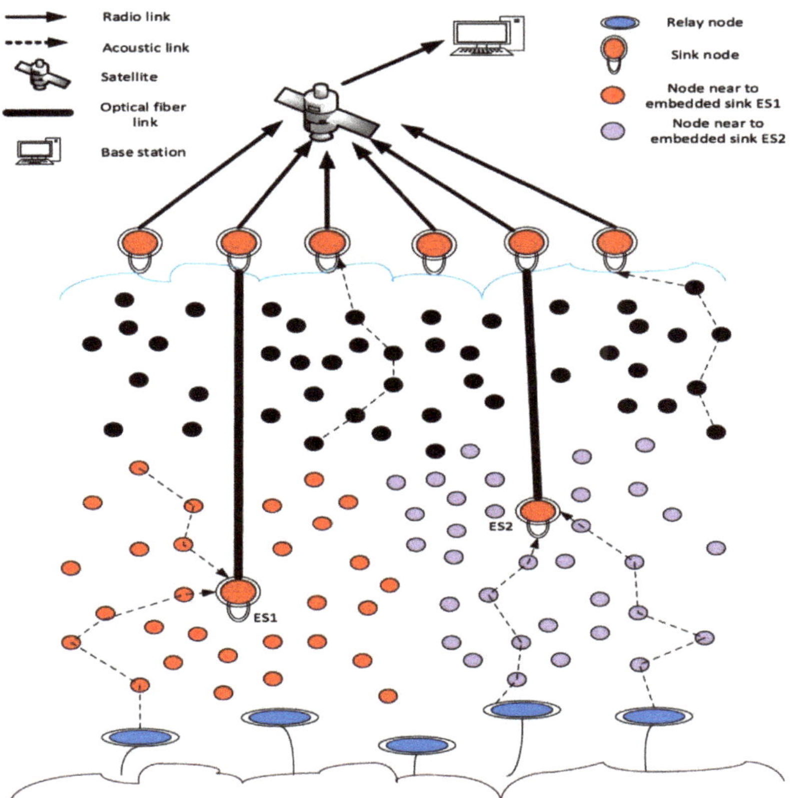

Figure 15. Network Topology with Two Embedded Sinks [54].

EDORQ starts the forwarding process using greedy mode, where the current forwarding node broadcasts the data packet to its neighbors. Each candidate neighbor extracted the depth (d) and void-flag information of the current node from the packet header after receiving the data packet and then compared d with its own depth. In order to ensure that the data packets are quickly sent in the sink's direction, the greedy mode helps locate a collection of candidate nodes closer to the water's surface. In order to achieve this, the void-flag field in the packet header is set to "0," indicating that only nodes whose depth is less than the current forwarder are eligible to be chosen as candidates. However, the protocol switches to void recovery mode when the packet is stuck in the void node as, illustrated in Figure 16.

The current node will retransmit the data packet in a void recovery mode, where the value "1" is entered in the void-flag field, allowing the neighbor nodes with the greatest depth to be chosen as candidate nodes. A node should only forward packets with the same ID once for a predetermined period of time in order to reduce duplicate transmissions, comparable to the DBR. As a result, the new candidate set of current nodes would continue the forwarding process. The next best packet forwarder from the current node will then transmit packets in a greedy manner toward the sink if no other the void node is reached.

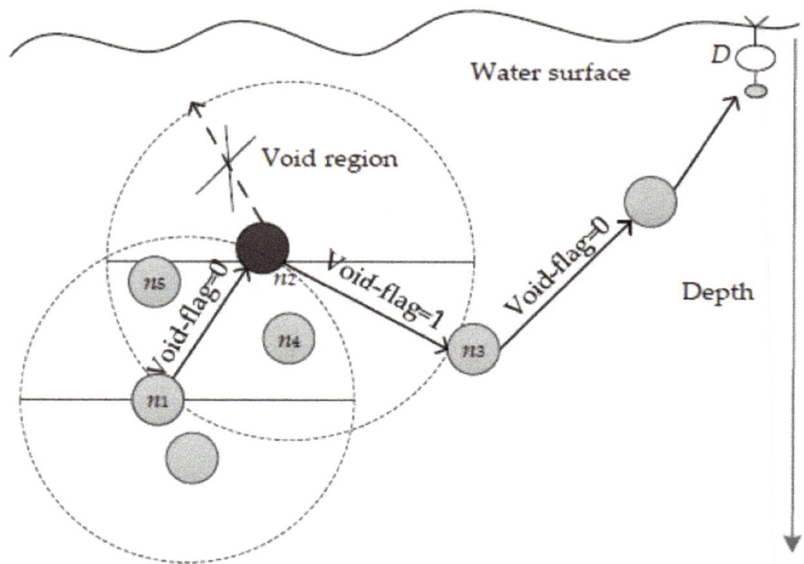

Figure 16. EDORQ void recovery mode [55].

RPSOR

In [56], another novel OR protocol called Reliable Path Selection and Opportunistic Routing (RPSOR) for UWSNs is presented to address the void area problem in UWSNs. It is an improved version of the WDFAD-DBR protocol. RPSOR operates in two stages: knowledge acquisition and packet forwarding. In the knowledge acquisition stage, nodes exchange their information through hello packets sent from surface sinks, neighbor request packets, and ACK packets generated by each sensor node. In addition, the node maintains three different tables, which are the source info table, the first hop info table, and the routing table. Furthermore, in the packet forwarding stage, the decision for PFN selection will be made based on the priority function, which is defined by three elements: the reliability index, the advancement factor, and the shortest path index. RPSOR only selects the nodes of the upper hemisphere as the forwarding neighbors. Therefore, nodes having higher a depth than the current node simply drop the packet.

In RPSOR, two sinks are mobile, as can be seen in Figure 17. Mobile sinks are utilized to travel to denser network areas that experience high traffic.

At the beginning of each simulation round, the network uses hello messages to assess the node density at various hops, and it then permits the sink to travel to any hops with a high node density. The nodes located at the following hop must transmit a large amount of load created by denser network locations. The majority of the packets are lost when this high traffic enters the network's sparse area since the network cannot handle such high traffic levels. Utilizing the position data of the denser hop, which was acquired by the greeting message, the sink determines the vertical trajectory.

PCR

Recently, a novel power control-based opportunistic (PCR) routing protocol for the Internet of Underwater Things (IoUTs) was proposed in [57]. They develop opportunistic routing and transmission power control methods in order to send data in IoUTs with the least amount of energy possible. Each node in PCR checks many transmission power levels before selecting its candidate set for the next-hop. The PCR protocol implements a periodic beaconing technique during the neighbor discovery phase for each transmission power level in order to gather information from neighbors and update the neighbors table. The candidate set will be expanded to include the neighbor node exhibiting positive packet progress. The appropriate transmission power level and the next-hop forwarding set are then computed based on the energy waste for each candidate set. Hence, the set of candidate nodes with the least energy waste is chosen as the best candidate set to continue forwarding the packet to the next hop until the packet reaches the destination. The nodes in the candidate set will then be sorted based on their normalized packet advancement to define the node's priority. Then, PCR applies a timer-based approach to manage the transmission coordination between the candidate nodes. Therefore, the candidate node's packet holding time decreases as its priority increases. Additionally, if a lower priority candidate node detects packet transmission from a higher priority candidate node, it will cancel its own transmission.

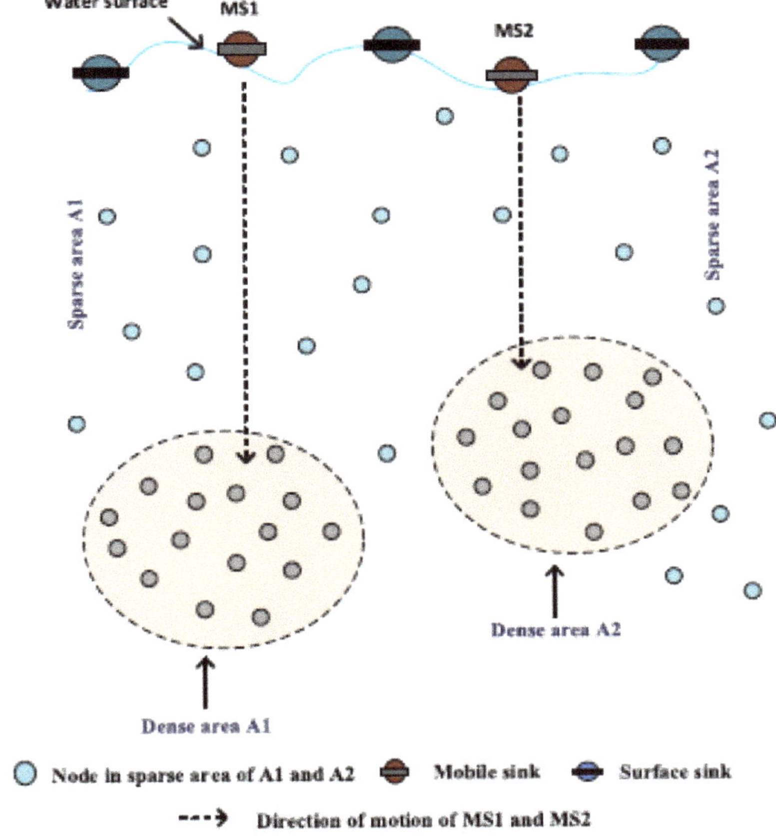

Figure 17. Sink mobility model [56].

By changing the transmission power level at each hop, the PCR packet delivery ratio was enhanced in order to select the most suitable candidate nodes from the sender neighbors to continue passing data packets to the sink(s) on the water's surface. In dense networks, PCR also lowers the node's transmission power level to lessen the need for retransmissions, which lowers energy usage in some cases. The energy consumption is still higher than the related works, as we can see from their data, and this will shorten the lifespan of the network.

SEEORVA

A secure and energy-efficient opportunistic routing protocol with void avoidance for underwater acoustic sensor networks, (SEEORVA) was presented in [58]. This protocol employs the OR strategy for reliable data delivery in UWSNs and uses energy thresholds in the forwarding process to give a priority to the forwarding nodes, which have energy above that particular threshold; in that way, energy efficiency and expanding network lifetime can be achieved. The protocol handles the communication void problem and encrypts transmitted packets using a secure, lightweight encryption technique for security.

In SEEORVA, the best forwarder selection was performed as follows: when a source node has packets to transmit, it creates a virtual vector pipe to the sink (as can be seen in Figure 18). The source then lists all the nodes that are detected within this pipe to be considered, calculating the highest energy of the nodes and the threshold energy value based on the calculated highest energy value.

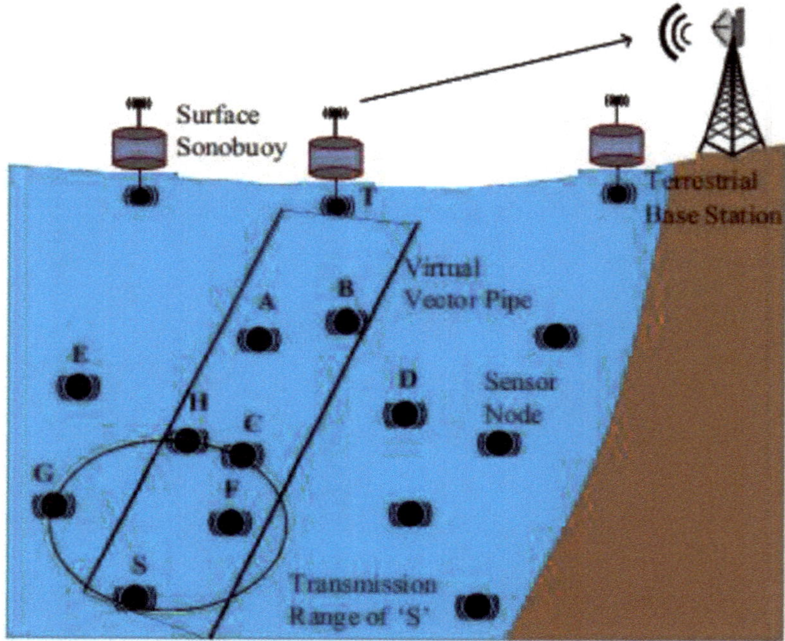

Figure 18. The best forwarder selection process in the SEEORVA protocol [58].

In the forwarding process, only candidate nodes within the source transmission range that have residual energy greater than the threshold and are making maximum progress to the sink will be given the highest priority and chosen as the best forwarder node. If this node could not forward the packet within the allocated transmission time, the next node in the list would forward the packet to the sink. Therefore, to ensure the security of transmitted data, SEEORVA uses a lightweight security protocol, the novel tiny symmetric encryption algorithm, to encrypt data packets before sending them through the network to the sink. These encrypted data packets can only be decrypted by the collection and processing centers at the water's surface.

Moreover, the proposed protocol addresses the void problem by encouraging the forwarder node to send a data packet void alert to its previous node if the forwarder node faces a communication void. The previous node searches for an alternative route to avoid the void and uses this alternate route to transmit the remaining data packets from the previous node to the sink.

The nodes' remaining energy is used as a significant factor to determine the priority of the next forwarder nodes in the forwarding process, thus extending the lifetime of each sensor node and the network overall. While the technique used to handle communication voids gives much better quality of service (QoS) results, it is also easy to implement with less overhead and delay. In addition, the encryption method ensures secure data packet transmission and avoids any leakage in data packets that can be harmful in any way.

EEDOR-VA

In [59], the most recently published routing protocol to address the void area issue is named energy efficient depth-based opportunistic routing with void avoidance protocol for UWSNs (EEDOR-VA). EEDOR-VA aims to improve network performance by developing a routing protocol that achieves a high packet delivery ratio while using less energy by choosing the shortest routing path. EEDOR-VA decides on routing based on the nodes' ability to reach the surface sink. This protocol introduces Hop Count Request (HCREQ) and Hop Count Reply (HCREP) messages to update the node's hop count to the nearest sink that can be approached. In the proposed protocol, data packets will not get stuck in any void and trapped nodes located in the transmission range of a source and/or relay node because these void and trapped nodes do not respond to the HCREQ message and are therefore removed from being one of the forwarding candidates. As a result, each P_{holder} can easily construct its forwarding set. That is, sensor nodes use the information from the hop-count discovery algorithm to update their hop count from the sink(s) and exclude void and trapped nodes in the P_{holder} nodes' transmission range from being included in the forwarding set. Periodic beaconing and its related costs is eliminated by the hop count discovery mechanism proposed in EEDOR-VA. The main goal of EEDOR-VA is to find as many loop-free paths as possible between a source node and a single or multiple sinks on the sea surface. The protocol can easily change the chosen route from one path to another by electing the next relay nodes from a different path if this relay node is the best choice in the next hop forwarding range. As a result, this technique prevents having to start the hop-count discovery process all over again. If all routes to all of the sinks fail, then a hop-count discovery is initiated. EEDOR-VA updates relay node information using route information and ensures that nodes responding to the P_{holder} have a path to one of the sink(s) in order to avoid the void nodes. The EEDOR-VA protocol's process is depicted in Figure 19. When a source node has a packet to send, it sends HCREQ first, which is received by all of its neighbors. Each of these neighbors sends out a rebroadcast of the appeal to their own neighbors.

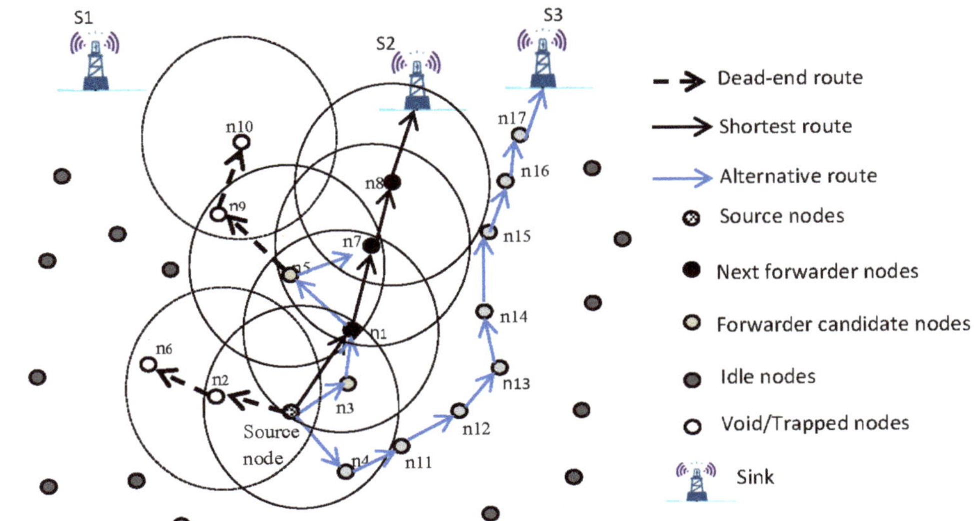

Figure 19. Underwater Network Architecture Model of the EEDOR-VA Protocol.

The EEDOR-VA protocol uses rounds; every round is comprised of three phases: a hop-count discovery phase, a forwarding set creation phase, and a data packet forwarding phase.

The hop-count discovery phase is in charge of determining the hop count of any source and/or relay nodes in the network to the sinks, whether the sink(s) are directly reachable within the transmission range of the source or reachable via one or more hops through relay nodes. Once the hop count of each of the route nodes is defined, the forwarding set formation phase is started. In each hop, the P_{holder} forms its next-hop forwarder set based on the extracted candidate information, and only a candidate with a hop count less than the P_{holder} hop count, no matter if it has less or more depth than the P_{holder}, will be added to that P_{holder} next-hop forwarder set. Finally, P_{holder} integrates the data packet with the sorted list of the selected forwarding candidate IDs and transmits it to its neighbors. Each neighbor checks the packet header and simply drops the packet if it cannot find its ID in the attached list or starts computing its holding time otherwise.

The EEDOR-VA protocol uses the node's hop count as the first metric to identify the best forwarding node and the node's depth as a secondary metric in the event of a tie. The best forwarding node will transmit the data packet immediately after receiving it to continue the forwarding process. Other forwarding candidates will drop the packet if they successfully hear the transmission from the most appropriate node. If not, the data packet will be transmitted by the following node in the sorted list, and so forth. These processes will be repeated hop by hop until the data packet reaches the sink or all the candidate nodes in the forwarding set fail to do so.

4. Comparison Study of OR Protocols for UWSNs

In the previous section, the literature review of the state-of-the-art of the OR protocols that are proposed for UWSNs to address the void area problem is presented. The main challenge in the protocols was to handle the void area problem by using different approaches. The occurrence of the void area in the routing path can significantly reduce network performance. In this section, the general comparison of these reviewed protocols based on their characteristics and features is summarized below in Table 1.

In Table 1, we can see that the existing void avoiding OR protocols for UWSNs are classified into two main classifications: geographic-based and pressure-based. In the first category, geographic-based, which includes [44,45,48,57,58], selecting the forwarding set candidates and making the forwarding packet decisions in OR requires information about the geographic position of sensor nodes. While in the pressure-based category, which includes [43,46,47,49–52,54,59], the depth information of nodes is needed to select forwarding set candidates and make forwarding packet decisions.

The reviewed protocols are divided into sender-side and receiver-side categories based on which node will decide if the candidate node can be added to the next hop forwarder set or not. A higher communication overhead is needed on the sender-side because the sensors frequently need to exchange node information in order to update their neighbors' tables. As a result, the limited resources of the node (i.e., battery and memory) are used up. On the receiver-side, the sender is unaware of its neighbors and is unaware of its forwarding set. This may result in a significant number of redundant broadcasts and raise the possibility of transmission collisions requiring reiterate transmissions. The entire network stability period may be shortened as a result of packet loss and sensor node energy consumption. A number of these protocols take advantage of the multi-sink architecture and consider the data packet as delivered if it reaches one of the deployed sinks on the water surface. This improves the network reliability.

Table 1. Comparison between void avoiding OR protocols.

Protocol	Category	Forwarding Set Selection Category	Sink(s)	Requirements	Knowledge Required/Maintained	Advantage	Disadvantage
HydroCast [43]	Pressure-based routing	Sender-side	Multi-sink	Nodes with special H/W	2-hop connectivity and the pairwise distances for the neighboring nodes	• Reduce end-to-end delay. • High delivery ratio. • Void handling technique by using recovery path.	• High energy consumption due to repeating the process of finding a detour path. • High overhead due to requiring 2-hop neighboring nodes information.
VAPR [44]	Geography-based routing	Sender-side	Multi-sink	SEA Swarm nodes	next-hop direction and hop distance-information at each node	• Reduce end-to-end delay. • Void handling technique by using directional opportunistic data forwarding algorithm. • Use multi-sink reduces the sensor node's battery drain and high traffic.	• High energy consumption because utilizing enhanced beaconing and measuring the distance to the neighboring nodes and broadcasting of the measured information.
GEDAR [45]	Geography-based routing	Sender-side	Multi-sink	Nodes with special H/W	Position information of its own, neighbors and sink	• Network topology control technique increases the connectivity of the network. • Reduce the number of packet retransmissions. • Void handling technique by utilizing a network topology control method.	• High physical energy consumption due to node movement to adjust their depth. • Ignore considering the sensor node energy level when selecting the forwarder node with high physical energy consumption may lead the protocol to be unable to select a forwarding node after a period of time due to exhausting their energy in physical movement.

Table 1. *Cont.*

Protocol	Category	Forwarding Set Selection Category	Sink(s)	Requirements	Knowledge Required/Maintained	Advantage	Disadvantage
IVAR [46]	Pressure-based routing	Receiver-side	Single-sink FIXED	Relay nodes and anchored nodes	Own depth, 1-hop neighbors and sink location	• Eliminates all the routes leading to a void area and therefore no need for a switch to recovery mode	• Redundant packet transmissions due to a hidden node problem. • Redundant packet transmissions increases energy consumption.
OVAR [47]	Pressure-based routing	Sender-side	Single-sink FIXED	Relay nodes and anchored nodes	Own depth, 1-hop neighbors and sink location info.	• No need for high overhead recovery mode for void handling since it ignores all the routes leading to a void area. • Hidden node problem addressed by selecting the candidate nodes near each other. • There is a trade-off between reliability and energy consumption.	• Modifying the number of nodes of the forwarding set affected the reliability of the network.
VHGOR [48]	Geography-based routing	Sender-side	Single-sink	Geo. location is available	Own location/neighboring table	• Void node handled in two ways (i) convex void handling and (ii) concave void handling (or) recovery mode.	• Consume restricted resources (memory through maintaining neighboring table, energy through nodes beacon).

Table 1. *Cont.*

Protocol	Category	Forwarding Set Selection Category	Sink(s)	Requirements	Knowledge Required/Maintained	Advantage	Disadvantage
WDFAD-DBR [49]	Pressure-based routing	Receiver-side	Multi-sinks	Anchored, relay and sink nodes.	Owen depth, 1-hop neighbor's information and 2-hop neighbor's depth.	• Duplicated packets were handled by dividing the forwarding area and neighbor node prediction mechanism which help to reduce energy consumption. • Sticking in void holes was reduced by using the depth of expected next hop.	• Broadcast control packets and ACKs periodically consume the node's battery and memory. • Retransmission is required if the best forwarding node failed to transmit the packet. • The flexibility of routing might be affected due to choosing a fixed primary forwarding area to form the forwarding set. • The void area is not handled completely since the trapped nodes are not eliminated from the forwarding set.
EVA-DBR [50] andSORP [51]	Pressure-based routing	Sender-side	Multi-sinks	Anchored, relay and sink nodes.	Owen depth, 1-hop neighbor's information and 2-hop neighbor's depth.	• By resizing the forwarding area the hidden problem is addressed in some cases. • A trade-off between the energy consumption and latency based on the predefined maximum delay. • Detect the void and trapped nodes before the data packet gets stuck in a void node.	• Periodically broadcasting neighbor's information consumes the node's resources. • Duplicated packets transmissions in spares network. • Hidden problem may appear if the forwarding range chosen to be more than half of the transmission range.

Table 1. *Cont.*

Protocol	Category	Forwarding Set Selection Category	Sink(s)	Requirements	Knowledge Required/Maintained	Advantage	Disadvantage
EDOVE [52]	Pressure-based routing	Receiver-side	Multi-sinks	Anchored, relay and sink nodes.	Owen depth, 1-hop neighbor's information and 2-hop neighbor's depth.	• Considering energy level as one of its parameters helps in reducing energy consumption and avoid energy holes.	• Exchange the neighbor's info, and maintains the neighbor's table consumes the nodes resources. • Duplicated packet transmissions increase the consumed energy. • The void area is not handled completely since the protocol only addressed energy void holes.
TORA [53]	Geography-based routing	Receiver-based	Multi-sinks	Sink nodes and ordinary nodes.	Sinks and ordinary position information.	• 2-hop Ack is used to improve data delivery ratios and handle the void node issue. • To reduce end-to-end delay and retransmission zero Ack is utilized.	• High-energy consumption due to periodic hello and Ack messages. • Extra cost due to equipping the ordinary nodes with a pressure sensor even though it is a geographic-based protocol.
EBER^2 [54]	Pressure-based routing	Sender-side	Multi-sink	Anchored, relay and underwater sink nodes.	Two-hop Potential Forwarding nodes.	• Residual energy of the nodes is used to reduce the duplicated packets and decreases the energy consumption. • Transmission energy adaptability supports reducing the void holes. • Embedded sinks used to increase the packet delivery ratio.	• Suffers from large end-to-end delay as well as accumulative propagation distance. • Communication between embedded sinks and on-surface sinks is costly. • Duplicate packets to surface due to the node's control power mechanism near the sinks

Table 1. *Cont.*

Protocol	Category	Forwarding Set Selection Category	Sink(s)	Requirements	Knowledge Required/Maintained	Advantage	Disadvantage
EDORQ [55]	Pressure-based routing	Receiver-side	Multi-sink	Relay nodes and anchored nodes	Own depth, 1-hop neighbors and sink location	• Minimizes the total energy consumption and achieves a high packet delivery ratio. • Handles the void area problem by switching to the void recovery mode. • Suppresses the duplicate packet transmission through assigning a holding time to each candidate based on its Q-value	• Recorded extra packet delay due to the holding time calculated based on Q-value. • Switching to void recovery mode because void nodes decreases network performance by increasing packet delay and energy consumption. • The protocol suffers from trapped nodes, which are not eliminated from the forwarding set. • High computational cost.
RPSOR [56]	Pressure-based routing	Receiver-side	Multi-sink	Mobile sinks, relay nodes and anchored nodes	Owen depth, 1-hop neighbor's information and 2-hop neighbor's depth	• Guarantee utilizing the shorts path. • Prevents/reduces void hole formation in the network. • Reduces duplicate packets	• Maintenance tables exhaust node resources. • Hello messages for exchanging global information increase the network overhead. • Data transmission delay. • Suffers from packet retransmission.

Table 1. *Cont.*

Protocol	Category	Forwarding Set Selection Category	Sink(s)	Requirements	Knowledge Required/Maintained	Advantage	Disadvantage
PCR [57]	Geography-based routing	Sender-side	Multi-sink	Nodes with power control mechanism.	Position information of its own, neighbors and sinks	• Joint design of OR and power control to improve the link quality at each hop. • Reduce the number of packet transmissions in the dense networks by reducing the transmission power level. • Void handling technique by exploiting a power control mechanism.	• Power control mechanise consumes more energy in forwarding set selection phase. • Communication overhead due to broadcasting the beacon messages using different power levels.
SEEORVA [58]	Geography-based routing	Sender-side	Multi-sink	Underwater sensor nodes (source/ relay).	Nodes located in the virtual vector pipe between source node and sink	• It uses energy threshold for energy conservation. • Handle void areas by sending a data packet void_alert to the previous node. • For security purposes, the protocol encrypts data packets before sending them through the network.	• Large amounts of energy consumed to make nodes' information global in the network • In sparse networks, it is possible that the pipe does not have sufficient nodes to forward messages. • The node density influences the pipe efficiency.
EEDOR-VA [59]	Pressure-based routing	Hybrid technique	Multi-sink	Nodes with special H/W (depth sensor).	Nodes' hop count and depth.The previous HCREQ sender to unicast the HCREP to that node.	• It proposed novel a technique to handle the void area problem. • It identifies all trapped and void nodes. • High PDR with energy efficiency maintaining. • Minimizes redundant and retransmissions. • Establishes a loop-free route between source and sink(s).	• End-to-end delay due to route establishing. • The protocol suffers from hidden terminal problem.

These state-of-the-art protocols employ various void area handling techniques to address the void area problem in order to increase network performance and deliver data packets properly.

Moreover, Table 1 also includes a brief summary of the benefits and drawbacks of the reviewed protocols in the two last fields. Additionally, most of the protocols deal with the void region by switching from the forwarding approach to the recovery mechanism, and the bulk of them have the stuck node issue. The IVAR, SEEORVA, and EEDOR-VA protocols are the only ones that address the void area issue and recognize every void/trapped node. However, both IVAR and SEEORVA implement periodic beaconing to provide sensor nodes with sink(s) reachability information. The network performance is significantly impacted by the beacon interval. Due to the prioritizing process, which depends on the depth that could be the same for more than one node, and the holding time, which depends on shared parameters between more nodes, both protocols still suffer from duplicate transmissions. While the EEDOR-VA protocol addresses the limitations of these two protocols through the novel hop-count discovery mechanism and the prioritizing technique.

5. Open Issues and Challenges in UWSNs

- Energy efficiency: Due to the harsh underwater environment restrictions and limitations on recharging or replacing the deployed sensor node, energy efficiency is a major constraint that can restrict many applications from achieving their goals. The current focus of study is on energy conservation and routing process energy optimization. Future studies will continue to focus heavily on this topic.
- Channel utilization: effective and efficient channel usage is a significant area of investigation in UWSNs that has attracted a lot of attention; it has an effect on energy consumption and void areas that form easily in UWSNs. The channel must be used to its full potential to overcome its limitations, like interference, high error rates, continual sensor node mobility, and propagation latency.
- Void areas: Due to the reasons listed in Section 2.2, void areas result in more frequent packet drops, decreasing the network QoS. To ensure a high QoS for various UWSN applications and increase network reliability, more mechanisms to handle void areas and void communications effectively and establish trust for user apps are required.
- Security: A significant area of concern is the security of data exchanged between sensor nodes. The security of data is the most crucial issue in many UWSN applications, especially the military ones. In such applications, any information leak could lead to harmful effects and severe repercussions. To secure the connection between the sensor nodes in UWSNs as attacks and threats grow, investigations that consider security and privacy will be a continuing and extremely difficult effort.

These are some of the most significant and active fields of research for UWSNs that need more investigation, and they will remain so in the upcoming years.

6. Conclusions

These days, opportunistic routing in UWSNs has drawn a lot of attention from researchers. OR has been shown to be more effective than the conventional routing strategy for wireless networks because it makes use of the broadcast nature of wireless networks. A number of factors have an impact on the performance and effectiveness of the UWSNs, including a shortage of resources (restricted battery power and memory), the harsh underwater environment, and a weak communication channel. The void area problem employing an OR approach is one of the significant concerns and research challenges in UWSNs. We have investigated in this paper the existing OR protocols proposed to address this problem.

First, we discussed the aspects of routing protocols for UWSNs covering the main challenges facing researchers when designing routing protocols, the concept of the void area problem in UWSNs, and reasons for this problem. OR and its key elements, including OR construction blocks and OR classification have been introduced. Second, the state-of-the-art void avoiding protocols that use the OR technique were investigated in depth.

The reviewed protocols have then been compared in many aspects, including the type of protocol, number of sinks, network topology requirements, and the special information required or needed to be maintained during the data packet routing. Their advantages and limitations were listed in the last two columns of Table 1. Moreover, we provided some of the open research issues in UWSNs that require further investigation.

Author Contributions: Conceptualization, R.M., W.P., F.C. and N.A.; writing—original draft preparation, R.M.; writing—review and editing, W.P., F.C. and N.A.; visualization, R.M.; supervision, W.P., F.C. and N.A.; All authors have read and agreed to the published version of the manuscript.

Funding: This research received no external funding.

Institutional Review Board Statement: Not applicable.

Informed Consent Statement: Not applicable.

Data Availability Statement: Not applicable.

Conflicts of Interest: The authors declare no conflict of interest.

Abbreviations

ACKs	Acknowledgment messages
ADCs	Analog to Digital Converters
ADV	Advancement Distance
AUV	Autonomous Underwater Vehicle
DTN	Double Transmission Node
EBER2	An Energy Balanced Efficient and Reliable Routing Protocol
EDORQ	Energy-efficient Depth-based Opportunistic Routing with Q-Learning
EDOVE	Energy and Depth variance-based Opportunistic Void avoidance
EEDOR-VA	Energy Efficient Depth-based Opportunistic Routing protocol
EVA-DBR	Energy-efficient and Void Avoidance Depth Based Routing
GEDAR	GEographic and opportunistic routing with Depth Adjustment-based topology control for communication Recovery
HCREP	Hop-Count Reply
HCREQ	Hop-Count Request
HydroCast	A Hydraulic Pressure Based Anycast Routing Protocol
IoUTs	Internet of Underwater Things
IVAR	An Inherently Void Avoidance Routing Protocol for Underwater Sensor Networks
OR	Opportunistic Routing
OREPP	Opportunistic Routing based Expected Packet Progress
PCR	Power Control-based opportunistic Routing protocol
PDP	Packet Delivery Probability
QoS	Quality of Service
RPSOR	Reliable Path Selection and Opportunistic Routing
SEEORVA	Energy-Efficient Opportunistic Routing Protocol with Void Avoidance For Underwater Acoustic Sensor Networks
STN	Single Transmission Node
TORA	Totally Opportunistic Routing Algorithm
TWSNs	Terrestrial Wireless Sensor Networks
uw-sinks	underwater sinks
UWSNs	Underwater Wireless Sensor Networks
VAPR	Void-Aware Pressure Routing
VHGOR	Void Handling using Geo-Opportunistic Routing in underwater wireless sensor networks
WDFAD-DBR	Weighting Depth and Forwarding Area Division DBR routing protocol

References

1. Coutinho, R.W.; Boukerche, A.; Vieira, L.F.; Loureiro, A.A. Design Guidelines for Opportunistic Routing in Underwater Networks. *IEEE Commun. Mag.* **2016**, *54*, 40–48. [CrossRef]
2. Shovon, I.I.; Shin, S. Survey on Multi-Path Routing Protocols of Underwater Wireless Sensor Networks: Advancement and Applications. *Electronics* **2022**, *11*, 3467. [CrossRef]
3. Akyildiz, I.F.; Pompili, D.; Melodia, T. Underwater acoustic sensor networks: Research challenges. *Ad Hoc Netw.* **2005**, *3*, 257–279. [CrossRef]
4. Allen, C.O. Wireless sensor network nodes: Security and deployment in the niger-delta oil and gas sector. *Int. J. Netw. Secur. Its Appl. (IJNSA)* **2011**, *3*, 68–79.
5. Du, X.; Liu, X.; Su, Y. Underwater acoustic networks testbed for ecological monitoring of Qinghai Lake. In *OCEANS 2016-Shanghai*; IEEE: Piscataway, NJ, USA, 2016.
6. Freitag, L.; Grund, M.; Von Alt, C.; Stokey, R.; Austin, T. A shallow water acoustic network for mine countermeasures operations with autonomous underwater vehicles. In Proceedings of the Undersea Defence Technology (UDT), Amsterdam, The Netherlands, 21–23 June 2005; pp. 1–6.
7. Henry, N.F.; Henry, O.N. Wireless sensor networks based pipeline vandalisation and oil spillage monitoring and detection: Main benefits for nigeria oil and gas sectors. *SIJ Trans. Comput. Sci. Eng. Its Appl.* **2015**, *3*, 1–6.
8. Khan, A.; Jenkins, L. Undersea wireless sensor network for ocean pollution prevention. In Proceedings of the 2008 3rd International Conference on Communication Systems Software and Middleware and Workshops (COMSWARE'08), Bangalore, India, 6–10 January 2008.
9. Kumar, O.; Priyadarshini, P. Underwater acoustic sensor network for early warning generation. In Proceedings of the 2012 Oceans, Hampton Roads, VA, USA, 14–19 October 2012; IEEE: Piscataway, NJ, USA, 2012; pp. 1–6.
10. Lloret, J.; Sendra, S.; Garcia, M.; Lloret, G. Group-based underwater wireless sensor network for marine fish farms. In Proceedings of the 2011 IEEE GLOBECOM Workshops (GC Wkshps), Houston, TX, USA, 5–9 December 2011; IEEE: Piscataway, NJ, USA, 2011.
11. Mohamed, N.; Jawhar, I.; Al-Jaroodi, J.; Zhang, L. Sensor network architectures for monitoring underwater pipelines. *Sensors* **2011**, *11*, 10738–10764. [CrossRef] [PubMed]
12. Casey, K.; Lim, A.; Dozier, G. A sensor network architecture for tsunami detection and response. *Int. J. Distrib. Sens. Netw.* **2008**, *4*, 27–42. [CrossRef]
13. Alsulami, M.; Elfouly, R.; Ammar, R. Underwater Wireless Sensor Networks: A Review. In Proceedings of the 11th International Conference on Sensor Networks (SENSORNETS 2022), Virtual Event, 7–8 February 2022.
14. Garg, P.; Waraich, S. Parametric Comparative Analysis of Underwater Wireless Sensor Networks Routing Protocols. *Int. J. Comput. Appl.* **2015**, *116*, 0975–8887. [CrossRef]
15. Khasawneh, A.; Abd Latiff, M.S.B.; Kaiwartya, O.; Chizari, H. Next forwarding node selection in underwater wireless sensor networks (UWSNs): Techniques and Challenges. *Information* **2017**, *8*, 3. [CrossRef]
16. Wosowei, J.; Shastry, C. Underwater Wireless Sensor Networks: Applications and Challenges in Offshore Operations. *Int. J. Curr. Adv. Res.* **2021**, *10*, 23729–23733.
17. Mittal, S.; Kumar, R.K.R. Different Communication Technologies and Challenges for Implementing UWSN. In Proceedings of the 2021 2nd International Conference on Advances in Computing, Communication, Embedded and Secure Systems (ACCESS), Ernakulam, India, 2–4 September 2021.
18. Nayyar, A.; Puri, V.; Le, D.N. Comprehensive Analysis of Routing Protocols Surrounding Underwater Sensor Networks (UWSNs). In *Data Management, Analytics and Innovation*; Springer: Singapore, 2019; pp. 435–450.
19. Solayappan, A.; Frej, M.B.H.; Rajan, S.N. Energy Efficient Routing Protocols and Efficient Bandwidth Techniques in Underwater Wireless Sensor Networks—A Survey. In Proceedings of the 2017 IEEE Long Island Systems, Applications and Technology Conference (LISAT), Farmingdale, NY, USA, 5 May 2017.
20. Yee, L.; Mehmood, S.; Almogren, A.; Ali, I.; Anisi, M.H. Improving the performance of opportunistic routing using min-max range and optimum energy level for relay node selection in wireless sensor networks. *PeerJ Comput. Sci.* **2020**, *6*, e326. [CrossRef] [PubMed]
21. Chen, D.; Varshney, P.K. A survey of void handling techniques for geographic routing in wireless networks. *IEEE Commun. Surv. Tutor.* **2007**, *9*, 50–67. [CrossRef]
22. Kavar, J.M.; Wandra, K.H. Survey paper on Underwater Wireless Sensor Network. *Int. J. Adv. Res. Electr. Electron. Instrum. Eng.* **2014**, *3*, 7294–7300.
23. Kiran, J.S.; Sunitha, L.; Rao, D.K.; Sooram, A. Review on Underwater Sensor Networks: Applications, Research Challenges and Time Synchronization. *Int. J. Eng. Res. Technol.* **2015**, *4*, 1329–1332.
24. Erol-Kantarci, M.; Mouftah, H.T.; Oktug, S. A Survey of Architectures and Localization Techniques for Underwater Acoustic Sensor Networks. *IEEE Commun. Surv. Tutor.* **2011**, *13*, 487–502. [CrossRef]
25. Darehshoorzadeh, A.; Boukerche, A. Underwater Sensor Networks: A New Challenge for Opportunistic Routing Protocols. *IEEE Commun. Mag.* **2015**, *53*, 98–107. [CrossRef]
26. Luo, J.; Fan, L.; Wu, S.; Yan, X. Research on Localization Algorithms Based on Acoustic Communication for Underwater Sensor Networks. *Sensors* **2018**, *18*, 67. [CrossRef]

27. Ayaz, M.; Baig, I.; Abdullah, A.; Faye, I. A Survey on Routing Techniques in Underwater Wireless Sensor Networks. *J. Netw. Comput. Appl.* **2011**, *34*, 1908–1927. [CrossRef]

28. Lu, Q.; Liu, F.; Zhang, Y.; Jiang, S. Routing Protocols for Underwater Acoustic Sensor Networks: A Survey from an Application Perspective. In *Advances in Underwater Acoustics*; InTech: London, UK, 2017.

29. Islam, T.; Lee, Y.K. A Comprehensive Survey of Recent Routing Protocols for Underwater Acoustic Sensor Networks. *Sensors* **2019**, *19*, 4256. [CrossRef]

30. Jadhav, P.; Satao, R. A Survey on Opportunistic Routing Protocols for Wireless Sensor Networks. *Procedia Comput. Sci.* **2016**, *79*, 603–609. [CrossRef]

31. Khan, H.; Hassan, S.A.; Jung, H. On Underwater Wireless Sensor Networks Routing Protocols: A Review. *IEEE Sens. J.* **2020**, *20*, 10371–10386. [CrossRef]

32. Felemban, E.; Shaikh, F.K.; Qureshi, U.M.; Sheikh, A.A.; Qaisar, S.B. Underwater Sensor Network Applications: A Comprehensive Survey. *Int. J. Distrib. Sens. Netw.* **2015**, *11*, 14. [CrossRef]

33. Akyildiz, I.F.; Pompili, D.; Melodia, T. Challenges for Efficient Communication in Underwater Acoustic Sensor Networks. *ACM Sigbed Rev.* **2004**, *1*, 3–8. [CrossRef]

34. Jain, U.; Hussain, M. UnderwaterWireless Sensor Networks. In *Handbook of Computer Networks and Cyber Security*; Springer: Cham, Switzerland, 2020; pp. 227–245.

35. Awan, K.M.; Shah, P.A.; Iqbal, K.; Gillani, S.; Ahmad, W.; Nam, Y. Underwater Wireless Sensor Networks: A Review of Recent Issues and Challenges. *Wirel. Commun. Mob. Comput.* **2019**, *2019*, 20. [CrossRef]

36. Mhemed, R. Energy Efficient and Void Avoiding Routing Protocols for Underwater Wireless Sensor Networks. Ph.D. Thesis, Dalhousie University, Halifax, NS, Canada, 2022.

37. Khalid, M.; Ullah, Z.; Ahmad, N.; Arshad, M.; Jan, B.; Cao, Y.; Adnan, A. A Survey of Routing Issues and Associated Protocols in Underwater Wireless Sensor Networks. *J. Sens.* **2017**, *2017*, 7539751. [CrossRef]

38. Vieira, L.F.M. Performance and Trade-off of Opportunistic Routing in Underwater Networks. In Proceedings of the IEEE Wireless Communications and Networking Conference: Mobile and Wireless Networks, Paris, France, 1–4 April 2012.

39. Jiang, J.; Han, G.; Guo, H.; Shu, L.; Rodrigues, J.J. Geographic multipath routing based on geospatial division in duty-cycled underwater wireless sensor networks. *J. Netw. Comput. Appl.* **2016**, *59*, 4–13. [CrossRef]

40. Liu, H.; Zhang, B.; Mouftah, H.T.; Shen, X.; Ma, J. Opportunistic routing for wireless ad hoc and sensor networks: Present and future directions. *IEEE Commun. Mag.* **2009**, *47*, 103–109. [CrossRef]

41. Fang, Q.; Gao, J.; Guibas, L.J. Locating and Bypassing Routing Holes in Sensor Networks. *Mob. Netw. Appl.* **2006**, *11*, 187–200. [CrossRef]

42. Al-Dharrab, S.; Uysal, M.; Duman, T.M. Cooperative Underwater Acoustic Communications. *IEEE Commun. Mag.* **2013**, *51*, 146–153. [CrossRef]

43. Lee, U.; Wang, P.; Noh, Y.; Vieira, L.F.; Gerla, M.; Cui, J.H. Pressure routing for underwater sensor networks. In Proceedings of the 2010 Proceedings IEEE INFOCOM, San Diego, CA, USA, 14–19 March 2010.

44. Noh, Y.; Lee, U.; Wang, P.; Choi, B.S.C.; Gerla, M. VAPR: Void-Aware Pressure Routing for Underwater Sensor Networks. *IEEE Trans. Mob. Comput.* **2013**, *12*, 895–908. [CrossRef]

45. Coutinho, R.W.; Boukerche, A.; Vieira, L.F.; Loureiro, A.A. Geographic and Opportunistic Routing for Underwater Sensor Networks. *IEEE Trans. Comput.* **2016**, *65*, 548–561. [CrossRef]

46. Ghoreyshi, S.M.; Shahrabi, A.; Boutaleb, T. An Inherently Void Avoidance Routing Protocol for Underwater Sensor Networks. In Proceedings of the 2015 International Symposium on Wireless Communication Systems (ISWCS), Brussels, Belgium, 25–28 August 2015. [CrossRef]

47. Ghoreyshi, S.M.; Shahrabi, A.; Boutaleb, T. A Novel Cooperative Opportunistic Routing Scheme for Underwater Sensor Networks. *Sensors* **2016**, *16*, 297. [CrossRef]

48. Kanthimathi, N. Void handling using Geo-Opportunistic Routing in underwater wireless sensor networks. *Comput. Electr. Eng.* **2017**, *64*, 365–379. [CrossRef]

49. Yu, H.; Yao, N.; Wang, T.; Li, G.; Gao, Z.; Tan, G. WDFAD-DBR:Weighting depth and forwarding area division DBR routing protocol for UASNs. *Ad Hoc Netw.* **2015**, *37*, 256–282. [CrossRef]

50. Ghoreyshi, S.M.; Shahrabi, A.; Boutaleb, T. An Underwater Routing Protocol with Void Detection and Bypassing Capability. In Proceedings of the 2017 IEEE 31st International Conference on Advanced Information Networking and Applications, Taipei, Taiwan, 27–29 March 2017.

51. Ghoreyshi, S.M.; Shahrabi, A.; Boutaleb, T. A Stateless Opportunistic Routing Protocol for Underwater Sensor Networks. *Wirel. Commun. Mob. Comput.* **2018**, *2018*, 18. [CrossRef]

52. Bouk, S.H.; Ahmed, S.H.; Park, K.J.; Eun, Y. EDOVE: Energy and Depth Variance-Based Opportunistic Void Avoidance Scheme forUnderwater Acoustic Sensor Networks. *Sensors* **2017**, *17*, 2212. [CrossRef]

53. Rahman, Z.; Hashim, F.; Rasid, M.F.A.; Othman, M. Totally opportunistic routing algorithm (TORA) for underwater wireless sensor network. *PLoS ONE* **2018**, *13*, 1–28. [CrossRef] [PubMed]

54. Wadud, Z.; Ismail, M.; Qazi, A.B.; Khan, F.A.; Derhab, A.; Ahmad, I.; Ahmad, A.M. An Energy Balanced Efficient and Reliable Routing Protocol for Underwater Wireless Sensor Networks. *IEEE Access* **2019**, *7*, 175980–175999. [CrossRef]

55. Lu, Y.; He, R.; Chen, X.; Lin, B.; Yu, C. Energy-Efficient Depth-Based Opportunistic Routing with Q-Learning for Underwater Wireless Sensor Networks. *Sensors* **2020**, *20*, 1025. [CrossRef]
56. Ismail, M.; Islam, M.; Ahmad, I.; Khan, F.A.; Qazi, A.B.; Khan, Z.H.; Wadud, Z.; Al-Rakhami, M. Reliable Path Selection and Opportunistic Routing Protocol for Underwater Wireless Sensor Networks. *IEEE Access* **2020**, *8*, 100346–100364. [CrossRef]
57. Coutinho, R.W.; Boukerche, A.; Loureiro, A.A. A novel opportunistic power controlled routing protocol for internet of underwater things. *Comput. Commun.* **2020**, *150*, 72–82. [CrossRef]
58. Menon, V.; Midhunchakkaravarthy, D.; John, S.; Jacob, S.; Mukherjee, A. A secure and energy-efficient opportunistic routing protocol with void avoidance for underwater acoustic sensor networks. *Turk. J. Electr. Eng. Comput. Sci.* **2020**, *28*, 2303–2315. [CrossRef]
59. Mhemed, R.; Comeau, F.; Phillips, W.; Aslam, N. Void Avoidance Opportunistic Routing Protocol for Underwater Wireless Sensor Networks. *Sensors* **2021**, *21*, 1942. [CrossRef] [PubMed]

Review

Effects and Prospects of the Vibration Isolation Methods for an Atomic Interference Gravimeter

Wenbin Gong [1], **An Li** [1], **Chunfu Huang** [1], **Hao Che** [1], **Chengxu Feng** [2] **and Fangjun Qin** [1,*]

[1] College of Electrical Engineering, Naval University of Engineering, Wuhan 430033, China; herwbin@yahoo.com (W.G.); lian196101@126.com (A.L.); wateriness_e@163.com (C.H.); hg15441@126.com (H.C.)

[2] College of Weapon Engineering, Naval University of Engineering, Wuhan 430033, China; kerryfengcx@126.com

* Correspondence: haig2005@126.com; Tel.: +86-189-7144-4646

Abstract: An atomic interference gravimeter (AIG) is of great value in underwater aided navigation, but one of the constraints on its accuracy is vibration noise. For this reason, technology must be developed for its vibration isolation. Up to now, three methods have mainly been employed to suppress the vibration noise of an AIG, including passive vibration isolation, active vibration isolation and vibration compensation. This paper presents a study on how vibration noise affects the measurement of an AIG, a review of the research findings regarding the reduction of its vibration, and the prospective development of vibration isolation technology for an AIG. Along with the development of small and movable AIGs, vibration isolation technology will be better adapted to the challenging environment and be strongly resistant to disturbance in the future.

Keywords: atomic interference gravimeter; active vibration isolation; vibration compensation; vibration noise; gravity measurement

Citation: Gong, W.; Li, A.; Huang, C.; Che, H.; Feng, C.; Qin, F. Effects and Prospects of the Vibration Isolation Methods for an Atomic Interference Gravimeter. *Sensors* **2022**, *22*, 583. https://doi.org/10.3390/s22020583

Academic Editors: Haixin Sun and Xuebo Zhang

Received: 8 December 2021
Accepted: 10 January 2022
Published: 13 January 2022

Publisher's Note: MDPI stays neutral with regard to jurisdictional claims in published maps and institutional affiliations.

1. Introduction

Gravity is the force applied by the attraction of the Earth to the objects on the ground. Under the effect of gravity, an object descends at an accelerated speed, so that its acceleration is called gravitational acceleration. Gravitational acceleration varies with time and space. At a place on the Earth, gravitational acceleration is dependent on such factors as local latitude, altitude, landform, density and distribution of underground matters. As an important parameter in describing the gravitational field of the earth, gravitational acceleration has been extensively applied in studies of inertial navigation, geological survey, geophysics and basic physics, etc. [1–3]. Today, high-precision gravity field maps, like remote sensing satellite images, SAR images and other satellite images, play an important role in the field of national economy and people's livelihood [4–7]. In the underwater navigation of submarines, the traditional sonar technology is unable to meet the requirements for high precision navigation at the seabed since it cannot receive any signal in deep waters, but a seabed high precision gravity map can be used to assist submarines in rapidly locating and avoiding the obstacles at seabed [1,8,9]. Hence, underwater gravity navigation entails high precision gravity measurement.

De Broglie claimed that physical particles had a wave particle duality. Like lights, atoms could be interfered by laser in beam splitting, reflection and combination because of their wave nature. In 1991, Kasevich and Chu et al. [10] utilized stimulated Raman transition in the coherent manipulation of cold atomic cloud for the first time, and implemented a cold atomic interferometer. Based on the measured gravity, the averaged resolution of the interferometer for 1000 s was 3×10^{-6} g. In 1992, Kasevich and Chu et al. [11] designed the world's first atomic fountain gravimeter based on the stimulated Raman transition, and achieved the gravity measurement resolution of 30 uGal (1uGal $= 10^{-8} \mathrm{m/s^2} \approx 10^{-9}$ g) within the integral time of 2000 s. Continual improvements were then made for eliminating the local gravity error driven by earth tides, so as to achieve the measurement resolution of 0.3 uGal within the integral time of 60 s, and realize more accurate gravitational acceleration measurement [12].

An atomic interference gravimeter (AIG), as a high precision measuring device for absolute gravity based on atomic interferometer, can be used to measure the absolute value of gravitational acceleration. At present, more than 50 research and development institutions are engaging in the exploration worldwide including LNE-SYRTE in France [13–15], Humboldt University in Germany [16,17], Stanford University in USA [18], Zhejiang University of Technology (ZJUT) [19,20], Huazhong University of Science and Technology (HUST) [21,22], University of Science and Technology of China (USTC) [23], and Wuhan Institute of Physics and Mathematics (WIPM) [24,25] in China, among others [26–31]. A short list is given in Table 1. In 2009, LNE-SYRTE in France developed the world's first movable atomic gravimeter [15]. While improving the sensitivity of atomic gravimeter, more efforts must be made to develop the atomic interference gravimeters of small size and good mobility. For this purpose, a number of dynamic experiments have been carried out ever since [20,32–38].

Table 1. Overview of studies on AIG.

Year	Institution	Interrogation Time	Sensitivity	References
2001	Stanford	160 ms	20 uGal/$\sqrt{\text{Hz}}$	Peters et al. [18]
2008	Stanford	400 ms	8 uGal/$\sqrt{\text{Hz}}$	Holger Müller [39]
2019	University of California	130 ms	37 uGal/$\sqrt{\text{Hz}}$	Wu et al. [32]
2008	LNE-SYRTE	50 ms	1.4 uGal/$\sqrt{\text{Hz}}$	Le Gouët J et al. [15]
2013	ONERA	48 ms	42 uGal/$\sqrt{\text{Hz}}$	Bidel et al. [36]
2018	Muquans	60 ms	50 uGal/$\sqrt{\text{Hz}}$	Menoret et al. [35]
2014	LENS	160 ms	30 uGal/$\sqrt{\text{Hz}}$	Rosi et al. [40]
2013	Humboldt University	260 s	3 uGal/$\sqrt{\text{Hz}}$	Hauth et al. [16]
2016	Humboldt University	0.26 s	9.6 uGal/$\sqrt{\text{Hz}}$	Freier et al. [17]
2013	ANU	60 ms	60 uGal/$\sqrt{\text{Hz}}$	Altin et al. [41]
2013	HUST	300 ms	4.2 uGal/$\sqrt{\text{Hz}}$	Hu et al. [42]
2019	HUST	160 ms	53 uGal/$\sqrt{\text{Hz}}$	Luo et al. [43]
2014	Zhejiang University	60 ms	10 uGal/$\sqrt{\text{Hz}}$	Wu et al. [44]
2018	National Institute of Metrology	70 ms	44 uGal/$\sqrt{\text{Hz}}$	Wang et al. [31]
2019	ZJUT	120 ms	300 uGal/$\sqrt{\text{Hz}}$	Fu et al. [19]
2019	WIPM	200 ms	30 uGal/$\sqrt{\text{Hz}}$	Huang et al. [45]

Compared with the traditional absolute gravimeter based on laser interferometer, an atomic gravimeter uses cold atomic cloud as a material to measure gravity. It can measure the gravity continuously for a long time and without mechanical wear, and achieve the sensitivity and accuracy as good as an FG5 gravimeter. Presently, the best combined standard uncertainty of atomic gravimeters can be up to 4.5 uGal in the world [46]. Due to its high measurement sensitivity and accuracy, an AIG has been widely applied in the accurate measurement of physical quantities including gravitational acceleration [37,47], gravity gradient [48], and universal gravitational constant [26]. Therefore, it is of great value in underwater navigation, resource scanning, and geologic monitoring, etc. [48–51].

The noise source of AIG is mainly composed of detection noise, vibration noise, Raman optical phase noise, optical frequency shift noise, etc. [52]. With the prior art, the detection noise, phase noise and frequency noise can be reduced to the level of mrad/single measurement; however, the vibration and noise can only be reduced to the level of 10–100 mrad/single measurement even if complex active and passive damping platforms are used, which makes the vibration and noise become the main noise source limiting the sensitivity of AIG.

Vibration, as a common phenomenon in the nature, often undermines the stability and reliability of various engineering instruments or equipment. Hence, vibration isolation is indispensable for these instruments and equipment [53,54]. In most engineering applications, attention has to be mainly paid to the vibration noise at 10–100 Hz or higher frequency, e.g., various means of transport and engines. The vibration at these frequency bands can be satisfactorily suppressed by simply applying the passive vibration technology. Nevertheless, more attention is paid to the vibration noise at low frequency and even very low frequency in some precision measurement experiments, e.g., microscopes, laser frequency stabilization systems and gravitational wave detection experiments [55–57]. For the purpose of higher measurement sensitivity, the good technology of low frequency vibration isolation should be adopted in these experiments [58–60]. When an AIG is used to measure gravitational acceleration, its measurement accuracy is dependent on a variety of factors with the increasing accuracy and the decreasing scale of measurement. Among these factors, the vibration of a Raman light reflection mirror significantly affects the measurement accuracy of an AIG, while vibration noise becomes a considerable restriction over its measurement accuracy and reliability [61,62]. Consequently, the technology for isolation and attenuation of vibration noise is crucial to accurately obtaining the atomic interference phase and implementing the accurate information detection of gravitational field [63,64].

The measurement error caused by Raman mirror vibration in the process of atomic interference can be effectively suppressed by differential measurement, such as atomic interference gradiometer [65]. Two sets of interference platforms at different positions in the atomic interference gravity gradiometer share the same optical and electronic system, and the two groups of atoms share the same pair of Raman light in the interference process, which can better suppress the Raman phase noise, vibration noise and other common mode noise in the measurement process without vibration isolation system, so as to obtain higher performance indexes than a single gravimeter. However, the measuring principle of AIG is different from that of gradiometer. There is only one interference platform, it is difficult to suppress or eliminate the vibration noise measured by a single atomic interferometer from the change of interferometer configuration, and can only be controlled from the vibration source or propagation path.

According to the noise transfer function of AIG, the low-frequency vibration noise has a great impact on AIG, while the high-frequency noise is attenuated by the mirror vibration noise transfer function. The response to the AIG to vibration and noise is like a low-pass filter. The cut-off frequency is the reciprocal of the time interval of the Raman light pulse of the atomic interferometer. The low-frequency vibration and noise has the greatest impact on the measurement. High frequency vibration and noise can be isolated by passive vibration isolation platform. For low-frequency vibration, the passive vibration isolation platform has no or little effect. In order to further improve the measurement accuracy of AIG, restraining low-frequency vibration and noise is the main research direction.

The other sections of this paper are organized as follows: the second section gives an introduction to the working principles of an AIG; the third section presents an analysis on how vibration noise affects the error of an AIG; the fourth section contains a review of the research findings regarding the vibration isolation technology for an AIG; and the fifth section gives a summary and discussion of the prospective development of the vibration isolation technology for an AIG.

2. Working Principles of an AIG

Like light wave interference, matter wave interference needs a beam splitter and combiner, which are often implemented by virtue of two-photon stimulated Raman transition in an atomic interferometer [10]. An AIG applies three Raman pulse beams onto an atomic wave packet to implement beam splitting, reflection and combination, so as to achieve atomic interference. Atoms are affected by gravity in the interference process. The gravity information can be obtained from atomic interference fringe to implement gravity measurement in the principles as shown in Figure 1a.

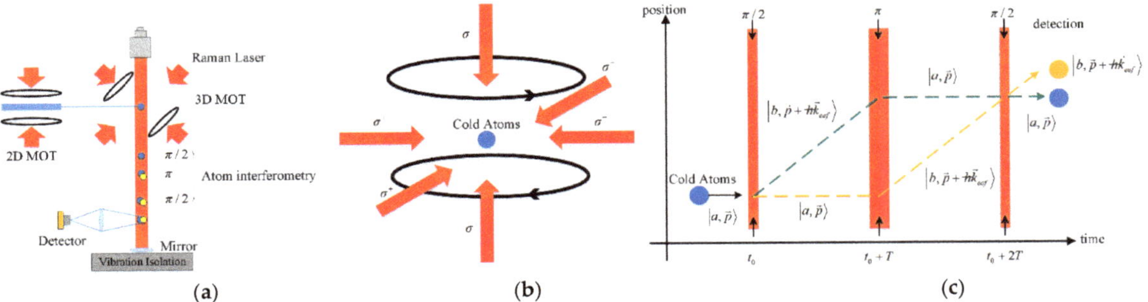

Figure 1. Principles of an AIG. (**a**) Operation diagram of an AIG; (**b**) Diagram of three-dimensional MOT atom cooling and trapping; (**c**) Diagram of Raman pulse atomic interference. Three Raman light pulse beams cause the beam splitting, reflection and combination of atomic wave packet to result in interference.

Under normal temperature, atoms are not easily manipulated since they move fast. Therefore, atoms must be slowed down first to generate cold atomic cloud. The dipole force and scattering force under the interaction of light and atoms are utilized for laser cooling and trapping of atoms in the vacuum cavity [66]. Rubidium atoms are decelerated by the Doppler cooling effect of resonant laser to generate cold atomic cloud in the end. Polarization gradient cooling is further implemented using the magnetic field generated by the Helmholtz coil in a magneto-optical trap (MOT) and a light field generated by three pairs of circular polarization lasers with propagation directions vertical to each other. This brings the temperature of cold atomic cloud to the uK level [67]. The decelerated cold atomic cloud is eventually trapped in the MOT as shown in Figure 1b. After shutting down the magnetic field, state preparation is conducted with the microwave state selection method when atoms move downward freely under the effect of gravity. The atoms at the level of basic state, which are sensitive to the magnetic field and have zero magnetic quantum in the atomic cloud, are therefore selected [68].

As shown in Figure 1c, MOT is utilized to prepare a cold atomic cloud. A sequence of three Raman pulse beams ($\pi/2 - \pi - \pi/2$) is then used and applied to cold atoms for the beam splitting, reflection and combination of the atomic wave packet. Meanwhile, chirped scanning is carried out for the frequency of Raman transition to regulate the interference phase. It is assumed that atoms are in the state $|a\rangle$ at the initial time. After interaction with the first pulse beam $\pi/2$, atoms are in the superposition state of $|a\rangle$ and $|b\rangle$ have different momentums, and their difference is $\hbar \vec{k}_{eff}$. The atomic wave packet experiences beam splitting, and then evolves along two paths in space. In the meanwhile, the phase of light field is also transferred to atoms. After the free evolution time T, the second pulse beam π interacts with atoms. At this time, the atoms in the state $|a\rangle$ transit to the state $|b\rangle$, while the atoms in the state $|b\rangle$ transit to the state $|a\rangle$. The momentum of atoms also changes correspondingly, causing the reorientation of the atomic wave packet. After the time T, the third pulse beam $\pi/2$ interacts with atoms. Under this circumstance, beam combination happens to the atomic wave packet, causing the interference.

In the detection area, the time of flight method (TOF) is employed to detect the fluorescence signal of atoms, and obtain the normalized atomic transition probability

P of atoms in two states [33,69,70]. In this case, P represents the probability of atomic interference in the macroscopic observable, and can be described by the following function:

$$P = P_0 + C \cdot \cos(\Delta\Phi)/2 \tag{1}$$

where P_0 is the average of atomic transition probability; C is the contrast of atomic interference fringe; $\Delta\Phi$ is the measured phase difference of interference fringe, which is originated from the accumulation of atomic interference in different paths, and expressed as

$$\Delta\Phi = \phi_1 - 2\phi_2 + \phi_2 = \left(k_{eff}g - 2\pi\alpha\right)T^2 \tag{2}$$

where ϕ_1, ϕ_2, ϕ_3 is the Raman light phase at the time of interaction with three Raman pulses, respectively; α is the chirp rate of Raman laser, which is used to compensate for the gravity-induced Doppler frequency shift by linearly scanning the Raman light frequency difference; k_{eff} is the equivalent wave vector; T is the time interval between two adjacent Raman light pulse beams, i.e., free evolution time of atoms.

Doppler frequency shift is positively correlated with the square of fall time [71]. In the implementation of an atomic interferometer, the Doppler frequency shift caused in the fall of atoms should be compensated to ensure the resonance of atoms with all three Raman pulse beams. The Raman light frequency difference is linearly scanned to compensate for the Doppler frequency shift. When the Doppler frequency shift can be perfectly compensated, cold atomic cloud always resonates with Raman light. The phase difference is $\Delta\Phi = \left(k_{eff}g - 2\pi\alpha\right)T^2 = 0$. The gravitational acceleration measured by a cold AIG is $g = 2\pi\alpha/k_{eff}$.

3. Influence of Vibration Noise on an AIG

The noise of AIG is mainly composed of detection noise, vibration noise, Raman optical phase noise, optical frequency shift noise, etc. With the prior art, the detection noise, phase noise and frequency noise can be reduced to the level of mrad/single measurement. However, the vibration and noise can only be reduced to the level of 10–100 mrad/single measurement even if complex active and passive damping platforms are used, which makes the vibration and noise become the main noise source limiting the sensitivity of cold atom gravimeter.

An AIG is presently one of the highest precision measuring devices for gravitational acceleration g in the world. In the measurement experiment, external vibration noise is very easily coupled into the total phase of inference fringe by virtue of the vibration of the Raman light reflection mirror in an atomic interferometer. This results in some measurement error. The vibration noise of most optical components in an atomic interferometer is in common mode during measurement. Therefore, phase noise can be caused to interference fringe only by the vertical vibration displacement of the Raman light reflection mirror placed at the bottom of the interferometer, and then undermine the sensitivity of the instrument [72].

The concept of sensitivity function is normally taken to describe the time domain atomic interferometer [63,73]. It is assumed that Raman light phase ϕ jumps $\delta\phi$ at the time t, causing the variation $\delta P(\delta\phi, t)$ of transition probability P. The sensitivity function $g_s(t)$ may be defined by

$$g_s(t) = 2 \lim_{\delta\phi \to 0} \frac{\delta P(\delta\phi, t)}{\delta\phi} \tag{3}$$

Interference signal has an approximately linear relationship with phase. The relationship between phase change and variation of transition probability is $\sigma_{\Delta\phi} = 2\sigma_P$. In terms of phase change, the sensitivity function is therefore expressed as

$$g_s(t) = \lim_{\delta\phi \to 0} \frac{\delta\Phi(\delta\phi, t)}{\delta\phi} \tag{4}$$

The variation of transition probability is solved by segmenting evolution matrix. The time at the center of the second pulse beam π is taken as the time origin to obtain the sensitivity function of atomic transition signal to Raman light as follows:

$$g(t) = \begin{cases} \sin \Omega_R(T+t) & -T \leq t < -T+\tau \\ 1 & -T+\tau \leq t < -\tau \\ -\sin \Omega_R t & -\tau \leq t < \tau \\ -1 & \tau \leq t < T-\tau \\ -\sin \Omega_R(T-t) & T-\tau \leq t < T \end{cases} \tag{5}$$

where Ω_R is the Rabi frequency; τ is the length of finite Raman pulse; and T is the Raman pulse time interval. Transfer function is a weighting function for the relationship between system input and output. It is the Fourier transform of the sensitivity function $G(\omega) = \int_{-\infty}^{\infty} g_s(t)e^{i\omega t}dt$. The transfer function between the interference phase of an interferometer and the Raman light modulated phase is $H_\phi(\omega) = \omega G(\omega)$.

In the measurement of atomic interference gravity, a reflection mirror is used to generate a pair of Raman lights in opposite directions, which jointly affect the atomic cloud. The effective laser phase perceived by atoms is originated from the phase difference between two Raman light beams that are transmitted downward and reflected by the reflection mirror, respectively. Hence, the change of laser phase is directly attributed to the motion of the Raman light reflection mirror at the stage of interference. Noise is then introduced. It is assumed that two Raman light beams in an atomic interferometer have the wave vector k_1, k_2, respectively. If the reflection mirror has the vertical vibration displacement $\delta_z(t)$, the phase noise introduced by the Raman light because of vibration is $k_{eff}\delta_z(t)$. The phase difference of interference fringe $\Delta\Phi = \phi_1 - 2\phi_2 + \phi_3 + k_{eff}\delta_z(t)$. When the vibration noise $k_{eff}\delta_z(t)$ is large enough to make the influence of vibration on $\Delta\Phi$ greater than π, the interference fringe of atoms will be entirely eliminated. It is therefore evident that the phase of an AIG is affected by the position change of the reflection mirror. Vibration isolation should be therefore provided for the reflection mirror to achieve more accurate measurement of gravitational acceleration. The power spectral density of the Raman reflection mirror phase can be expressed as

$$s_\phi(\omega) = k_{eff}^2 s_a(\omega)/\omega^4 \tag{6}$$

where ω is the angular frequency of vibration; and $s_a(\omega)$ is the noise power spectrum of vibration acceleration. The phase variance of an AIG can be expressed as

$$\sigma_\phi^2 = \int_0^{\infty} H_\phi^2(\omega)k_{eff}^2 s_a(\omega)/\omega^4 d\omega \tag{7}$$

The influence of the reflection mirror vibration noise on gravity is defined by

$$\sigma_g^2 = \int_0^{\infty} H_a^2(\omega)s_a(\omega)d\omega \tag{8}$$

where $H_a^2(\omega)$ is the transfer function of the reflection mirror vibration noise to the gravimeter. While satisfying $\omega \ll \Omega_R$ and $\tau \ll T$, there is

$$|H_a(\omega)|^2 = \frac{k_{eff}^2}{\omega^4}|H_\phi(\omega)|^2 = \frac{16\sin^4(\omega T/2)}{T^4\omega^4} \tag{9}$$

The transfer function of the reflection mirror vibration noise for an AIG can be used to create the transfer function curve of the vibration noise at different Raman light time intervals T as shown in Figure 2. As shown in the figure, when interrogation time T is 50 ms, 90 ms, 120 ms, respectively, the noises with the frequency lower than 1 Hz are all transferred nearly at the ratio of 1:1 to the interference phase. When the frequency is greater than 1 Hz, the transfer rate of vibration begins to attenuate. Additionally, when the frequency exceeds 10 Hz, the transfer rate has attenuated by five orders of magnitude. Evidently, the vibration noises with the frequency lower than $f = 1/T$ has the highest influence on measurement. The increasing frequency causes the transfer rate of vibration to attenuate at the rate f^2. Hence, the transfer function of vibration noise is typical of low pass just as a low-pass filter. In other words, it is more sensitive to low frequency vibration. The vibration noises with the frequency lower than 10 Hz affect a cold atomic gravimeter most significantly. For this reason, special attention must be paid to the vibration of low frequency bands for the vibration isolation system of an atomic gravimeter.

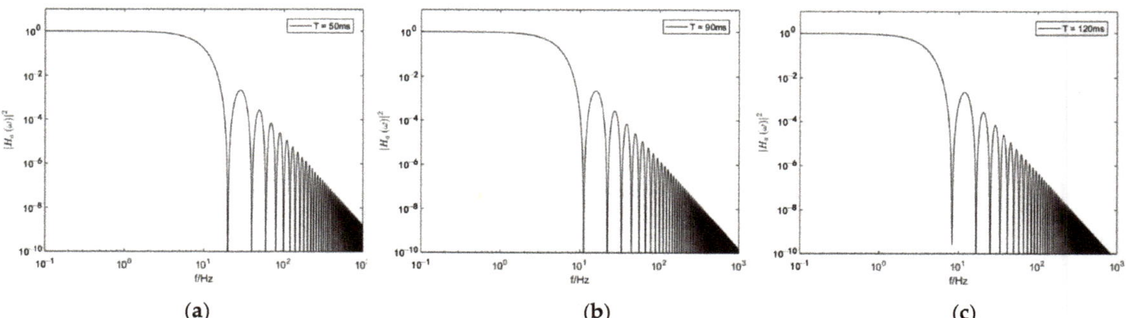

(a) (b) (c)

Figure 2. Transfer function spectrum of the reflection mirror vibration noise in an AIG at different time intervals under the effect of Raman light. (**a**) $T = 50$ ms, (**b**) $T = 90$ ms, and (**c**) $T = 120$ ms.

4. Research Status of Vibration Isolation Technology for an AIG

Vibration noise may be reduced by causing the attenuation of vibration and lowering the motion of the reflection mirror as much as possible. This can be implemented in a low noise environment, e.g., cold atomic gravimeter [13] (CAG), which had been used for gravity measurement in the Walferdange Underground Laboratory for Geodynamics. Nevertheless, a gravimeter will be significantly limited to a laboratory and not applied extensively if the measurement of gravitational acceleration is performed only in a low noise environment [74].

Vibration isolation is one of the major methods for vibration control. In this method, a vibration isolation system is used to isolate a vibration source from precision instruments. At present, vibration noise is mainly suppressed in three ways, including passive vibration isolation, active vibration isolation and vibration compensation. Among them, the last two ways are mainly applied for an AIG.

4.1. Passive Vibration Isolation

A passive vibration isolation system relies on an elastic damping material and mechanical structure to absorb or attenuate the mechanical waves of vibration. It is normally a mass-spring-damping system as shown in Figure 3. The attenuation of vibration is mainly achieved by such devices as a coil spring, elastomer pad, and air spring. The advantage of this system is the realization of the best vibration isolation with a simple structure, but not relying on any external energy, sensor, actuator or control system. Nevertheless, this system is troubled by very poor isolation of vibration at low frequency bands, very long time needed for stabilization, and low operability. It is often applied in industrial equipment, civil engineering structure, precision instrument and equipment. The common devices of passive vibration isolation include pneumatic vibration isolator [75], zero-length spring vibration isolator [76], and negative stiffness spring vibration isolator [77–79].

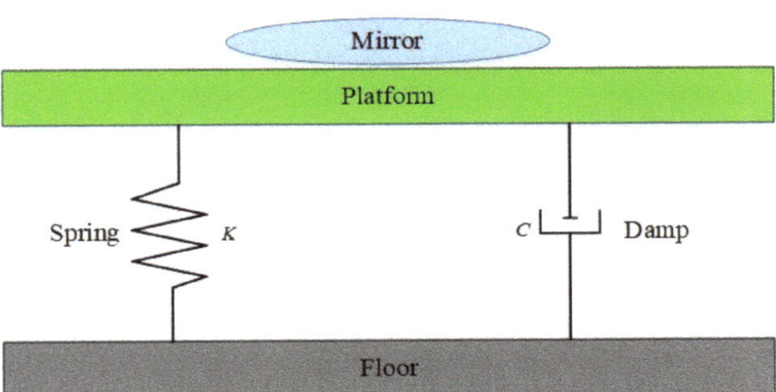

Figure 3. Composition of a passive vibration isolation system.

A zero-length spring based on long period is one of the main passive vibration isolation system designs for an AIG, IMGC-02 absolute gravimeter adopts this passive vibration isolation method [80,81]. It mainly consists of zero-length spring structure, geometric reverse spring structure, Euler column spring structure, and torsion balance spring structure, etc. In [76], Li et al., of Tsinghua University utilized feedback control to improve the zero-initial-length spring structure, devised and manufactured an ultralow frequency vertical vibration isolator based on spring link. The vibration isolation system employed the optical lever to detect the angular displacement of swing link. The feedback circuit controlled the voice coil motor to drive the swing link based on the detected displacement signal, which compensated for the influence of creep and temperature shift on the spring. After careful modulation, the system could achieve the stable oscillation within the natural cycle of 32 s, and constantly operate for more than one year. The system had been tested in T^{-1} absolute gravimeter, and realized the uncertainty of 2 μGal in 12 h measurement. At present, compared with other types of passive vibration isolation structures, zero-length spring structure is widely used.

However, because the spring needs a large volume, the volume of the zero-length spring structure is large, and will continue to accumulate due to the influence of temperature drift and creep. The requirements of miniaturization and mobility are difficult to meet the requirements of vibration isolation.

The negative stiffness spring can locally reduce the overall stiffness of the spring. The greater the spring stiffness, the stronger the bearing capacity and the greater the natural frequency. The negative stiffness spring is connected in parallel with the positive and negative stiffness springs, so that the whole has nonlinear characteristics near the equilibrium position, and the stiffness is close to 0. Negative stiffness vibration isolation system is widely used in gravity measurement experiment based on atomic interference, which can realize low vibration environment [79].

The research and development cost of high-precision passive vibration isolation platform based on complex spring and support structure is high, while the passive vibration isolation platform based on negative stiffness is simple and efficient, and can provide multi degree of freedom vibration isolation. The negative rigid commercial vibration isolation platform developed by Minus K company has good vibration isolation performance, small volume, simple operation, and it is easy to retrofit and install a voice coil motor [32]. It is widely used in the vibration isolation of the reflector of atomic gravimeter with good effect. The 100BM-10 commercial passive vibration isolation platform produced by the company is only in size 310 mm × 310 mm × 117 mm, with a payload range of 34–50 kg. It can provide 0.5 Hz vertical natural frequency and 1.5 Hz horizontal natural frequency, which is suitable for miniaturized system applications.

The Müller research team of the University of California at Bernoulli [32] used the passive damping platform (25BM-10, Minus K) to carry out vehicle flow static gravity measurement, and obtained the measurement sensitivity of 500 uGal/$\sqrt{\text{Hz}}$ and the measurement accuracy of 40 μGal.

A passive vibration isolation system can bear very high loads regardless of its simple structure, but depends much on the assembly and debugging accuracy of structure. It is essentially dependent on the elastic components of special structure for good vibration isolation, so that it is easily affected by the creep and temperature shift of elastic materials. For this reason, the system cannot maintain its good vibration isolation for a long time, and has poor resistance to disturbance. Its performance of vibration isolation is normally inferior to that of an active vibration isolation system, so that the active isolation is more common.

The passive vibration isolation platform can suppress high-frequency vibration and noise, but it has a poor suppression effect on low-frequency vibration and noise below 0.5 Hz, and even resonance will occur and increase vibration and noise. According to the transfer function of vibration, low-frequency vibration noise has a greater impact on atomic interference than high-frequency noise. Therefore, the passive vibration isolation system cannot meet the high-precision requirements of atomic gravimeter.

4.2. Active Vibration Isolation

An active vibration isolation system is equipped with a vibration sensor and actuator. By virtue of feedback control, it can make an effective improvement to the poor vibration isolation of a passive vibration isolation system at low frequency bands. This system is normally consisted of spring, sensor (accelerometer or seismograph), actuator and control system [82], as shown in Figure 4. Vibration signal is converted into the output signal of a brake through amplifier and control circuit. The feedback control is imposed on the vibration isolation platform to effectively control vibration. As the active vibration isolation can use a certain control algorithm according to the vibration signal of the sensor and use feedforward or feedback to achieve the control effect, compared with the passive vibration isolation platform, it can produce lower resonance frequency and achieve a stronger effect of vibration suppression. An active vibration isolation system has been widely applied in metrological, photoetching and medical fields as well as semiconductor industry. In the study of the active vibration isolation system for an AIG, research institutions have developed a variety of vibration isolation systems and achieved good effective improvements to measurement accuracy and sensitivity [83–85].

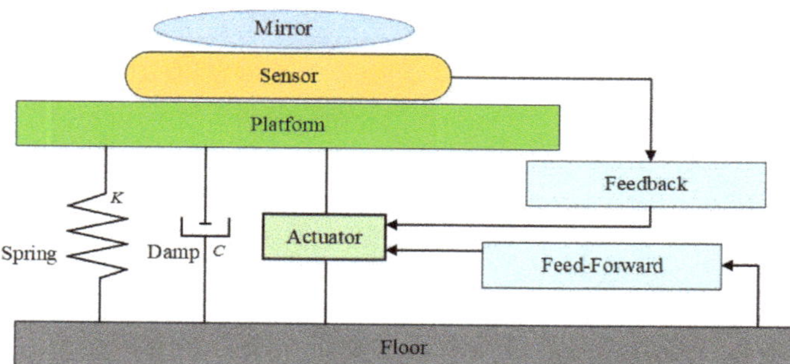

Figure 4. Composition of an active vibration isolation system.

Hensley et al. of the Stanford University, USA designed an experimental system with vertical ground motion and atomic gravimeter isolation as given in [86]. The system combined an active system with a passive system formed by a mechanical spring and an optical workstation suspended in the compressed air. The active system was used to measure the acceleration of an object to be isolated, and then fed it back to an electromagnetic actuator as an offset against the motion. Eventually, an active spring vibration isolation system was developed with the effective natural resonance frequency of 0.033 Hz. It could lower the vibration noise to 10^{-8} g/$\sqrt{\text{Hz}}$ from 0.1–20 Hz. The vibration noise of the system was reduced by 300 times. The system was tested in an atomic interference measurement experiment to obtain the uncertainty 3×10^{-9} g. In the meanwhile, a comparative experiment was carried out to prove that acceleration error signal could be lowered by 30–1000 times when noise was at 10 Hz to 100 Hz, and by 1000 times when noise was above 100 Hz.

In [87], Freier of the Humboldt University of Berlin simplified the structure of active vibration isolation, and devised a single rate spring active vibration isolation system. The system followed the basic principle that an isolated platform was supported by a principal spring to isolate high frequency vibration noise and form a passive vibration isolation system. An accelerometer placed on the isolated platform was employed to detect the vibration of the platform. Through active feedback compensation, feedback force was then applied to compensate for the influence of such vibration, so as to achieve the equivalent ultra-long period and isolate the influence of low frequency vibration noise. The vibration noises within the range from 0.03 Hz to 5 Hz were suppressed by 200 times. The natural resonance frequency of the system was 0.025 Hz. The sensitivity of an atomic gravimeter reached 7×10^{-8} g/$\sqrt{\text{Hz}}$.

Schmidt et al. developed a gravimetric atom interferometer (GAIN) as detailed in [88]. To reduce the mechanical vibration noise, a vibration isolation system with active feedback was devised by placing a reflection mirror on a retrofitted commercial passive vibration isolation platform (Minus K50BM-10). A feedback circuit was implemented by measuring the residual vibration of the vibration isolation platform with a Guralp CMG-3VL uniaxial force feedback accelerometer, so that a voice coil motor was used to feed back the vibration to a vibration isolator. An electronic feedback device was installed in a detached control unit to keep the small size of the sensor, which lowered the effective resonance frequency from 0.50 Hz to 0.025 Hz. Hauth et al. [16] further optimized the active vibration isolation system in GAIN given in [15], and achieved the same low resonance frequency for both horizontal and vertical axes by remodeling the passive vibration isolation platform on which the reflection mirror was placed. Additionally, an inclined workstation was also devised to obtain the atomic interference fringe with the pulse interval T = 230 ms and realize the measurement sensitivity of 3×10^{-8} g/$\sqrt{\text{Hz}}$.

Following the vibration isolation scheme proposed by Schmidt, Zhou et al., of HUST made an improvement to the active vibration isolation system of a cold atomic gravimeter in 2012 as detailed in [22]. In this improved vibration isolation system, the uniaxial accelerator in the original design was replaced by a triaxial commercial seismograph (Guralp CMG-3ESP) to achieve the positive correlation of the system's damping force with the output of sensor. In this way, the stability of the feedback system was further improved to better obtain the data of horizontal vibration. The improved system could suppress the vibration noise at 0.1–1 Hz by 100 times, and realize the sensitivity less than 1×10^{-9} g/$\sqrt{\text{Hz}}$ to the vibration noise at lower than 2 Hz, and its natural resonance frequency 0.016 Hz. When the vibration isolation system was applied in an AIG for experimental gravity measurement, its sensitivity reached 5.5×10^{-8} g/$\sqrt{\text{Hz}}$, and its resolution was 6.5×10^{-9} g within the integral time of 60 s, which was comparable to that of the most advanced atomic gravimeter.

In [89], Zhou et al., designed an ultralow frequency active vibration isolator that was able to suppress the vibration noise in three-dimensional directions simultaneously. In the system, a passive vibration isolation platform suppressed the vibration noise on the ground in three directions. A triaxial microseismograph was introduced to detect the residual vibration noise on the passive vibration isolation platform, and then the vibration signal was converted into a control signal for a voice coil motor. Feedback force was applied in the vertical direction and two horizontal axial directions to further attenuate the vibration noise on the ground. This could also lower the vertical vibration acceleration caused by coupling with the horizontal vibration, so as to realize the optimal vibration isolation. After the vibration isolation system formed a feedback circuit, the equivalent resonance frequency in the vertical direction was 0.01 Hz, while the equivalent resonance frequency in the horizontal direction was 0.083 Hz. The vibration noise in the vertical direction was suppressed by approximately 50 times within the frequency range of 0.2 Hz to 2 Hz, but the vibration noise in the horizontal direction was suppressed by around

5 times. When there was not any active vibration isolation in the horizontal direction, the vibration noise was 1.8×10^{-9} g/$\sqrt{\text{Hz}}$. When active vibration isolation was added in the horizontal direction, the performance of vibration isolation was mainly restricted by the self-noise of the sensor and the electronic noise of electronic devices, but the vibration noise was lowered to 7.5×10^{-10} g/$\sqrt{\text{Hz}}$. Hence, the atomic interferometer achieved the sub-microgal level of gravity measurement accuracy.

As detailed in [90], Tang et al. of University of Chinese Academy of Sciences designed and implemented a compact and stable active low frequency vibration isolation system to improve the active vibration isolation system devised by Freier [87]. In the system, the vibration signals detected by a seismograph (CMG-3VL) were processed in a digital control system and then fed back to a voice coil motor, which could control and suppress the vibration of the passive vibration isolation platform. The natural frequency of the system decreased from 0.8 Hz to 0.015 Hz, so that the vibration in the vertical direction was effectively suppressed. The vibration noise at the frequency of around 1 Hz was attenuated to 1×10^{-9} g/$\sqrt{\text{Hz}}$. The vibration noise was reduced by 100 times on the whole. Therefore, its measurement accuracy was considerably improved.

Luo et al. of ZJUT developed a compact low frequency active vibration isolation system as presented in [91]. A sliding mode robust control system was utilized to process and feedback the vibration signals detected by a seismograph, while a voice coil motor was employed to control and eliminate the motion of the passive vibration isolation platform. Within the frequency range of 0.1–10 Hz, the sliding mode robust control system achieved the power spectral density of residual vibration noise 99.9% lower than that of the passive vibration isolation platform to the maximum, and 83.3% lower than that of the lead-lag compensation control method to the maximum. Apart from better performance of vibration isolation, it needed a setting of only three parameters, and offered a strong resistance to disturbance.

Chen et al. of USTC constructed an easily hauled three-dimensional active vibration isolation system for a movable atomic gravimeter as detailed in [92]. The system could effectively isolate the motion on the ground to enhance the measurement sensitivity of a movable atomic gravimeter. With a devised comprehensive feedback algorithm, it could isolate the vertical vibration on the ground by three orders of magnitude and the horizontal vibration on the ground by one order of magnitude. At the frequency of below 10 Hz to which an atomic gravimeter was sensitive, the vibration noise in the vertical direction was suppressed to 4.8×10^{-9} g/$\sqrt{\text{Hz}}$, while the vibration noise in the horizontal direction was lowered to 2.3×10^{-7} g/$\sqrt{\text{Hz}}$. The influence of vibration noise on the sensitivity of an interferometer was reduced to below 2 uGal/$\sqrt{\text{Hz}}$, which was two orders of magnitude lower than that of the interferometer without a vibration isolation system.

To measure the vibration noise of a reflection mirror in an atomic gravimeter, Zhang et al. [93] introduced an evaluation scheme for measuring the mirror vibration noise of an atomic gravimeter with a Michelson Raman laser (MIRL) interferometer. The MIRL interferometer was composed of the intrinsic Raman beam of the atomic gravimeter and a four-channel phase shift detector. The scheme presented an approach of using an improved "AI-MI-AI" three-interference system with a "three-cornered hat" to measure the contribution of mirror vibration to measurement instability. Restricted by the equivalence principle, the approach could not give the absolute vibration of the reflection mirror, but it was very simple, inexpensive, efficient and accurate to apply. Therefore, it offered another way to evaluate the contribution of vibration noise to measurement instability apart from commercial seismograph or accelerometer.

In [94], Zhou et al. put forward a cold atomic interference active vibration isolation system based on linear auto disturbance rejection control (LADRC) algorithm to implement the effective isolation for the low frequency noise at 0.1 Hz and below. The active vibration isolation system was compact and stable while offering low frequency and good performance. The LADRC controller involved two parts, i.e., extended state observer (ESO) and control law state equation. The ESO in the LADRC controller could directly observe the total disturbance of the system and make a timely compensation for such disturbance. As an active vibration isolation system combining a commercial passive vibration isolation system with an electronic feedback circuit, the system achieved the effective resonance frequency of 0.0152 Hz. As a spring-mass system that could generate nearly critical damping, it could significantly reduce the influence of frequency vibration at 0.1–10 Hz. Within the frequency range of 0.1–5 Hz, the system reduced the vertical vibration by 1000 times. Within the frequency range of 0.1–2 Hz, it could suppress the noise to 2×10^{-9} g/$\sqrt{\text{Hz}}$. Meanwhile, the system had a stable oscillation and a natural period of 66 s. Its performance was better than that of a classic lag compensation filter. The system required the adjustment to only a few parameters, and could be applied by simply adjusting the feedback gain of system state error.

Regardless of good performance, an active vibration isolation system still has some limitations especially when an AIG is used in a noisy environment. A compact vibration isolation system may only cause the vibration to attenuate by several orders of magnitude. A large vibration attenuation system may achieve better vibration attenuation, but it weighs more than one ton, and is not easy to carry because of its large size. Moreover, it has a very complex structure. Additionally, the system is often limited to a very small dynamic range, and requires the adjustment and consideration of environmental conditions including temperature. At the time of resonance, the system could not attenuate or properly suppress, but actually increase noise.

4.3. Vibration Compensation

The cold atomic interference gravity measurement based on vibration compensation satisfies the urgent need for the integrated and small weight measuring system [95]. It is applicable to measuring absolute gravity in the field and on a movable platform. Vibration compensation is to measure vibrations at the same time AIG is performed, and then after the measurement adjust the AIG data to compensate for the vibrations. The advantage is that it is necessary only to measure the vibrations accurately, but not to mechanically correct for them [62,96].

As shown in Figure 5, the Raman light reflection mirror at the bottom of an AIG in a vibration compensation scheme is normally installed above a sensor (accelerometer or seismograph) to monitor the vibration of the platform and evaluate the vibration noise. The sensor should be installed as close to the reflection mirror as possible, and also accurately leveled. After passing an analog-to-digital converter and a digital filter, the longitudinal output signal of a sensor is approximate to the vibration signal of the reflection mirror. The sensitivity function convolving with the vibration acceleration is used to calculate the phase shift resulted from vibration within a measurement period. In the end, the phase shift is compensated in the phase-transition probability curve to reconstruct the fringe and obtain the actual gravitational acceleration.

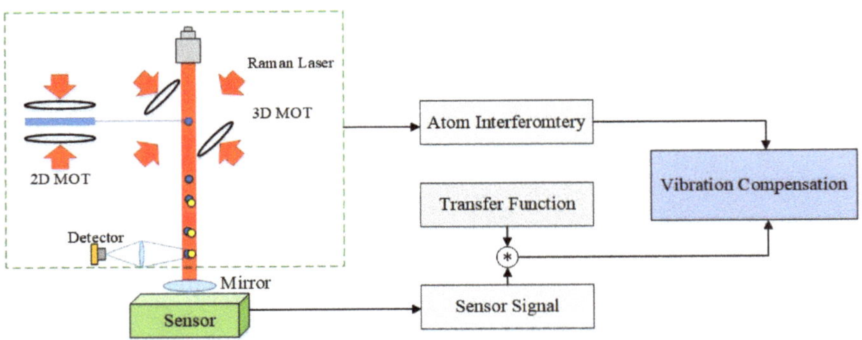

Figure 5. Vibration compensation scheme for an AIG.

In 2009, Merle et al. utilized a low noise seismograph to independently measure the vibration noise of a portable atomic gravimeter on the ground as detailed in [97]. After comparing two measurement schemes, i.e., fringe fitting and nonlinear locking, they found that the average phase of an interferometer could be determined in the phase measurement even if its phase noise exceeded 2 π. The scanned fringe of vibration noise was fitted when the interaction happened in a very short period (2T = 100 ms). The sensitivity of measurement at night reached 5.5×10^{-8} g. In the experiment, they explored the sensitivity limit of an atomic gravimeter without vibration isolation. This provided an idea for vibration compensation, so that it was of great significance to the study of portable atomic gravimeter.

The Le Gouët group of the Observatory of Paris designed the M-Z atomic interference in [15,98] using vibration compensation. The acceleration signal of a seismograph test device on a passive vibration isolation platform was utilized to calculate the phase shift of interference fringe, which was caused by vibration noise in different time periods. In the data processing, the phase noise caused by vibration noise was deducted. By virtue of vibration compensation, an atomic gravimeter achieved the sensitivity 1.4×10^{-8} g. Moreover, the group proposed to compensate the phase of a laser device using the vibration noise detected by the seismograph, so that the influence of vibration noise on measurement was reflected in the phase shift of interference fringe in the interaction of atoms and laser.

As described in [99], Barret et al. adopted the complementary working mode of "cold atomic gravimeter/accelerometer + classic accelerometer" and the vibration noise correction technology in the gravity measurement experiment in a parabolic flight micro-gravity environment. They analyzed the probability density of atomic interference signals to obtain the displacement P_0 and amplitude A of atomic interference fringe and solve the phase of atoms. Subsequently, an accelerometer fixed to the Raman reflection mirror was employed to measure the vibration acceleration and estimate the fringe period in which the phase of atoms fell into. The phase of atoms was then corrected. The corrected phase of atoms corresponded to the resultant acceleration of vehicle acceleration and vibration acceleration. Therefore, the vehicle acceleration could be directly determined when the vibration acceleration was known. This experiment had realized a dynamic gravitational acceleration measurement in a real sense for the first time.

Muquans, a French company, also successfully realized the real-time vibration compensation in its absolute quantum gravimeter (AQG) as presented in [35]. A highly sensitive accelerometer (Nanometrics Titan) was utilized to measure the vibration of a gravity measurement system. The acceleration data was then filtered and digitalized to compensate for the phase change caused by vibration, so that AQG could ensure the gravity measurement of high sensitivity even under the effect of severe vibration. In the experiment, the long-term stability of AQG could be lower than 1 µGal for absolute gravity measurement.

Lautier et al. applied the signal of a classic accelerometer in the real-time phase correction of an atomic gravimeter, so that it could operate with the best performance when there was not any isolation platform [100]. Moreover, it overcame the dead time problem in continuous measurement. Therefore, it was made ready for applications in geophysical and inertial navigation.

In [52], a rubidium AIG designed by Logan Latham Richardson of the University of Hannover was tested for low noise environment and simulated high noise environment in a vibration compensation experiment. A low noise ultra-wide-band seismograph (Trillium 240) was employed for vibration compensation. When the free evolution time was set to 78 ms, its short-term uncertainty increased from 4.4×10^{-6} g/$\sqrt{\text{Hz}}$ to 9.2×10^{-7} g/$\sqrt{\text{Hz}}$. In the meanwhile, a relatively high noise accelerometer (Nanometrics Titan) was used for vibration compensation. When the free evolution time was set to 10 ms, its short-term uncertainty was enhanced from 7.4×10^{-3} g/$\sqrt{\text{Hz}}$ to 1.0×10^{-4} g/$\sqrt{\text{Hz}}$.

In [62], a seismograph, a reflection mirror and a vibration source were fixed to a large vibration isolation platform to implement the calibration of transfer function. Subsequently, the Raman light reflection mirror was placed against the seismograph (CMG-3ESP-C) to generate a vibration signal. After filtering and integral compensation, the vibration signal was aggregated with the original interference fringe for correction. The gravitational acceleration was eventually obtained after correction. The correction was implemented in the following way: commercial software was used to reconstruct and simulate the waveform by inversing the transfer function of the seismograph to the reflection mirror. The Fourier transform was applied to the collected waveform. After correcting the spectrum, the inverse Fourier transfer was utilized to inverse back the time domain waveform. When the free evolution time was set to 60 ms, the resolution of the system could reach 32 μGal after 25.6 s integration. As effectively proved in an experiment, the passive vibration isolation platform could be used to suppress the vibration of most frequencies. The vibration isolation method could be applied in compensation for the vibration of large amplitude at given frequencies, and could achieve better vibration isolation than the passive vibration isolation platform when there was severe vibration in the environment.

At present, studies have been gradually conducted on applying an AIG in dynamic measurement. As presented in [33], Cheng et al., conducted the absolute gravity measurement of a ship in moored condition. A high precision accelerometer was placed below a Raman reflection mirror to measure the power spectrum of hull acceleration noise. After correction by vibration compensation, the sensitivity of gravity measurement reached 16.6 mGal/$\sqrt{\text{Hz}}$, and the resolution of gravity measurement within the integral time of 1000 s was 0.7 mGal. Li et al. [101] carried out a lake navigation test for an AIG. Based on the preliminary vibration isolation with an inertial stabilization platform, an accelerometer was employed to measure the vibration residual error of the gravimeter. After correction by vibration compensation, the measurement accuracy reached the mGal level.

The vibration compensation method is technically difficult to implement because of heavy computing workload. With regard to transfer function, it is easy to inverse the transfer function of a seismograph or an accelerometer, but difficult to determine the transfer function of the seismometer/accelerometer to the reflection mirror. In addition, the inverse Fourier transform is very complex for signal filtering, and very difficult to implement. The vibration compensation method may achieve the same effect as the passive vibration isolation method. Compared with the vibration isolation method based on mechanical structure, the vibration compensation method has a great advantage, that is, better realizing the high precision gravitational acceleration measurement when there is any strong external disturbance. Additionally, the vibration compensation method can satisfactorily optimize the vibration in all frequency bands, and is of great practical significance to measure the gravitational acceleration in the harsh field environment. The advantages and disadvantages of the vibration isolation system of the AIG are shown in Table 2.

Table 2. Comparison of vibration isolation systems for an AIG.

Methods	Previous Studies	Advantages	Disadvantages
Passive vibration isolation based	Li et al. [76] Refs. [80,81] Wu et al. [32]	It does not require external energy, sensor, actuator or control system, and can achieve the greatest vibration isolation with a simple structure.	It has very poor performance for the resonance and vibration isolation at low frequency bands, needs a very long stabilization time, and is poorly operable.

Table 2. *Cont.*

Methods	Previous Studies	Advantages	Disadvantages
Active vibration isolation based	Hensley J. M. et al. [86] Freier [87] M. Schmidt et al. [88] M. Hauth et al. [16] Zhou et al. [22] Zhou et al. [89] Tang et al. [90] Luo et al. [91] Chen et al. [92] Zhang et al. [93] Zhou et al. [94]	It may generate lower resonance frequency, better vibration suppression effect, and realize the high precision of static measurement.	It has a complex system structural design, is difficult to eliminate own noise, and offers the low precision of dynamic measurement.
Vibration compensation based	S Merlet et al. [97] Refs. [15,98] Brynle Barrett et al. [99] Menoret, V. et al. [35] Lautier et al. [100] Logan Latham Richardson [52] Xu et al. [62] Cheng et al. [33] Li et al. [101]	It can effectively isolate vibration at all frequency bands, be strongly resistant to disturbance, adapt to the harsh field environment in applications, and meet the needs for small and movable atomic gravimeters in applications.	It causes a heavy computing workload, and is difficult to determine the transfer function between the measuring device and equipment, and to achieve high precision vibration isolation.

5. Conclusions and Outlook

An AIG has developed into a significant tool for precision gravity measurement. Its measurement accuracy, sensitivity and applicability have improved. However, vibration isolation is one of its key technical problems in the development of an absolute gravimeter. The performance of vibration isolation system directly affects the measuring results and observation accuracy of an absolute gravimeter, even up to the mGal level. For this reason, it becomes a key and difficult point in the development of an absolute gravimeter. Up to now, an atomic gravimeter has been gradually expanded from static measurement in a laboratory to dynamic measurement in the field. For the development of small atomic gravimeters, vibration isolation system is urgently needed to eliminate vibration noise and enhance measurement accuracy.

The vibration of Raman mirror in AIG and the vibration of reference prism in falling angle cubic gravimeter are the reasons for their measurement errors, respectively. When active vibration isolation and passive vibration isolation methods are used, their vibration isolation principle is the same, that is, to suppress the vibration of Raman optical mirror/reference prism. However, the principle of vibration compensation is different. AIG compensates the phase, while the falling angle cube gravimeter compensates the interference trajectory. Due to the long research time of the vibration isolation method of a falling angle cubic gravimeter, the method theory is relatively mature, which can provide a reference for AIG vibration isolation.

Active vibration isolation is currently a mainstream vibration isolation method for a high precision AIG. It is employed by most atomic interference gravimeters that can achieve the accuracy up to 10^{-9} g. To apply the method, many parameters should be adjusted in the control system. However, an AIG can achieve very high measurement accuracy if these parameters are properly adjusted. The method is restricted by its complex system design and higher requirement for small size of gravimeter. Additionally, the noise of instruments in the active feedback system may introduce an error, which is a major contributor to inaccurate measurement. For the purpose of higher measurement accuracy, it is very important to improve the performance of sensor and satisfactorily process vibration signal.

In the vibration compensation method, compensation is directly made for the calculated interference fringe of vibration signal. When external vibration is more noticeable and there are many violent disturbances, the vibration compensation method can achieve better results than other vibration isolation methods. Presently, vibration compensation is mainly dependent on the limited sensitivity of an accelerometer used to measure vibration noise. It still needs further improvement to implement high accuracy measurement of absolute gravity. Moreover, the accelerometer and the transfer function between accelerometer and Raman light reflection mirror were not taken into account in the vibration compensation. The cold atomic interference gravity measurement based on vibration compensation satisfies the urgent need for an integrated and small gravity measurement system, and applies to the measurement of absolute gravity in the field and on a movable platform. It is the trend of future development.

Author Contributions: Conceptualization, W.G. and A.L.; methodology, W.G.; validation, A.L., F.Q. and C.H.; formal analysis, A.L. and C.F.; data curation, H.C.; writing—original draft preparation, W.G. and F.Q.; writing—review and editing, A.L. and C.H.; visualization, W.G.; project administration, F.Q. All authors have read and agreed to the published version of the manuscript.

Funding: This research was funded by the National Natural Science Foundation of China under Grants (61873275), and this work was supported by the Foundation of Basic Strengthening Technology of the Military Science and Technology Commission (2019JCJQJJ047), this research is also funded by Natural Science Foundation of Hubei Provincial of China (2017CFB377).

Institutional Review Board Statement: Not applicable.

Informed Consent Statement: Not applicable.

Data Availability Statement: Not applicable.

Conflicts of Interest: The authors declare no conflict of interest.

References

1. Cheiney, P.; Fouche, L.; Templier, S.; Napolitano, F.; Battelier, B.; Bouyer, P.; Barrett, B. Navigation-Compatible Hybrid Quantum Accelerometer Using a Kalman Filter. *Phys. Rev. Appl.* **2018**, *10*, 034030. [CrossRef]
2. Welker, T.C.; Pachter, M.; Huffman, R.E. Gravity gradiometer integrated inertial navigation. In Proceedings of the 2013 European Control Conference (ECC), Zurich, Switzerland, 17–19 July 2013; pp. 846–851.
3. Bong, K.; Holynski, M.; Vovrosh, J.; Bouyer, P.; Condon, G.; Rasel, E.; Schubert, C.; Schleich, W.P.; Roura, A. Taking atom interferometric quantum sensors from the laboratory to real-world applications. *Nat. Rev. Phys.* **2019**, *1*, 731–739. [CrossRef]
4. Zhang, X.B.; Yang, P.X.; Huang, P.; Sun, H.X.; Ying, W.W. Wide-bandwidth signal-based multireceiver SAS imagery using extended chirp scaling algorithm. *IET Radar Sonar Navig.* **2021**, 1–11. [CrossRef]
5. Gong, W.B.; Shi, Z.S.; Wu, Z.H.; Luo, J.R. Arbitrary-Oriented Ship Detection via Feature Fusion and Visual Attention for High-Resolution Optical Remote Sensing Imagery. *Int. J. Remote Sens.* **2021**, *42*, 2622–2640. [CrossRef]
6. Zhang, X.B.; Wu, H.R.; Sun, H.X.; Ying, W.W. Multireceiver SAS Imagery Based on Monostatic Conversion. *IEEE J. Sel. Top. Appl. Earth Observ. Remote Sens.* **2021**, *14*, 10835–10853. [CrossRef]
7. Zhang, X.; Yang, P. An Improved Imaging Algorithm for Multi-Receiver SAS System with Wide-Bandwidth Signal. *Remote Sens.* **2021**, *13*, 5008. [CrossRef]
8. Musso, C.; Bresson, A.; Bidel, Y.; Zahzam, N.; Dahia, K.; Allard, J.-M.; Sacleux, B. Absolute gravimeter for terrain-aided navigation. In Proceedings of the 20th International Conference on Information Fusion, Xi'an, China, 10–13 July 2017; pp. 1326–1332.
9. Zhang, X.B.; Ying, W.W.; Yang, P.X.; Sun, M. Parameter estimation of underwater impulsive noise with the Class B model. *Iet Radar Sonar Navig.* **2020**, *14*, 1055–1060. [CrossRef]
10. Kasevich, M.; Chu, S. Atomic interferometry using stimulated Raman transitions. *Phys. Rev. Lett.* **1991**, *67*, 181–184. [CrossRef]
11. Kasevich, M.; Chu, S. Measurement of the gravitational acceleration of an atom with a light-pulse atom interferometer. *Appl. Phys. B* **1992**, *54*, 321–332. [CrossRef]
12. Peters, A.; Chung, K.Y.; Chu, S. Measurement of gravitational acceleration by dropping atoms. *Nature* **1999**, *400*, 849–852. [CrossRef]
13. Gillot, P.; Cheng, B.; Imanaliev, A.; Merlet, S.; Santos, F.P.D. The LNE-SYRTE cold atom gravimeter. In Proceedings of the 2016 European Frequency and Time Forum (EFTF), York, UK, 4–7 April 2016; pp. 1–3.
14. Merlet, S.; Karcher, R.; Imanaliev, A.; Dos Santos, F.P. Ultracold Atom Gravimeter. In Proceedings of the 2018 Conference on Precision Electromagnetic Measurements, Paris, France, 8–13 July 2018.
15. Le Gouët, J.; Mehlstäubler, T.E.; Kim, J.; Merlet, S.; Clairon, A.; Landragin, A.; Pereira Dos Santos, F. Limits to the sensitivity of a low noise compact atomic gravimeter. *Appl. Phys. B* **2008**, *92*, 133–144. [CrossRef]

16. Hauth, M.; Freier, C.; Schkolnik, V.; Senger, A.; Schmidt, M.; Peters, A. First gravity measurements using the mobile atom interferometer GAIN. *Appl. Phys. B* **2013**, *113*, 49–55. [CrossRef]

17. Freier, C.; Hauth, M.; Schkolnik, V.; Leykauf, B.; Schilling, M.; Wziontek, H.; Scherneck, H.-G.; Müller, J.; Peters, A. Mobile quantum gravity sensor with unprecedented stability. In *Proceedings of the Journal of Physics: Conference Series*; IOP Publishing: Bristol, UK, 2016; p. 012050.

18. Peters, A.; Chung, K.Y.; Chu, S. High-precision gravity measurements using atom interferometry. *Metrologia* **2001**, *38*, 25–61. [CrossRef]

19. Fu, Z.; Wu, B.; Cheng, B.; Zhou, Y.; Weng, K.; Zhu, D.; Wang, Z.; Lin, Q. A new type of compact gravimeter for long-term absolute gravity monitoring. *Metrologia* **2019**, *56*, 025001. [CrossRef]

20. Bin, W.; Yin, Z.; Bing, C.; Dong, Z.; Nan, W.K.; Xin, Z.X.; Jun, C.P.; Xing, W.K.; Hai, Y.Q.; Hong, L.J.; et al. Static measurement of absolute gravity in truck based on atomic gravimeter. *Acta Phys. Sin.* **2020**, *69*, 25–32. [CrossRef]

21. Zhang, J.Y.; Chen, L.L.; Cheng, Y.; Luo, Q.; Shu, Y.B.; Duan, X.C.; Zhou, M.K.; Hu, Z.K. Movable precision gravimeters based on cold atom interferometry. *Chin. Phys. B* **2020**, *29*, 88–96. [CrossRef]

22. Zhou, M.-K.; Hu, Z.-K.; Duan, X.-C.; Sun, B.-L.; Chen, L.-L.; Zhang, Q.-Z.; Luo, J. Performance of a cold-atom gravimeter with an active vibration isolator. *Phys. Rev. A* **2012**, *86*, 043630. [CrossRef]

23. Xie, H.T.; Chen, B.; Long, J.B.; Xue, C.; Chen, L.K.; Chen, S. Calibration of a compact absolute atomic gravimeter. *Chin. Phys. B* **2020**, *29*, 093701. [CrossRef]

24. Jin, W. Precision measurement with atom interferometry. *Chin. Phys. B* **2015**, *24*, 053702. [CrossRef]

25. Zhou, L.; Xiong, Z.-Y.; Yang, W.; Tang, B.; Peng, W.-C.; Wang, Y.-B.; Xu, P.; Wang, J.; Zhan, M.-S. Measurement of Local Gravity via a Cold Atom Interferometer. *Chin. Phys. Lett.* **2011**, *28*, 013701. [CrossRef]

26. Geiger, R.; Landragin, A.; Merlet, S.; Santos, F.P.D. High-accuracy inertial measurements with cold-atom sensors. *AVS Quantum Sci.* **2020**, *2*, 024702. [CrossRef]

27. Cooke, A.K.; Champollion, C.; Le Moigne, N. First evaluation of an absolute quantum gravimeter (AQG#B01) for future field experiments. *Geosci. Instrum. Method. Data Syst.* **2021**, *10*, 65–79. [CrossRef]

28. Zhang, L.; Zhou, Y.; Weng, K.; Cheng, B.; Wu, B.; Lin, Q.; Hu, Z. The gravity estimation with square-root unscented Kalman filter in the cold atom gravimeter. *Eur. Phys. J. D* **2020**, *74*, 147. [CrossRef]

29. Wu, X.; Weiner, S.; Pagel, Z.; Malek, B.S.; Muller, H. Mobile quantum gravimeter with a novel pyramidal magneto-optical trap. In Proceedings of the Conference on Lasers and Electro-Optics, Washington, DC, USA, 10–15 May 2020; p. JW2C.5.

30. Chen, B.; Long, J.; Xie, H.; Li, C.; Chen, L.; Jiang, B.; Chen, S. Portable atomic gravimeter operating in noisy urban environments. *Chin. Opt. Lett.* **2020**, *18*, 12–17. [CrossRef]

31. Wang, S.-K.; Zhao, Y.; Zhuang, W.; Li, T.-C.; Wu, S.-Q.; Feng, J.-Y.; Li, C.-J. Shift evaluation of the atomic gravimeter NIM-AGRb-1 and its comparison with FG5X. *Metrologia* **2018**, *55*, 360–365. [CrossRef]

32. Wu, X.; Pagel, Z.; Malek, B.S.; Nguyen, T.H.; Zi, F.; Scheirer, D.S.; Müller, H. Gravity surveys using a mobile atom interferometer. *Sci. Adv.* **2019**, *5*, eaax0800. [CrossRef] [PubMed]

33. Bing, C.; Yin, Z.; Jun, C.P.; Jun, Z.K.; Dong, Z.; Nan, W.K.; Xing, W.K.; Lin, W.H.; Ping, P.S.; Long, W.X.; et al. Absolute gravity measurement based on atomic gravimeter under mooring state of a ship. *Acta Phys. Sin.* **2021**, *70*, 040304. [CrossRef]

34. Desruelle, B.; Vermeulen, P.; Menoret, V.; Landragin, A.; Bouyer, P.; Le Moigne, N.; Gabalda, G.; Bonvalot, S. Sub µGal Absolute Gravity Measurements with a Transportable Quantum Gravimeter. In Proceedings of the AGU Fall Meeting Abstracts, New Orleans, LA, USA, 11–15 December 2017; p. G54A-05.

35. Menoret, V.; Vermeulen, P.; Le Moigne, N.; Bonvalot, S.; Bouyer, P.; Landragin, A.; Desruelle, B. Gravity measurements below 10(-9) g with a transportable absolute quantum gravimeter. *Sci. Rep.* **2018**, *8*, 12300. [CrossRef]

36. Bidel, Y.; Carraz, O.; Charriere, R.; Cadoret, M.; Zahzam, N.; Bresson, A. Compact cold atom gravimeter for field applications. *Appl. Phys. Lett.* **2013**, *102*, 144107. [CrossRef]

37. Bidel, Y.; Zahzam, N.; Blanchard, C.; Bonnin, A.; Cadoret, M.; Bresson, A.; Rouxel, D.; Lequentrec-Lalancette, M.F. Absolute marine gravimetry with matter-wave interferometry. *Nat. Commun.* **2018**, *9*, 9. [CrossRef]

38. Bidel, Y.; Zahzam, N.; Bresson, A.; Blanchard, C.; Cadoret, M.; Olesen, A.V.; Forsberg, R. Absolute airborne gravimetry with a cold atom sensor. *J. Geod.* **2020**, *94*, 20. [CrossRef]

39. Müller, H.; Chiow, S.W.; Herrmann, S.; Chu, S.; Chung, K.Y. Atom-interferometry tests of the isotropy of post-Newtonian gravity. *Phys. Rev. Lett.* **2008**, *100*, 031101. [CrossRef]

40. Rosi, G.; Sorrentino, F.; Cacciapuoti, L.; Prevedelli, M.; Tino, G.M. Precision measurement of the Newtonian gravitational constant using cold atoms. *Nature* **2014**, *510*, 518–521. [CrossRef] [PubMed]

41. Altin, P.A.; Johnsson, M.T.; Negnevitsky, V.; Dennis, G.R.; Anderson, R.P.; Debs, J.E.; Szigeti, S.S.; Hardman, K.S.; Bennetts, S.; McDonald, G.D.; et al. Precision atomic gravimeter based on Bragg diffraction. *New J. Phys.* **2013**, *15*, 023009. [CrossRef]

42. Hu, Z.-K.; Sun, B.-L.; Duan, X.-C.; Zhou, M.-K.; Chen, L.-L.; Zhan, S.; Zhang, Q.-Z.; Luo, J. Demonstration of an ultrahigh-sensitivity atom-interferometry absolute gravimeter. *Phys. Rev. A* **2013**, *88*, 043610. [CrossRef]

43. Luo, Q.; Zhang, H.; Zhang, K.; Duan, X.-C.; Hu, Z.-K.; Chen, L.-L.; Zhou, M.-K. A compact laser system for a portable atom interferometry gravimeter. *Rev. Sci. Instrum.* **2019**, *90*, 043104. [CrossRef]

44. Wu, B.; Wang, Z.; Cheng, B.; Wang, Q.; Xu, A.; Lin, Q. The investigation of a µGal-level cold atom gravimeter for field applications. *Metrologia* **2014**, *51*, 452–458. [CrossRef]

45. Huang, P.W.; Tang, B.; Chen, X.; Zhong, J.Q.; Xiong, Z.Y.; Zhou, L.; Wang, J.; Zhan, M.S. Accuracy and stability evaluation of the Rb-85 atom gravimeter WAG-H5-1 at the 2017 International Comparison of Absolute Gravimeters. *Metrologia* **2019**, *56*, 7. [CrossRef]

46. Shuqing, W.; Tianchu, L. Technical Development of Absolute Gravimeter: Laser Interferometry and Atom Interferometry. *Acta Opt. Sin.* **2021**, *41*, 44–59. [CrossRef]

47. Tino, G.M. Testing gravity with cold atom interferometry: Results and prospects. *Quantum Sci. Technol.* **2021**, *6*, 024014. [CrossRef]

48. Savoie, D.; Altorio, M.; Fang, B.; Sidorenkov, L.A.; Geiger, R.; Landragin, A. Interleaved atom interferometry for high-sensitivity inertial measurements. *Sci. Adv.* **2018**, *4*, eaau7948. [CrossRef]

49. Musso, C.; Sacleux, B.; Bresson, A.; Allard, J.M.; Dahia, K.; Bidel, Y.; Zahzam, N.; Palmier, C. Terrain-aided navigation with an atomic gravimeter. In Proceedings of the 22nd International Conference on Information Fusion, Ottawa, ON, Canada, 2–5 July 2019.

50. Battelier, B.; Barrett, B.; Fouché, L.; Chichet, L.; Antoni-Micollier, L.; Porte, H.; Napolitano, F.; Lautier, J.; Landragin, A.; Bouyer, P. Development of compact cold-atom sensors for inertial navigation. In *Quantum Optics*; International Society for Optics and Photonics: Bellingham, WA, USA, 2016; p. 990004.

51. Rice, H.F.; Benischek, V.; Sczaniecki, L. Application of Atom Interferometric Technology For GPS Independent Navigation and Time Solutions. In Proceedings of the 2018 IEEE/Ion Position, Location And Navigation Symposium, Monterey, CA, USA, 23–26 April 2018; pp. 1097–1106.

52. Richardson, L.L. Inertial Noise Post-Correction in Atom Interferometers Measuring the Local Gravitational Acceleration. Ph.D. Dissertation, Gottfried Wilhelm Leibniz Universität Hannover, Hannover, Germany, 2019.

53. Lee, D.-H.; Kim, Y.-B.; Chakir, S.; Huynh, T.; Park, H.-C. Noninteracting Control Design for 6-DoF Active Vibration Isolation Table with LMI Approach. *Appl. Sci.* **2021**, *11*, 7693. [CrossRef]

54. Wu, K.; Li, G.; Hu, H.; Wang, L. Active low-frequency vertical vibration isolation system for precision measurements. *Chin. J. Mech. Eng.* **2017**, *30*, 164–169. [CrossRef]

55. Wu, Q.; Teng, Y.; Wang, X.; Wu, Y.; Zhang, Y. Vibration error compensation algorithm in the development of laser interference absolute gravimeters. *Geosci. Instrum. Method. Data Syst.* **2021**, *10*, 113–122. [CrossRef]

56. Yao, J.; Wu, K.; Guo, M.; Wang, L. Improvement of the ultra-low-frequency active vertical vibration isolator with geometric anti-spring structure for absolute gravimetry. *Rev. Sci. Instrum.* **2021**, *92*, 054503. [CrossRef] [PubMed]

57. Wen, Y.; Wu, K.; Guo, M.; Wang, L. An Optimized Vibration Correction Method for Absolute Gravimetry. *IEEE Trans. Instrum. Meas.* **2021**, *70*, 1003607. [CrossRef]

58. Yao, J.; Wu, K.; Guo, M.; Wang, G.; Wang, L. An Ultralow-Frequency Active Vertical Vibration Isolator With Geometric Antispring Structure for Absolute Gravimetry. *IEEE Trans. Instrum. Meas.* **2020**, *69*, 2670–2677. [CrossRef]

59. Wang, G.; Wu, K.; Hu, H.; Li, G.; Wang, L.J. Ultra-low-frequency vertical vibration isolator based on a two-stage beam structure for absolute gravimetry. *Rev. Sci. Instrum.* **2016**, *87*, 7. [CrossRef]

60. Wang, G.; Hu, H.; Wu, K.; Wang, L.J. Correction of vibration for classical free-fall gravimeters with correlation-analysis. *Meas. Sci. Technol.* **2017**, *28*, 6. [CrossRef]

61. Jipeng, W.; Dong, H.; Jinhai, B.; Hao, G. Review on Vibration Isolation Method of Atom-interferometric Gravimeter. *Metrol. Meas. Technol.* **2020**, *40*, 15–20. [CrossRef]

62. Aopeng, X.; Delong, K.; Zhijie, F.; Zhaoying, W.; Qiang, L. Vibration compensation of an atom gravimeter. *Chin. Opt. Lett.* **2019**, *17*, 070201. [CrossRef]

63. Cheinet, P.; Canuel, B.; Dos Santos, F.P.; Gauguet, A.; Yver-Leduc, F.; Landragin, A. Measurement of the sensitivity function in a time-domain atomic interferometer. *IEEE Trans. Instrum. Meas.* **2008**, *57*, 1141–1148. [CrossRef]

64. LeLe, C.; Qin, L.; XiaoBing, D.; YuJie, T.; DeKai, M.; Heng, Z.; MinKang, Z.; XiaoChun, D.; ChengGang, S.; ZhongKun, H. Precision gravity measurements with cold atom interferometer. *Sci. Sin.-Phys. Mech. Astron.* **2016**, *46*, 21–38. [CrossRef]

65. Weiner, S.; Wu, X.; Pagel, Z.; Li, D.; Sleczkowski, J.; Ketcham, F.; Mueller, H. A Flight Capable Atomic Gravity Gradiometer With a Single Laser. In Proceedings of the 7th IEEE International Symposium on Inertial Sensors and Systems, Hiroshima, Japan, 23–26 March 2020.

66. Ming-Sheng, Z. Progress in laser manipulating single neutral atoms. *Physics* **2015**, *44*, 518–526. [CrossRef]

67. Jin, W.; Xiaojun, L.; Jiaomei, L.; Hongtai, Z.; Kaijun, J.; Mingsheng, Z. Laser Trapping of Rubidium Atoms. *Acta Opt. Sin.* **2000**, *20*, 862–864.

68. Carraz, O.; Lienhart, F.; Charriere, R.; Cadoret, M.; Zahzam, N.; Bidel, Y.; Bresson, A. Compact and robust laser system for onboard atom interferometry. *Appl. Phys. B-Lasers Opt.* **2009**, *97*, 405–411. [CrossRef]

69. Zhijie, F.; Qiyu, W.; Zhaoying, W.; Bin, W.; Bing, C.; Qiang, L. Participation in the absolute gravity comparison with a compact cold atom gravimeter. *Chin. Opt. Lett.* **2019**, *17*, 011204. [CrossRef]

70. Dong, X.; Jin, S.; Shui, H.; Peng, P.; Zhou, X. Improve the performance of interferometer with ultra-cold atoms. *Chin. Phys. B* **2021**, *30*, 58–71. [CrossRef]

71. Zhang, X.B.; Huang, H.; Ying, W.W.; Wang, H.K.; Xiao, J. An Indirect Range-Doppler Algorithm for Multireceiver Synthetic Aperture Sonar Based on Lagrange Inversion Theorem. *IEEE Trans. Geosci. Remote Sens.* **2017**, *55*, 3572–3587. [CrossRef]

72. Tang, B. *Experimental Investigation of Active Vibration Isolation System for High Precision Cold Atom Interferometers*; Wuhan Institute of Physics and Mathematics Chinese Academy of Sciences: Wuhan, China, 2014.

73. Aleynikov, M.S.; Baryshev, V.N.; Blinov, I.Y.; Kupalov, D.S.; Osipenko, G.V. Prospects for the Development of a Sensitive Atomic Interferometer Based on Cold Rubidium Atoms. *Meas. Tech.* **2020**, *63*, 520–523. [CrossRef]
74. Fang, B.; Dutta, I.; Gillot, P.; Savoie, D.; Lautier, J.; Cheng, B.; Garrido Alzar, C.L.; Geiger, R.; Merlet, S.; Pereira Dos Santos, F.; et al Metrology with Atom Interferometry: Inertial Sensors from Laboratory to Field Applications. *J. Phys. Conf. Ser.* **2016**, *723*, 012049 [CrossRef]
75. Erin, C.; Wilson, B.; Zapfe, J. An improved model of a pneumatic vibration isolator: Theory and experiment. *J. Sound Vib.* **1998**, *218*, 81–101. [CrossRef]
76. Li, G.; Hu, H.; Wu, K.; Wang, G.; Wang, L.J. Ultra-low frequency vertical vibration isolator based on LaCoste spring linkage. *Rev. Sci. Instrum.* **2014**, *85*, 104502. [CrossRef]
77. Mizuno, T.; Toumiya, T.; Takasaki, M. Vibration isolation system using negative stiffness. *JSME Int. J. Ser. C Mech. Syst. Mach. Elem. Manuf.* **2003**, *46*, 807–812. [CrossRef]
78. Platus, D.L. Negative-stiffness-mechanism vibration isolation systems. In Proceedings of the Vibration Control in Microelectronics, Optics, and Metrology, San Jose, CA, USA, 1 February 1992; pp. 44–54.
79. Li, H.; Li, Y.; Li, J. Negative stiffness devices for vibration isolation applications: A review. *Adv. Struct. Eng.* **2020**, *23*, 1739–1755. [CrossRef]
80. D'Agostino, G.; Desogus, S.; Germak, A.; Origlia, C. The new IMGC-02 transportable absolute gravimeter: Measurement apparatus and applications in geophysics and volcanology. *Ann. Geophys.* **2008**, *51*, 39–49. [CrossRef]
81. Berrino, G. The state of the art of gravimetry in Italy. *Rend. Lincei. Sci. Fis. E Nat.* **2020**, *31*, 35–48. [CrossRef]
82. Bazinenkov, A.M.; Mikhailov, V.P. Active and Semi Active Vibration Isolation Systems Based on Magnetorheological Materials. *Procedia Eng.* **2015**, *106*, 170–174. [CrossRef]
83. Verma, M.; Collette, C. Active Vibration Isolation System for Drone Cameras. In Proceedings of the 14th International Conference on Vibration Problems, Crete, Greece, 1–4 September 2019; 2021; pp. 1067–1084.
84. Aggogeri, F.; Merlo, A.; Pellegrini, N. Active vibration control development in ultra-precision machining. *J. Vib. Control* **2021**, *27*, 790–801. [CrossRef]
85. Guo, M.; Wu, K.; Yao, J.; Wen, Y.; Wang, L. A Vertical Seismometer With Built-In Retroreflector for Absolute Gravimetry. *IEEE Trans. Instrum. Meas.* **2021**, *70*, 1004310. [CrossRef]
86. Hensley, J.M.; Peters, A.; Chu, S. Active low frequency vertical vibration isolation. *Rev. Sci. Instrum.* **1999**, *70*, 2735–2741 [CrossRef]
87. Freier, C. Measurement of Local Gravity Using Atom Interferometry. Ph.D. Thesis, Humboldt-Universität zu Berlin, Berlin, Germany, 2010.
88. Schmidt, M.; Senger, A.; Hauth, M.; Freier, C.; Schkolnik, V.; Peters, A. A mobile high-precision absolute gravimeter based on atom interferometry. *Gyroscopy Navig.* **2011**, *2*, 170. [CrossRef]
89. Zhou, M.-K.; Xiong, X.; Chen, L.-L.; Cui, J.-F.; Duan, X.-C.; Hu, Z.-K. Note: A three-dimension active vibration isolator for precision atom gravimeters. *Rev. Sci. Instrum.* **2015**, *86*, 046108. [CrossRef]
90. Tang, B.; Zhou, L.; Xiong, Z.; Wang, J.; Zhan, M. A programmable broadband low frequency active vibration isolation system for atom interferometry. *Rev. Sci. Instrum.* **2014**, *85*, 093109. [CrossRef] [PubMed]
91. Luo, D.-Y.; Cheng, B.; Zhou, Y.; Wu, B.; Wang, X.-L.; Lin, Q. Ultra-low frequency active vibration control for cold atom gravimeter based on sliding-mode robust algorithm. *Acta Phys. Sin.* **2018**, *67*, 82–87. [CrossRef]
92. Bin, C.; Bao, L.J.; Tai, X.H.; Kan, C.L.; Shuai, C. A mobile three-dimensional active vibration isolator and its application to cold atom interferometry. *Acta Phys. Sin.* **2019**, *68*, 117–125. [CrossRef]
93. Zhang, N.; Hu, Q.; Wang, Q.; Ji, Q.; Zhao, W.; Wei, R.; Wang, Y. Michelson laser interferometer-based vibration noise contribution measurement method for cold atom interferometry gravimeter. *Chin. Phys. B* **2020**, *29*, 70601. [CrossRef]
94. Zhou, Y.; Luo, D.; Wu, B.; Cheng, B.; Lin, Q. Active vibration isolation system based on the LADRC algorithm for atom interferometry. *Appl. Opt.* **2020**, *59*, 3487–3493. [CrossRef]
95. Weng, K.; Zhou, Y.; Zhu, D.; Wang, K.; Wu, B.; Cheng, B.; Lin, Q. High-accuracy gravity measurement with miniaturized quantum gravimeter. *Sci. Sin. Phys. Mech. Astron.* **2021**, *51*, 074204. [CrossRef]
96. Xie, H.T.; Chen, B.; Long, J.B.; Xue, C.; Chen, L.K.; Chen, S. Tilt adjustment for a portable absolute atomic gravimeter. *Chin. Phys. B* **2020**, *29*, 73701. [CrossRef]
97. Merlet, S.; Le Gouët, J.; Bodart, Q.; Clairon, A.; Landragin, A.; Pereira Dos Santos, F.; Rouchon, P. Operating an atom interferometer beyond its linear range. *Metrologia* **2009**, *46*, 87–94. [CrossRef]
98. Le Gouët, J. Étude des Performances d'Un Gravimètre Atomique Absolu: Sensibilité Limite et Exactitude Préliminaire. Ph.D. Dissertation, Université Paris Sud-Paris XI, Orsay, France, 2008.
99. Barrett, B.; Antoni-Micollier, L.; Chichet, L.; Battelier, B.; Lévèque, T.; Landragin, A.; Bouyer, P. Dual matter-wave inertial sensors in weightlessness. *Nat. Commun.* **2016**, *7*, 13786. [CrossRef] [PubMed]
100. Lautier, J.; Volodimer, L.; Hardin, T.; Merlet, S.; Lours, M.; Dos Santos, F.P.; Landragin, A. Hybridizing matter-wave and classical accelerometers. *Appl. Phys. Lett.* **2014**, *105*, 144102. [CrossRef]
101. An, L.; Hao, C.; Fangjun, Q.; Chunfu, H.; Wenbin, G. Development and prospect of cold atom interferometry gravimetry measurement. *J. Nav. Univ. Eng.* **2021**, *33*, 1–7. [CrossRef]

MDPI
St. Alban-Anlage 66
4052 Basel
Switzerland
Tel. +41 61 683 77 34
Fax +41 61 302 89 18
www.mdpi.com

Sensors Editorial Office
E-mail: sensors@mdpi.com
www.mdpi.com/journal/sensors